RELIGION
—— and ——
SCIENTIFIC
NATURALISM

To JACK,

With appreciation not only for the rave report & the great quote, but, more importantly, for your person & your own great contributions.

David

May 22, 2000

SUNY Series in
Constructive Postmodern Thought
David Ray Griffin, Editor

David Ray Griffin, editor, *The Reenchantment of Science: Postmodern Proposals*

David Ray Griffin, editor, *Spirituality and Society: Postmodern Visions*

David Ray Griffin, *God and Religion in the Postmodern World: Essays in Postmodern Theology*

David Ray Griffin, William A. Beardslee, and Joe Holland, *Varieties of Postmodern Theology*

David Ray Griffin and Huston Smith, *Primordial Truth and Postmodern Theology*

David Ray Griffin, editor, *Sacred Interconnections: Postmodern Spirituality, Political Economy, and Art*

Robert Inchausti, *The Ignorant Perfection of Ordinary People*

David W. Orr, *Ecological Literacy: Education and the Transition to a Postmodern World*

David Ray Griffin, John B. Cobb, Jr., Marcus P. Ford, Pete A. Y. Gunter, and Peter Ochs, *Founders of Constructive Postmodern Philosophy: Peirce, James, Bergson, Whitehead, and Hartshorne*

David Ray Griffin and Richard A. Falk, editors, *Postmodern Politics for a Planet in Crisis: Policy, Process, and Presidential Vision*

Steve Odin, *The Social Self in Zen and American Pragmatism*

Frederick Ferré, *Being and Value: Toward a Constructive Postmodern Metaphysics*

Sandra B. Lubarsky and David Ray Griffin, editors, *Jewish Theology and Process Thought*

J. Baird Callicott and Fernando J. R. de Rocha, editors, *Earth Summit Ethics: Toward a Reconstructive Postmodern Philosophy of Environmental Education*

David Ray Griffin, *Parapsychology, Philosophy, and Spirituality: A Postmodern Exploration*

Jay Earley, *Transforming Human Culture: Social Evolution and the Planetary Crisis*

Daniel A. Dombrowski, *Kazantzakis and God*

E. M. Adams, *A Society Fit for Human Beings*

Frederick Ferré, *Knowing and Value: Toward a Constructive Postmodern Epistemology*

Jerry H. Gill, *The Tacit Mode: Michael Polanyi's Postmodern Philosophy*

Nicholas F. Gier, *Spiritual Titanism: Indian, Chinese, Western Perspectives*

David Ray Griffin, *Religion and Scientific Naturalism: Overcoming the Conflicts*

RELIGION

——— and ———

SCIENTIFIC NATURALISM

Overcoming the Conflicts

DAVID RAY GRIFFIN

STATE UNIVERSITY OF NEW YORK PRESS

Published by
State University of New York Press, Albany

For information, address State University of New York Press,
State University Plaza, Albany, N.Y. 12246

Production by Michael Haggett
Marketing by Patrick Durocher

Library of Congress Cataloging-in-Publication Data

Griffin, David Ray, 1939–
    Religion and scientific naturalism : overcoming the conflicts /
David Ray Griffin.
        p. cm. — (SUNY series in constructive postmodern thought)
    Includes bibliographical references and index.
    ISBN 0-7914-4563-1 (alk. paper). — ISBN 0-7914-4564-X (pbk. :
alk. paper)
    1. Religion and science.   2. Naturalism—Religious aspects.
I. Title. II. Series.
BL240.2.G75   2000
215—dc21   99-37758
                                                                    CIP

10 9 8 7 6 5 4 3 2 1

To Ian Barbour and Robert John Russell

CONTENTS

INTRODUCTION TO SUNY SERIES IN CONSTRUCTIVE POSTMODERN THOUGHT*

The rapid spread of the term *postmodern* in recent years witnesses to a growing dissatisfaction with modernity and to an increasing sense that the modern age not only had a beginning but can have an end as well. Whereas the word *modern* was almost always used until quite recently as a word of praise and as a synonym for *contemporary*, a growing sense is now evidenced that we can and should leave modernity behind—in fact, that we must if we are to avoid destroying ourselves and most of the life on our planet.

Modernity, rather than being regarded as the norm for human society toward which all history has been aiming and into which all societies should be ushered—forcibly if necessary—is instead increasingly seen as an aberration. A new respect for the wisdom of traditional societies is growing as we realize that they have endured for thousands of years and that, by contrast, the existence of modern civilization for even another century seems doubtful. Likewise, *modernism* as a worldview is less and less seen as the Final Truth, in comparison with which all divergent worldviews are automatically regarded as "superstitious." The modern worldview is increasingly relativized to the status of one among many, useful for some purposes, inadequate for others.

Although there have been antimodern movements before, beginning perhaps near the outset of the nineteenth century with the Romanticists and the Luddites, the rapidity with which the term *postmodern* has become widespread

---

*The present version of this introduction is slightly different from the first version, which was contained in the volumes that appeared prior to 2000. My thanks to Catherine Keller and Edward Carlos Munn for helpful suggestions.

in our time suggests that the antimodern sentiment is more extensive and intense than before, and also that it includes the sense that modernity can be successfully overcome only by going beyond it, not by attempting to return to a premodern form of existence. Insofar as a common element is found in the various ways in which the term is used, *postmodernism* refers to a diffuse sentiment rather than to any common set of doctrines—the sentiment that humanity can and must go beyond the modern.

Beyond connoting this sentiment, the term *postmodern* is used in a confusing variety of ways, some of them contradictory to others. In artistic and literary circles, for example, postmodernism shares in this general sentiment but also involves a specific reaction against "modernism" in the narrow sense of a movement in artistic-literary circles in the late nineteenth and early twentieth centuries. Postmodern architecture is very different from postmodern literary criticism. In some circles, the term *postmodern* is used in reference to that potpourri of ideas and systems sometimes called *new age metaphysics*, although many of these ideas and systems are more premodern than postmodern. Even in philosophical and theological circles, the term *postmodern* refers to two quite different positions, one of which is reflected in this series. Each position seeks to transcend both *modernism,* in the sense of the worldview that has developed out of the seventeenth-century Galilean-Cartesian-Baconian-Newtonian science, and *modernity*, in the sense of the world order that both conditioned and was conditioned by this worldview. But the two positions seek to transcend the modern in different ways.

Closely related to literary-artistic postmodernism is a philosophical postmodernism inspired variously by physicalism, Ludwig Wittgenstein, Martin Heidegger, a cluster of French thinkers—including Jacques Derrida, Michel Foucault, Gilles Deleuze, and Julia Kristeva—and certain features of American pragmatism.* By the use of terms that arise out of particular segments of this movement, it can be called *deconstructive, relativistic,* or *eliminative* postmodernism. It overcomes the modern worldview through an antiworldview, deconstructing or even entirely eliminating various concepts that have generally been thought necessary for a worldview, such as self, purpose, meaning, a real world, givenness, reason, truth as correspondence, universally valid norms, and

---

*The fact that the thinkers and movements named here are said to have inspired the deconstructive type of postmodernism should not be taken, of course, to imply that they have nothing in common with constructive postmodernists. For example, Wittgenstein, Heidegger, Derrida, and Deleuze share many points and concerns with Alfred North Whitehead, the chief inspiration behind the present series. Furthermore, the actual positions of the founders of pragmatism, especially William James and Charles Peirce, are much closer to Whitehead's philosophical position—see the volume in this series entitled *The Founders of Constructive Postmodern Philosophy: Peirce, James, Bergson, Whitehead, and Hartshorne*—than they are to Richard Rorty's so-called neopragmatism, which reflects many ideas from Rorty's explicitly physicalistic period.

divinity. While motivated by ethical and emancipatory concerns, this type of postmodern thought tends to issue in relativism. Indeed, it seems to many thinkers to imply nihilism.* It could, paradoxically, also be called *ultramodernism*, in that its eliminations result from carrying certain modern premises—such as the sensationist doctrine of perception, the mechanistic doctrine of nature, and the resulting denial of divine presence in the world—to their logical conclusions. Some critics see its deconstructions or eliminations as leading to self-referential inconsistencies, such as "performative self-contradictions" between what is said and what is presupposed in the saying.

The postmodernism of this series can, by contrast, be called *revisionary, constructive*, or—perhaps best—*reconstructive*. It seeks to overcome the modern worldview not by eliminating the possibility of worldviews (or "metanarratives") as such, but by constructing a postmodern worldview through a revision of modern premises and traditional concepts in the light of inescapable presuppositions of our various modes of practice. That is, it agrees with deconstructive postmodernists that a massive deconstruction of many received concepts is needed. But its deconstructive moment, carried out for the sake of the presuppositions of practice, does not result in self-referential inconsistency. It also is not so totalizing as to prevent reconstruction. The reconstruction carried out by this type of postmodernism involves a new unity of scientific, ethical, aesthetic, and religious intuitions (whereas post-structuralists tend to reject all such unitive projects as "totalizing modern metanarratives"). While critical of many ideas often associated with modern science, it rejects not science as such but only that *scientism* in which the data of the modern natural sciences alone are allowed to contribute to the construction of our public worldview.

The reconstructive activity of this type of postmodern thought is not limited to a revised worldview. It is equally concerned with a postmodern *world* that will both support and be supported by the new worldview. A postmodern world will involve postmodern persons, with a postmodern spirituality, on the one hand, and a postmodern society, ultimately a postmodern global order, on the other. Going beyond the modern world will involve transcending its individualism, anthropocentrism, patriarchy, economism, consumerism, nationalism, and militarism. Reconstructive postmodern thought provides support for the ethnic, ecological, feminist, peace, and other emancipatory movements of our time,

---

*Peter Dews says that although Derrida's early work was "driven by profound ethical impulses," its insistence that no concepts were immune to deconstruction "drove its own ethical presuppositions into a penumbra of inarticulacy" (*The Limits of Disenchantment: Essays on Contemporary European Culture* [London, New York: Verso, 1995], 5). In his more recent thought, Derrida has declared an "emancipatory promise" and an "idea of justice" to be "irreducible to any deconstruction." Although this "ethical turn" in deconstruction implies its pulling back from a completely disenchanted universe, it also, Dews points out (6–7), implies the need to renounce "the unconditionality of its own earlier dismantling of the unconditional."

while stressing that the inclusive emancipation must be from the destructive features of modernity itself. However, the term *postmodern*, by contrast with *premodern*, is here meant to emphasize that the modern world has produced unparalleled advances, as Critical Theorists have emphasized, which must not be devalued in a general revulsion against modernity's negative features.

From the point of view of deconstructive postmodernists, this reconstructive postmodernism will seem hopelessly wedded to outdated concepts, because it wishes to salvage a positive meaning not only for the notions of selfhood, historical meaning, reason, and truth as correspondence, which were central to modernity, but also for notions of divinity, cosmic meaning, and an enchanted nature, which were central to premodern modes of thought. From the point of view of its advocates, however, this revisionary postmodernism is not only more adequate to our experience but also more genuinely postmodern. It does not simply carry the premises of modernity through to their logical conclusions, but criticizes and revises those premises. By virtue of its return to organicism and its acceptance of nonsensory perception, it opens itself to the recovery of truths and values from various forms of premodern thought and practice that had been dogmatically rejected, or at lease restricted to "practice," by modern thought. This reconstructive postmodernism involves a creative synthesis of modern and premodern truths and values.

This series does not seek to create a movement so much as to help shape and support an already existing movement convinced that modernity can and must be transcended. But in light of the fact that those antimodern movements that arose in the past failed to deflect or even retard the onslaught of modernity, what reasons are there for expecting the current movement to be more successful? First, the previous antimodern movements were primarily calls to return to a premodern form of life and thought rather than calls to advance, and the human spirit does not rally to calls to turn back. Second, the previous antimodern movements either rejected modern science, reduced it to a description of mere appearances, or assumed its adequacy in principle. They could, therefore, base their calls only on the negative social and spiritual effects of modernity. The current movement draws on natural science itself as a witness against the adequacy of the modern worldview. In the third place, the present movement has even more evidence than did previous movements of the ways in which modernity and its worldview are socially and spiritually destructive. The fourth and probably most decisive difference is that the present movement is based on the awareness that *the continuation of modernity threatens the very survival of life on our planet.* This awareness, combined with the growing knowledge of the interdependence of the modern worldview with modernity's militarism, nuclearism, patriarchy, global apartheid, and ecological devastation, is providing an unprecedented impetus for people to see the evidence for a postmodern worldview and to envisage postmodern ways of relating to each other, the rest of nature, and the cosmos as a whole. For these reasons, the failure of the previous antimodern movements says little about the possible success of the current movement.

Advocates of this movement do not hold the naively utopian belief that the success of this movement would bring about a global society of universal and lasting peace, harmony and happiness, in which all spiritual problems, social conflicts, ecological destruction, and hard choices would vanish. There is, after all, surely a deep truth in the testimony of the world's religions to the presence of a transcultural proclivity to evil deep within the human heart, which no new paradigm, combined with a new economic order, new childrearing practices, or any other social arrangements, will suddenly eliminate. Furthermore, it has correctly been said that "life is robbery": a strong element of competition is inherent within finite existence, which no social-political-economic-ecological order can overcome. These two truths, especially when contemplated together, should caution us against unrealistic hopes.

No such appeal to "universal constants," however, should reconcile us to the present order, as if it were thereby uniquely legitimated. The human proclivity to evil in general, and to conflictual competition and ecological destruction in particular, can be greatly exacerbated or greatly mitigated by a world order and its worldview. Modernity exacerbates it about as much as imaginable. We can therefore envision, without being naively utopian, a far better world order, with a far less dangerous trajectory, than the one we now have.

This series, making no pretense of neutrality, is dedicated to the success of this movement toward a postmodern world.

David Ray Griffin
Series Editor

PREFACE

The central question of this book is simply whether there is anything essential to science that is in conflict with any beliefs essential to vital religion, especially theistic religion. My answer is No, but the dominant answer has been Yes. In many cases, to be sure, this Yes is merely implicit. That is, whereas many modern intellectuals have declared the scientific worldview to have rendered religious beliefs incredible, many others, including many theologians, have avoided this conclusion only by redefining the religion in question, such as Christianity, out of all recognition. Besides being unrecognizable, the remaining beliefs could not serve as the nucleus of a vital spirituality.

The main reason for the conviction, whether announced or camouflaged, that science conflicts with essential religious beliefs is a twofold equation: the equation of religion with supernaturalism and the equation of science, since about the middle of the nineteenth century, with a materialistic version of scientific naturalism. Given this twofold equation, the "scientific worldview" necessarily conflicts, in various ways, with the worldview presupposed by religious believers. Unless these perceived conflicts can be overcome, the division in our culture between religious and anti-religious forces will not be healed. There are, to be sure, many other important issues to be discussed under the heading of "science and religion." But unless they contribute to overcoming the perceived conflicts between the worldview presupposed by the scientific community and that presupposed by the religious communities, they are largely irrelevant. One reason I wrote this book was to try to drive home this point—that the overriding issue is that of *worldview*.

My aim, however, is not simply to draw attention to this formal point, but also to suggest how the perceived conflicts can be overcome. This suggestion is oriented around a distinction between the "minimal" and the "maximal" construals of *scientific naturalism*. Although the scientific community really only requires naturalism in the minimal sense, I claim, almost everyone has understood the term in the maximal sense. One result has been that theologians and constructive philosophers of religion have thereby been confronted with a lose-lose proposition: If they wanted to reconcile religion with the worldview of the scientific community, they had to reconceive religious belief so drastically as to make their "reconciliation" irrelevant to their religious community. If they wanted to reaffirm the beliefs vital to their religious community, they had to articulate a worldview at odds with that of the scientific community. The difference between these two responses has largely defined the differences between liberal and conservative theologies.

Once it is seen that science requires naturalism only in the minimal sense, however, the apparent conflicts can be overcome, because naturalism in the minimal sense is fully compatible with theistic religion. In fact, given a wider, more open form of naturalism, such as that provided by the philosophy of Alfred North Whitehead, the change from a supernaturalistic to a naturalistic form of theism need not lead to a weaker, less vital spirituality. It can, in fact, have the opposite effect. When this is realized, our culture may overcome not only the perceived conflicts between science and religion, but also the antithesis between liberal and conservative forms of religion.

If the fundamental problem is that of worldview, the solution must be primarily philosophical. What is needed is a philosophical cosmology that, besides commending itself in terms of the normal criteria of adequacy and self-consistency, can also be seen as meeting the respective needs of the scientific and religious communities. The recognition of the capacity of Whitehead's philosophy to play this role has, I suggest, been blocked on both sides. On the one hand, the widespread equation of naturalism with its maximal form has made it difficult to recognize that Whitehead's philosophy *is* a form of naturalism. On the other hand, the suspicion in the scientific community that Whitehead's philosophy was not fully naturalistic was matched by suspicions in many religious circles that it was not fully religious. To some extent, this was due to the fact that the conclusions of some of Whitehead's most prominent advocates made his philosophy appear to be more in line with the maximal version of naturalism than is really the case. Partly, however, the view that a Whiteheadian worldview could not be religiously adequate has been due to the assumption that theistic religions, such as Christianity and Judaism, require a supernaturalistic version of theism. This book seeks to show this assumption not to be true.

The immediate stimulation for this book came from my participation in a seminar on science and religion sponsored by the John Templeton Foundation and conducted by the Center for Theology and the Natural Sciences (CTNS), which is located in Berkeley at the Graduate Theological Union. I am deeply indebted to both the John Templeton Foundation and to Robert Russell's CTNS, both of which are playing very important roles in promoting new reflection and dialogue on this crucially important topic. My dedication of this book to Bob reflects my deep appreciation for him as a person and for the enormous contribution he has made to increasing both the visibility and the level of the science-and-religion discussion.[1] In also dedicating this book to Ian Barbour, I mean to pay tribute to the other constructive thinker who has done the most in recent times to elevate the level of the discussion with regard to the relations between science and religious thought. Besides being widely recognized as the "dean" of the recent science-and-religion discussion in general—to whom Bob Russell, I, and countless others are deeply indebted—Ian Barbour has also been the pioneer in relating Whiteheadian process philosophy to this discussion. To the extent that process philosophy is already recognized as providing a viable basis for integrating science and religion, this fact is due primarily to Ian's work. The significance of Ian's contributions over many decades has recently been recognized, of course, by his being awarded the 1999 Templeton Prize for Progress in Religion.

In coming to my present position on these issues, I have, of course, learned from many prior and contemporary thinkers. Most of those to whom I am directly indebted for the argument in this book are named in the list of references at the end. I am especially indebted to Jack Haught, Ted Peters,[2] and an anonymous SUNY Press reader for many helpful suggestions.

I am also deeply grateful to my primary institution, the Claremont School of Theology, which recognizes that it is part of the task of theological schools to provide informed reflection, for the churches and for the wider culture, about crucial religious issues on which there has been confusion and conflict.

I am delighted again to express publicly my gratitude to my multi-talented wife, Ann Jaqua (perhaps known to some readers through Nora Gallagher's *Things Seen and Unseen*), for the many ways in which she provides support in general, for her enthusiasm for this book in particular, and for her goodness, wisdom, dedication, and many contributions to our world.

---

1. Among those contributions have been a series of conferences at the Vatican Observatory resulting in a number of excellent volumes on a topic central to the present volume, divine influence in the world: *Quantum Cosmology and the Laws of Nature*, ed. Robert John Russell, Nancey Murphy, and C. J. Isham (1993), *Chaos, Complexity and Self-Organization*, ed. Robert John Russell, Nancey Murphy, and Arthur Peacocke (1995), and *Evolutionary and Molecular Biology*, ed. Robert John Russell, William R. Stoeger, S.J., and Francisco J. Ayala (1998), all with the subtitle *Scientific Perspectives on Divine Action* and all co-published by the Vatican Observatory and CTNS.

2. Ted Peters, who is a central figure at CTNS, probably has done more than anyone else to promote the term *consonance* for describing the relation between science and theology. This term usually connotes a view that, by emphasizing the degree to which science and theology are independent enterprises even while looking for parallels between them, is in tension with the full-scale *integration* advocated by process thinkers such as myself. I am pleased to see, however, that Peters' own construal of "consonance" in his edited volume on the subject, *Science and Theology: The New Consonance*, is the same as the position I suggest in the present volume, which I call "possible harmony." For example, he says that, "In the strongest sense of the word, 'consonance' means full accord or harmony" (ST 1). Pointing out that no such harmony exists today, he says that we now have only "consonance in a weak sense," which involves the hypothesis that, because there is only one reality and science and theology both pursue truth about it, we should "expect that sooner or later shared understandings will develop" (18, 1). Far from denying that consonance in the strong sense is desirable and possible, Peters simply says that "[a]ccord or harmony might be a treasure we hope to find, but we have not found it yet" (18).

Part I

SCIENCE, RELIGION, AND NATURALISM

1

SCIENCE, RELIGION, AND WORLDVIEW

Alfred North Whitehead once wrote that, when we consider what science and religion are, "it is no exaggeration to say that the future course of history depends [upon our decision] as to the relations between them." Lying behind this statement was his view of science and religion not as two bodies of doctrine but as two *forces*—"the force of our religious intuitions, and the force of our impulse to accurate observation and logical deduction"—and his judgment that they are "the two strongest general forces" influencing us (SMW, 181).[1] When this statement was written, in 1925, it would have evoked much dissent: The modern view that religion was atavistic, a superstitious relic from bygone times soon destined to disappear, was widely held. For Whitehead to regard religion as a force to be mentioned in the same breath with science would have seemed strange. In the latter part of the twentieth century, however, the various religions of the world revealed that they are indeed sources of tremendous power. This resurgence of religion has been surprising to the modern secular mind. This resurgence is, indeed, one of the main reasons for saying that we live in distinctively postmodern times. In our day, we can appreciate, perhaps

---

1. All books are cited by abbreviations, which are listed directly after the author's name in the bibliography at the end of the book. For the present citation, for example, locate "Whitehead, Alfred North (SMW)," which refers to his *Science and the Modern World.*

better than could most of Whitehead's contemporaries, his conclusion that it is of utmost importance to overcome the fact that these two forces "seem to be set one against the other" (SMW, 182).

This conviction lies behind the present book. If our religious impulses and our scientific impulses are indeed the two strongest general forces on our thought and behavior, and yet these two forces appear to be opposed to each other, then we are drawn in opposite directions. If we are thus divided, it will be difficult to motivate and organize ourselves to take the kind of concerted action that will be necessary if we are to meet the unprecedented challenges of our day, such as political and economic injustice, domestic and international insecurity, over-population and ecological deterioration.[2] This internal division has become per-haps the central phenomenon of our political and cultural life, as those representing resurgent "religion" usually have radically different agendas from those repre-senting "scientific rationality."

The conflict between scientific and religious impulses, however, does not occur simply between two kinds of people, as if some were purely "scientific types" and others purely "religious types." Rather, the conflict means that we as individuals are ourselves internally torn, drawn in opposing directions. For ex-ample, we read that the "fate of the earth" is in danger and that it is up to us to prevent its destruction by nuclear weapons or polluting technology, but then we hear from other voices that the fate of the world is entirely in God's hands, that it is human hubris to try to "save the world." Or, agreeing with the book's author that the fate of the world is really in human hands, we then read his paralyzing question: If life is ultimately meaningless, what difference does it really make if life on Earth is extinguished millions or even billions of years prematurely?

---

2. In reflecting on a meeting that issued a "Joint Appeal by Science and Religion on the Environ-ment," John Haught brings out the importance of consensus on the nature of reality by the scientific and religious communities if they are to reach meaningful consensus on the moral and practical level. Among the participants in the meeting, he reports, were Carl Sagan, E. O. Wilson, and Stephen Jay Gould, all of whom are well known for saying that religions, insofar as they purport to reflect the truth about reality, are essentially illusory. At this meeting, however, they each spoke "very favorably of religion's possible role in alleviating the ecological crisis," with each giving the same rationale—that although science alone gives us the facts about the cosmos, "religions can foster the kind of moral fervor that the environmental movement sorely needs" (PN 8). As Haught points out, however, this attempt to benefit from the moral passion generated by religion while denying any cognitive truth to religious worldviews is self-defeating: "[I]t is only because believers take their religious symbols and ideas to be disclosive of the *truth* of reality that they are aroused to moral passion in the first place. If devotees thought that their religions were *not* representative of the way things *really* are, then the religions would be ethically impotent" (PN 9). If Haught is right, and I am deeply convinced that he is, then the environmental crisis should provide many people today an especially urgent motive to ask whether, beneath the obviously false and mythological elements in the traditional religions, there are contained some basic truths about the nature of the universe.

How, then, are we to conceive the relation between science and religion? It might appear that Whitehead has expressed a version of the "conflict thesis" as to their relation. That thesis, however, regards science and religion to be in *essential* conflict, meaning that the very essence of what "religion" is conflicts with the very essence of what "science" is. Whitehead's statement, by contrast, says only that science and religion *seem* to be in conflict with each other. This statement leaves open the possibility that the conflict between them is merely apparent, or at least merely temporary, being based upon contingent, accidental factors that can be overcome. That, in fact, was Whitehead's view, and his own philosophical writings were devoted primarily to overcoming this apparent conflict. Philosophy, he said, "attains its chief importance by fusing the two, namely, religion and science, into one rational scheme of thought" (PR, 15).

With that statement, Whitehead pointed to one of the major ways of understanding the "relation" between science and religion. According to this view, although science and religion may at times be in conflict, it is possible for them to be in harmony. This harmony is to be effected *by integrating them into a philosophical worldview.* The distinctiveness of this position can be seen by showing its place in a typology of the three major ways of understanding the relation between science and religion: independence, conflict, and possible harmony.[3]

The Relation Between Science and Religion: Three Views

Speaking of the relation between "science" and "religion" can, of course, be confusing, because both of these terms can be taken to refer to types of *activities,* which as such cannot conflict. What is really at issue, of course, is whether some *beliefs* that are well-supported by science are in conflict with some *beliefs* that are taken to be essential to religion. Theology is the attempt to state, reformulate, and systematize the beliefs of a religious community. Most precisely put, therefore, the question is whether there is any conflict between scientific

---

3. This typology is the same as Ian Barbour's (*RAS,* Ch. 1), with two exceptions. First, Barbour adds a fourth position, called "dialogue," whereas I regard those discussed under this heading as proposing either a type of integration (such as Ernan McMullin, discussed in Ch. 3) or the modified version of independence that Ted Peters calls "weak consonance." Second, whereas he calls the third position "integration," I call it "possible harmony" while considering integration to be the ideal. Besides agreeing with Barbour on this formal ideal, I also agree with him on most substantive issues. Insofar as my work differs from his, it does so by focusing on the issue of "scientific naturalism," by working out an integrated position from a Whiteheadian stance more fully on a few issues, by showing, partly through the use of parapsychological evidence, how this stance can provide a more religiously robust theology (as illustrated in Chapters 7 and 8), and by emphasizing how this type of "naturalistic theism" differs from other stances to which this label is sometimes applied.

beliefs and *theology*. I will sometimes use this term. When I speak simply of the relation between science and *religion,* the referent is to the ideational or theological aspect of religion. The present question, then, concerns the three major ways of understanding the relation between scientific beliefs, on the one hand, and religious beliefs or theology, on the other.

One popular view in recent times has been idea that scientific and religious beliefs are independent from each other in such a way that they cannot possibly come into conflict. There is, therefore, no need to try to bring them into harmony. One way to declare them thus independent is simply to hold that truth is not one: Theology tells us one set of truths, science tells us another. This two-truth solution, however, is difficult for most of us to accept. We feel that truth, ultimately, must all be of a piece.

Various attempts have been made, accordingly, to provide a more palatable version of the two-truth solution. One such position holds that, although truth is one, we are, at least in this life, incapable of seeing this unity. We must simply, as religious persons, accept some truths on the basis of revelation, without being able to see how these truths are compatible with the results of scientific investigations. Science, for example, may show that all events in the world, including those events in which we make conscious decisions, are fully enmeshed in a deterministic nexus of causes and effects; from revelation, however, we know that we are responsible for our actions, which implies that we have a significant degree of freedom. As scientists, accordingly, we affirm determinism, while as religious persons we affirm freedom. The same duality may obtain with regard to the existence of God, divine providence in the world, the objective existence of ethical norms, and immortality. In the middle part of the twentieth century, this two-truth solution was phrased in terms of two languages: When speaking scientific language, we speak in terms of determinism, relativism, and nihilism, as befitting a godless universe; when speaking religiously, however, we speak in terms of God, freedom, ethical norms, and immortality. Many exponents of the independence thesis have, however, said that science and religion cannot conflict because they deal with *different domains.* René Descartes' dualism between soul and body was used, for example, as a line of demarcation: Science was to pronounce on the physical world, including the human body, while the human soul, with its relation to God, was allocated to religion. Albert Einstein's version of the independence thesis, urged more recently by Stephen Jay Gould (RA), says that science deals with facts, religion with values. Another distinction, used especially in discussions of evolution and creation, says that science asks *how* while theology asks *why.*

According to those who hold one of the other two views—the conflict thesis or the thesis of possible harmony—the independence thesis fails in all its forms. Although theology and science differ in important ways and to some extent deal with different domains, they also overlap significantly, and in this area of overlap there is the potential for conflict. The attempt to divide theology and science along the lines of soul and body, for example, breaks down by virtue of the fact that soul and body interact: This interaction prevents a neat division between a free mind and fully deterministic bodily behavior. The same is true of the attempted division in terms of facts and values: Science involves not only facts but also values, such as the value of knowing the truth in spite of possible undesirable consequences from a religious or moral viewpoint, while the articulation of religious belief inevitably asserts, or at least presupposes, various factual claims, such as the claim that human behavior is partly free and that the world is God's creation. No more successful is the distinction between how and why, or method and purpose: The claim that the world exists because it is God's creation, for example, cannot intelligibly be made without implying something about how the world has been created that might conflict with the scientific community's attempt to state how our world has come about. And, if these various attempts to divide religion and science into separate domains all fail, it is also true, advocates of both conflict and possible harmony insist, that we cannot rest content with two irreconcilable sets of alleged truths. We need, as Whitehead put it, "a vision of the harmony of truth" (SMW, 185).

Turning now to the conflict thesis: It is important to understand that it says not merely that science and religion *have been* in conflict, but also that there is an *essential* and therefore *permanent* conflict between them. This thesis may be held by representatives of either religion or science. When held by representatives of religion, it maintains that the only reliable guide to truth is provided by a distinctively religious way of knowing. In theistic religions, this has been understood as a "revelation" that has been authoritatively transmitted. All true knowledge, at least about matters of ultimate concern, is based upon this revelation. The attempt to develop autonomous scientific knowledge, based upon experience and reason not subordinated to the truths of revelation, will inevitably lead to error. The classic statement of this position is Tertullian's rhetorical question, "What has Athens to do with Jerusalem?", with "Jerusalem" standing for the biblical revelation and "Athens" for Greek philosophy, which contained the beginnings of what is today called "natural science." Although the view that science and religion are necessarily in conflict is still held by some advocates of religious revelation, since the middle of the nineteenth century it has more commonly been associated with advocates of the scientific way of knowing, some examples of which will be provided in Chapter 2.

Sometimes science is said not to be in essential conflict with religion as such but only with theology. This was the thesis of one of the most well-known books promulgating the "conflict" or "warfare" thesis, Andrew Dickson White's *A History of the Warfare of Science with Theology in Christendom* (1896). White's polemic presupposed the equation of theology with "revealed religion" based on taking the Bible to be a scientific text. White's understanding of the nature of theology, however, is not the only one. Indeed, that understanding of theology has been rejected by the tradition of "liberal theology," which has existed in various forms since the eighteenth century and to which the present book belongs.

The third basic way of relating science and religion is that of seeking to show that, although conflict certainly can occur between scientific and religious beliefs, this conflict is not necessary, which means that harmony is possible. Those who accept this third way agree that harmony is possible between theology, properly construed, and science, properly construed. Within this agreement, however, there are many different ways of understanding the proper construal of theology and the proper construal of science. There are, accordingly, many different understandings of the best way to overcome the apparent conflicts between them so as to demonstrate the essential harmony, or at least the absence of essential conflict, between them. At one extreme is the view that scientific beliefs are to be adjusted so as to be harmonious with traditional theological beliefs. This approach, which is that taken by "creation scientists," results in a "science" that is unrecognizable as such to most members of the scientific community—a "science" that is inconsistent with various empirical data as well as with the naturalistic presupposition of today's scientific community, according to which all occurrences, without exception, are explainable in principle without appeal to supernatural interventions. At the other extreme is the view that theology is simply to be accommodated to the views that are dominant within the contemporary scientific community. Given the reductionistic perspective that has been dominant within the scientific community, this way of reaching harmony has resulted in theologies that are unrecognizable as such by the members of the religious communities that they are supposed to represent.

Most attempts to demonstrate or bring about harmony between science and theology, however, avoid these two extremes, seeking to do justice both to historic religious beliefs and to science's basic assumptions and established facts. These middle positions involve a twofold critique. Unlike "creation science," which accepts historic Christian doctrines virtually wholesale and modifies scientific doctrines accordingly, these middle positions modify inherited religious doctrines in the light of scientifically established facts. And, unlike the tendency of modern liberal theologies to accept contemporary scientific ideology virtually wholesale, these middle positions distinguish between science as such and the worldview or ideology with which it has been associated in recent times. These middle positions, in other words, think of the "relation" between scientific and religious beliefs as going both ways, so that the resulting harmony results from a *mutual modification*. One of the differences among these various middle positions is in terms of where they think the greater modifications are needed—on the side of historic religious doctrines or on the side of doctrines with which late modern science has been associated. Another crucial issue is whether they think of the two-way interaction as occurring directly between science and theology as such, or more indirectly, through the mediation of philosophy.

Whitehead's Position

Given this framework, we can characterize Whitehead's position. He believed that the apparent conflicts between science and religion have been due about equally to inherited religious ideas and to the worldview with which science has recently been associated (which he called "scientific materialism"). And he believed that the needed modifications on both sides could only be achieved by means of philosophy, with "philosophy" understood primarily as metaphysical cosmology, the attempt to create an all-inclusive worldview in which scientific facts and inescapable religious intuitions can be harmonized. Like those who speak of the mutual independence of religious and scientific beliefs, Whitehead recognized that they originate in very different types of experience.

> The dogmas of religion are the attempts to formulate in precise terms the truths disclosed in the religious experience of mankind. In exactly the same way the dogmas of physical science are the attempts to formulate in precise terms the truths disclosed in the sense-perception of mankind. (RM, 57)

Whereas scientific beliefs are based primarily on sensory perceptions, religious beliefs are based primarily on nonsensory perceptions. Unlike advocates of the independence thesis, however, Whitehead did not believe that the different roots of scientific and religious beliefs meant that they could remain unreconciled.

One reason why Whitehead believed reconciliation to be necessary is that, unlike most other science-based philosophers of recent times, he thought that truths are disclosed by nonsensory as well as sensory perceptions. He did not, therefore, think that philosophy, as the attempt to formulate an inclusive worldview, is to be based solely upon the systematization of truths derived from sensory perceptions. In fact, Whitehead's first metaphysical book, *Science and the Modern World,* begins with a repudiation of that view: "The various human interests which suggest cosmologies," he says, "are science, aesthetics, ethics, religion." Since the seventeenth century, however, "the cosmology derived from science has been asserting itself at the expense of older points of view with their origins elsewhere." It is the task of philosophy in our time to overcome this one-sidedness, which means that it is the task of philosophy "to harmonise, re-fashion, and to justify divergent intuitions as to the nature of things. It has to insist on . . . the retention of the whole of the evidence in shaping our cosmological scheme" (SMW, vii).

In speaking of "the whole of the evidence," Whitehead had in mind especially the "truths disclosed in the religious experience of mankind" referred to in the indented quotation, which came from his next book, *Religion in the Making.* Just as his former book had focused on the contribution to metaphysics made by recent developments in science, this second book contains a section titled "The Contribution of Religion to Metaphysics" (RM, 84). It also contains a succinct statement of Whitehead's conviction as to the way science, theology, and metaphysics are interrelated in the search for truth: "You cannot shelter theology from science, or science from theology; nor can you shelter either of them from metaphysics, or metaphysics from either of them. There is no short cut to truth" (RM, 76–77). We cannot, in other words, regard either theology or science as an autonomous discipline with truths to be protected from the encroachment of the other. We also are not to think of metaphysical philosophy as independent of experience. It is nothing but the attempt to think consistently and comprehensively about the whole range of evidence supplied by our sensory and nonsensory experiences, reconciling doctrines with origins in one type of experience with those originating in the other.

A second reason why it is necessary to reconcile science and religion is a twofold human tendency (1) to take a given method of obtaining truth as the only valid method, and (2) to exaggerate the truths obtained from that method. Of the first of these tendencies, which Whitehead called "obscurantism"—but equally well could have called "intellectual original sin"—he says:

This obscurantism is rooted in human nature more deeply than any particular subject of interest. It is just as strong among men of science as among the clergy. . . . A few generations ago the clergy, or to speak more accurately, large sections of the clergy were the standing examples of obscurantism. Today their place has been taken by scientists—

By merit raised to that bad eminence.

The obscurantists of any generation are in the main constituted by the greater part of the practitioners of the dominant methodology. Today scientific methods are dominant, and scientists are the obscurantists. (FR, 43–44)

The dominant form of obscurantism in earlier ages, in other words, was *revelationism*, the belief that the learned interpretation of divinely given revelation provided a self-sufficient path to truth; the dominant form of obscurantism today, especially in intellectual circles, is *scientism*, the belief that the scientific method (as hitherto employed) is the only way to discover truth.

Obscurantism is made all the worse by the second tendency, which is to exaggerate the truths that are found through one's preferred method, formulating them in such a way as to exclude complementary truths. "Thought is abstract," Whitehead says, "and the intolerant use of abstractions is the major vice of the intellect" (SMW, 18). Religious and scientific doctrines, in other words, both involve abstractions, and theologians and scientific philosophers both tend to exaggerate the truth of their respective abstractions, then to use these exaggerated doctrines to deny the truth of the abstractions from the other side. For example, the theologian or religion-based philosopher, being impressed by the evidence for divine providence in the world, may define this providence as all-determining causality, thereby ruling out the complementary truth that events in the world are causally conditioned by antecedent events. The scientist, or science-based philosopher, being impressed instead by this latter fact, may construe the world's cause-effect nexus so that divine providence is totally ruled out. Again, the theologian or philosopher of religion, in the interest of stressing human responsibility, may portray human freedom so as virtually to ignore the degree to which our freedom is often limited by a wide range of causal conditions beyond our control (such as genetic inheritance and early childhood experiences). The science-based philosopher, by contrast, may construe these causal constraints as total determination, thereby ruling out responsible freedom altogether.

The task of philosophy in all these conflicts is to be the "critic of abstractions" (SMW, 59), showing how the abstractions of religious thought and of scientific thought are compatible by overcoming their respective exaggerations and placing them within a larger, more inclusive worldview than either had provided by itself.

Scientific Naturalism

Today, the discussion of the apparent conflicts between science and religion has increasingly been stated in terms of the issue of "scientific naturalism." Science, it is widely agreed in scientific, philosophical, and liberal religious circles, necessarily presupposes naturalism. Given this assumption, having an integrated worldview with no conflict between scientific and religious beliefs would require a religion devoid of supernaturalism. Some liberal theologians have suggested the possibility of a worldview that is religious while being naturalistic. Most philosophers, theologians, and scientists, however, believe that scientific naturalism is incompatible with any significantly religious view of reality.

To a great extent, this difference revolves around an ambiguity in the idea of "scientific naturalism," which can be understood either in a minimal or a maximal sense. In the *minimal* sense, scientific naturalism is simply a rejection of supernatural interventions in the world, meaning interventions that interrupt the world's most fundamental pattern of causal relations. Understood *maximally*, by contrast, scientific naturalism is equated with sensationism, atheism, materialism, determinism, and reductionism. Thus construed, scientific naturalism rules out not only supernatural interventions, as just defined, but also much more, such as human freedom, variable divine influence in the world, and any ultimate meaning to life. If scientific naturalism is understood in this maximal sense, those who say that it rules out a significantly religious worldview are right. If, however, science is understood only to require naturalism in the minimal sense, the quest for a worldview that is fully religious while being fully naturalistic may not be quixotic.

The conflicts revolving around naturalism can be understood in terms of the tendency, discussed above, of parties on both sides of a conflict to become fixated on an exaggeration. Although each position is giving witness to a truth, in its exaggerated form this truth is a falsehood, because it excludes the element of truth in the other position. The suggestion of the present book is that both "scientific naturalism," as usually understood, and "supernaturalism," as usually understood, are falsifying exaggerations. On the one hand, the minimal form of scientific naturalism is true, but the maximal form, with its sensationism, atheism, and materialism, is false. On the other hand, theism, with its notion that a divine reality exerts variable influence in the world, is true, but it is a falsifying exaggeration to think that this influence can be all-determining, so that it could interrupt the causal powers and principles of the world.

Four Kinds of Conflict

Given the above understanding of the relation between science and religion and of the distinction between the two versions of scientific naturalism, the possible conflicts between science and religion are of four basic kinds. *Conflict of the first kind* results when a religious community is committed to a supernaturalist worldview, according to which God is understood to be a being outside the world who can and perhaps does supernaturally intervene in it, interrupting the causal powers of the creatures. This belief, which can be called *ontological supernaturalism,* conflicts with the naturalistic worldview, which has been increasingly presupposed and confirmed by the scientific community during the past two centuries. This minimal naturalism holds that all events are enmeshed in a universal cause-effect nexus, so that all events have natural causes and effects: There could be no events devoid of natural causes and no events devoid of natural effects.

Naturalism in this minimal sense can be identified with what has historically been called "uniformitarianism," which is the assumption that the same general causal principles obtain for all events. Naturalism in this (minimal) sense does not necessarily rule out many things that "scientific naturalism" is usually thought to rule out (such as divine influence, freedom, and paranormal events). But it does rule out the reality and even the possibility of occasional supernatural interruptions of the most fundamental causal principles of the world. Insofar as theology asserts or presupposes ontological supernaturalism, it necessarily stands in conflict with the most fundamental assumption of the contemporary scientific worldview.

Many features of traditional Christian belief have exemplified this supernaturalistic worldview. The most obvious example is the belief in "miracles" defined as supernatural interruptions of the normal cause-effect relationships. A miracle, in fact, has usually been *defined* as an event that is caused totally and directly by God, without the use of any natural ("secondary") causes. Another example of supernaturalism is provided by the traditional christology, according to which God's activity or presence in Jesus was metaphysically different in kind from God's activity or presence in all other human beings. A third example is the belief that the rise of life and then the rise of the human mind required supernatural interventions, in which God's activity was different in kind from the providential activity that God exerts always and everywhere. A fourth example is provided by the belief that life after death will be brought about by a supernatural act of God, such as a resurrection of our physical bodies. In many ways the most important example, however, is the belief in an infallible revelation or inspiration, according to which divine causation overruled the normal human thought processes, with their usual fallibility, so as to produce statements that directly express the divine knowledge and will. This last example is especially important because it provides the transition from ontological to *epistemic* supernaturalism, according to which some ideas are to be accepted not because of their intrinsic merits, but solely because of their alleged origin in an infallible revelation—which leads to the next category.

Conflict of the second kind occurs when a religious community remains committed to beliefs about particular facts after science has demonstrated them to be almost certainly false. Among the most notorious examples have been the geocentric view of the universe, the idea that the universe was created only a few thousand years ago, and the idea that Moses wrote the first five books of the Bible. These and other beliefs have been held onto, even after being shown by good evidence to be almost certainly false, because they were thought to have been infallibly revealed, so that reasoning on the basis of empirical evidence was not allowed to disconfirm them. This epistemic supernaturalism, therefore, depends upon ontological supernaturalism, according to which God can, through supernatural intervention, annul the fallibility that normally characterizes human ideas.

Conflicts of these first two kinds conform to the picture of the conflict between science and religion prevalent in intellectual circles today, according to which the conflict is the result of a dogmatic religious mentality, unwilling to accept the results of the scientific method and the naturalistic worldview upon which it is based. Conflicts of the third and fourth kinds, however, suggest a more complex picture.

Conflict of the third kind occurs insofar as the scientific community is committed not only to minimal naturalism, but also to the maximal form of naturalism, which rules out beliefs vital to religion. This has largely been true since the middle of the nineteenth century, as the scientific community became increasingly committed to a sensationist, mechanistic, materialistic, deterministic, reductionistic, relativistic, nihilistic worldview, which rules out not only super-naturalistic religious belief but also any significantly religious interpretation of reality whatsoever.

To see why this is so, we can begin with the epistemic side of this worldview, its *sensationism,* which says that we have no mode of perception except sensory perception. This doctrine rules out any theistic religious experience understood as a direct, nonsensory apprehension of a Divine Actuality distinct from oneself. It also rules out moral and aesthetic expe-rience, understood as a direct nonsensory apprehension of normative ideals or values. This sensationist doctrine often leads to relativism, according to which all value-judgments are purely subjective preferences, with no possi-bility of being true, in the sense of corresponding with any normative ideals in the nature of things.

The ontological dimensions of this worldview also rule out a significantly religious interpretation of the universe. Its *mechanism* forbids any purposive, teleological causation. Its *materialism* forbids any distinction between the mind or soul and the brain, thereby ruling out life after death (apart from a supernatu-ral resurrection of the body, which is, of course, also ruled out). The mechanism and materialism, taken together, imply determinism, thereby ruling out human freedom. This worldview's *reductionism,* according to which all vertical causa-tion goes upward, from the simpler to the more complex, reinforces the denial of freedom (which would require "downward causation" from the mind to the body). The *atheism* of this worldview, besides denying any transcendent source of religious experiences, combines with the reductionism to rule out the idea of a divine creation of the world and even any divine influence in the world. This atheism, especially when combined with relativism, leads to nihilism, according to which life has no ultimate meaning.

The *basic* ideas of this worldview, then, are its sensationism, mechanism, materialism, reductionism, and atheism, with its determinism, relativism, and nihilism being implications (which exponents may seek to deny or at least qualify). Also, the mechanism and reductionism of this worldview can be regarded as implicit in its materialism. The basic ideas, accordingly, can be reduced to three: sensationism, atheism, and materialism. Then, using "*s*" for sensationism, "*a*" for atheism, and "*m*" for materialism, we can refer to this maximal naturalism as *naturalism$_{sam}$.* This maximal naturalism is also called "scientific materialism," "reductionistic naturalism," "materialistic naturalism," and "atheistic naturalism." In any case, if science appears to be committed to *this* type of naturalistic worldview, then it is necessarily in conflict not only with supernaturalistic the-ology but with any significant religious belief whatsoever.

This scientific materialism is often thought to be part and parcel of the naturalistic worldview required, and increasingly confirmed, by science. This maximal naturalism, or naturalism$_{sam}$, however, goes far beyond the scientific naturalism discussed in relation to conflict of the first kind. That minimal naturalism insists upon nothing but the rejection of supernaturalism. Beyond presupposing the uniformitarian belief that the basic causal processes of the world are never violated, science need not be committed to any more restrictive dogma as to the nature of these processes. Scientific naturalism need not and should not be committed, for example, to the idea that the basic causal processes are all mechanistic, so that all apparently purposive causation must be illusory. Also, not being wedded to materialism and reductionism, scientific naturalism need not be committed to the identity of mind and brain but should leave that question open, to be decided by empirical and theoretical considerations. Scientific naturalism need not be closed, therefore, to genuine freedom or even the possibility of life after death. The naturalism required by science also requires no commitment to the idea that all the basic causal processes of the world are between contiguous things or events. It can be relaxed, therefore, about claims about so-called paranormal interactions, letting these claims be settled by the evidence. Scientific naturalism, not being committed to the sensationist view of perception, also need not rule out *a priori* the possibility of genuine moral, aesthetic, and religious experience. The rejection of supernaturalism, finally, does not even rule out theism of all forms. It rules out only the supernaturalistic form of theism, according to which God can interrupt the basic causal processes of the world. Scientific naturalism could, therefore, be compatible with a naturalistic theism, or theistic naturalism, if such there can be.

This nonreductionistic naturalism, accordingly, could be compatible with a significantly religious interpretation of the world. Scientific naturalism in this sense, therefore, is not necessarily in conflict with theology, at least if, as I will argue, a significant theology is possible without supernaturalism. This prospect of harmony is ruled out, however, if scientific naturalism is equated with naturalism$_{sam}$. Because this equation has usually been made, and because this equation is the most important source of conflict between science and religion in intellectual circles today, most of this book is directed against this equation.

Conflict of the fourth kind occurs when, on the basis of naturalism$_{sam}$— *rather than on the basis of factual evidence*—scientists make judgments about particular events that conflict with beliefs that are vital to religious communities. For example, scientists may declare that our universe's origin was in no way influenced by a purposive creator, a statement for which there is obviously no hard evidence—in fact, no evidence whatsoever. Another example is provided by the claim that a purposive creator in no way influenced the rise of life in general and the rise of human life in particular. Needless conflict of this fourth type also occurs if scientists go beyond the historical evidence, which suggests that the prophets, Jesus, and biblical authors were, like the rest of us, fallible human beings, to declare that they were *in no way* inspired by a Divine Reality. No science, be it physics, biology, psychology, or archeology, provides evidence to support such a claim. Another example is the claim that "science" implies that the reported post-crucifixion appearances of Jesus were either fabrications or purely subjective hallucinations, involving no influence from any continuing experiential activity of Jesus.

Conflict of this fourth kind, it should be evident, is parallel to conflict of the second kind, which arises from the epistemic supernaturalism of some forms of religion. Devotees of naturalism$_{sam}$ do not, of course, explicitly believe in a revelation from on high that authorizes them to make pronouncements not based upon empirical evidence. But they often allow their worldview to function in an analogous fashion. Just as religious supernaturalists may hold that their interpretation of various facts "must" be true, regardless of the empirical data, reductionistic naturalists may be so convinced that their interpretations "must" be true that they claim the authority of science for these interpretations, even if they are empirically groundless. When scientists engage in this kind of speculation while labeling it "science," they violate their own commitment to epistemic naturalism, according to which beliefs are supposed to be based upon the rational-empirical method rather than upon some presupposed revelation, whether explicit or implicit, which exempts their knowledge-claims from the need to be supported by evidence.

Although any of these four kinds of conflict by itself can be serious, the most dramatic and most publicized kinds of conflict between science and religion occur when reductionistic naturalism, with its sensationistic, atheistic, materialistic worldview, encounters religious supernaturalism. The best-known confrontation of this kind in our time is that between the neo-Darwinian theory of evolution and "scientific creationism," which claims to confirm on scientific grounds the idea, derived from a literalistic reading of Genesis, that the Earth was created only a few thousand years ago. Insofar as the "relation between science and religion" is equated with this conflict—an equation that the mass media, with their love for extreme confrontations, tend to promote—a harmonious relation seems impossible. This is because the confrontation is doubly extreme, with all four kinds of conflict occurring at once. On the one hand, "the religious view" is equated with a doctrine that, because of its supernaturalism, is incompatible with even the most open form of scientific naturalism and insists upon a wildly implausible reading of the empirical data. On the other hand, "the scientific view" is equated with a doctrine that, because of its materialistic atheism, is incompatible not only with supernaturalism but with *any* idea of theistic guidance of the evolutionary process.

To put the issue in terms of Whitehead's statement that "the intolerant use of abstractions is the major vice of the intellect": In conflicts of the first and second kinds, theologians are using their abstractions intolerantly. In conflicts of the third and fourth kinds, scientists (or science-based philosophers) are using *their* abstractions intolerantly. Confrontations involving all four kinds of conflict are particularly vicious because each side is using its abstractions intolerantly against the abstractions of the other side. However, although these confrontations are especially dramatic, they present no fundamentally new problem for the prospect of harmony. Their resolution would follow simply from the solutions implicit in the description of the four kinds of conflict. As that description showed, these four kinds ultimately reduce to two. That is, conflicts of the second kind, being based upon epistemic supernaturalism, are derivative from conflict of the first kind, which results when a religious community is committed to ontological supernaturalism. Conflicts of the fourth kind are equally derivative from conflict of the third kind, which is rooted in the identification of science with naturalism$_{sam}$. Conflicts of the first and third kinds, accordingly, are fundamental.

The Road to Harmony

On the basis of this analysis, we can see that the relation between science and religion could move from a relation of conflict to one of harmony if and only if two things were to occur: if the theistic religious communities were to give up all remnants of supernaturalism in favor of a theistic naturalism, and if the scientific community were to give up reductionistic naturalism, with its sensationism, atheism, and materialism, in favor of a naturalism restricted to the rejection of supernatural interruptions of the world's basic causal processes. However, the scientific community will not change its worldview simply for the sake of effecting this harmony, and the same is true of most theistic religious communities. Whether this dual development will appear conceivable, therefore, depends upon the answer to two fundamental questions. The first question is whether there is a form of theistic naturalism that can provide the basis for a theology or religious philosophy with sufficient robustness and continuity with historic Christian faith for it to be widely accepted by the coming generations of Christian thinkers. (A similar question would obtain for other theistic religious faiths, such as Judaism, Islam, and theistic forms of Hinduism.) The second question is whether there is a version of scientific naturalism that provides a more adequate basis for science than does the materialistic version. The harmony between the scientific and religious communities would be complete, of course, only if one and the same naturalistic worldview were to be accepted by both.

The twofold question, accordingly, is whether there is a naturalistic worldview that, besides being adequate to the various beliefs presupposed by religious faith, also provides a better context for science than the materialistic, sensationistic naturalism with which science has been associated since about the middle of the nineteenth century. If so, we could have an integrated worldview that is at once scientific and religious. The present book is based on the conviction that Whitehead's worldview, which is a theistic naturalism based upon a nonmaterialistic ontology and nonsensationist epistemology, provides the basis for giving an affirmative answer to this twofold question.

At the root of the distinctive aspects of the Whiteheadian worldview, which is sometimes called "process philosophy," is the ultimate reality of time, or temporal process. From the perspective of this worldview, most of the problems of philosophy and theology, including the conflicts between science and religion, have been finally rooted in the neglect of time or process—of the fundamentally temporal nature of existence. From this perspective, to be *actual* is to be a process with temporal duration. This seemingly tiny modification provides a new perspective on virtually every issue, from the reality of time for physics to the reality of time for God, from the mind-body relation to the God-world relation, and from the reality of human freedom to the possibility of speaking of teleological causation and progress in the evolutionary process. By providing the basis for affirming theistic naturalism and for rejecting materialism without returning to dualism, this modification gives us a way to integrate our religious intuitions and our scientific convictions into a unified vision.

Preview of the Following Chapters

The remaining chapters of this first part of the book continue the focus on science, religion, and naturalism. Chapter 2 traces the development through which the materialistic version of scientific naturalism arose out of the first version of the modern scientific worldview, with its supernaturalistic dualism, then illustrates the ways in which this materialistic naturalism leads to conflict with any significantly religious outlook. Crucial to this whole issue, I suggest, is the development of modern thinking about divine action in terms of the scheme of primary and secondary causation.

In the third chapter, I look at three approaches to harmonizing science and religion that challenge my central theses. One of these approaches rejects both of my theses—that religion can thrive without supernaturalism and that science rightly presupposes naturalism in the minimal sense. I examine this approach, which argues that harmony can result only if science is dissociated from even minimal naturalism, in terms of the writings of Alvin Plantinga and Phillip Johnson. A second approach accepts the view that science rightly presupposes naturalism, but rejects my view that harmony with religion can be achieved only if the religious community also accepts a form of naturalism. This approach, which places a purely methodological scientific naturalism within a supernaturalistic framework, will first be examined in terms of three of its contemporary representatives—William Hasker, Ernan McMullin, and Howard Van Till—then in terms of Rudolf Otto's more consistent version of it. A third approach accepts my view that harmony requires the rejection of supernaturalism by the religious community, but rejects my contention that harmony also requires the materialistic form of naturalism to be rejected by the scientific community. I examine this position in terms of its recent advocacy by Willem Drees.

In the fourth chapter, I begin the exposition of the approach advocated in this book: harmonizing science and religion in terms of a richer, more open version of naturalism. After looking briefly at the attempt to effect this harmony on the basis of John Dewey's version of naturalism, I summarize Whitehead's version, showing how he simultaneously rejected both religious supernaturalism and scientific materialism. Then, after indicating some ways in which Whitehead's alternative to materialism results in a naturalism that is more open to religiously important experiences and beliefs, I point briefly to some scientific developments that undermine the determinism, reductionism, and sensationism of the materialistic version of scientific naturalism.

Because the mechanistic view of nature, which is the one ontological doctrine held in common by the first and the second versions of the "modern scientific worldview," has been so strongly identified with *the* scientific view of nature, Chapter 5 backs up to look at the origin of this mechanistic view, showing that it was based less on scientific (rational and empirical) considerations than on theological and sociological interests of the time. Another point of this chapter is that the Neoplatonic-magical-spiritualist tradition, which was the main opponent of the mechanical philosophy in the seventeenth-century battle of the worldviews, can, as a religious and scientific naturalism, be regarded as a premodern precursor to the *postmodern* naturalistic worldview developed by Whitehead.[4]

Part II shows how Whitehead's naturalism provides the basis for harmonizing religious and scientific beliefs in relation to some issues on which there has been considerable conflict during the late modern period. Chapter 6 addresses the mind-body problem as it occurs in the current science-based philosophical discussion, showing how the retention of the Cartesian view of matter has increasingly led to the conclusion, in both materialistic and dualistic thinkers, that the relation of the brain to conscious experience is an inexplicable mystery. I then argue that Whitehead's panexperientialism not only can provide a naturalistic account of the rise of consciousness but also can account for the reality of the freedom that we all presuppose—in our scientific as well as our religious activities.

Chapter 7 then explores the relations between religious belief and the most controversial of all the sciences, parapsychology. I suggest that the prejudice

---

4. For the way in which the term "postmodern" is used here, in contrast with its more well-known usage, see my series introduction at the outset of this book.

against parapsychology has been just that—a prejudgment, made almost solely on *a priori*, philosophical grounds—and that religious philosophers and theologians, most of all, should not succumb to this prejudice, because the assumptions of the reigning orthodoxy in the scientific community that have created this prejudice are the same assumptions that have created a similar prejudgment against religious beliefs. I suggest, further, that it is finally time to respond appreciatively to the claim by some of the advocates of parapsychology that it is "religion's basic science," the one that most clearly provides positive support for a religious interpretation of human experience in particular and reality in general. I conclude by looking at the importance of parapsychological findings for various issues of religious importance, including the possibility of genuine religious experience and life after death. This chapter is crucial, because one of this book's main theses, that a robust religious worldview is possible within a naturalistic framework, depends heavily upon the evidence from parapsychology—this being one of the reasons that I did not heed the advice that, for the sake of "credibility," I should delete this chapter.

In Chapter 8, we come to the issue that has been at the very heart of the conflict between science and theistic religious belief since the time of Charles Darwin: the idea that our world, rather than being created *ex nihilo* by a supernatural creator, has come about by a naturalistic, evolutionary process. After reviewing various facts showing that the neo-Darwinian version of naturalistic evolutionism seems to be almost as far from the truth as supernaturalistic creationism, I suggest that Whitehead's theistic naturalism provides resources for developing a position that combines the strengths of each of these views while avoiding their problems. The resulting view, building on the idea of "punctuated equilibria," shows how the idea of rather radical jumps, which seems demanded both by conceptual considerations and the empirical evidence, is supported by Whitehead's theistic naturalism.

A comment about the difficulty of some of the chapters: The task that Whitehead took on, that of trying to integrate the truths of science, religion, ethics, and aesthetics into a comprehensive worldview, is not easy. It is a task that most of the philosophical movements of the twentieth century, such as logical positivism, linguistic philosophy, phenomenology, existentialism, and most types of philosophy called "postmodern," have tried to avoid. But if Whitehead is right in holding that there is no shortcut to truth—and he surely is—then the task of trying to integrate the truths from our various types of experience cannot be avoided.

The understanding of the relation of science, theology, and philosophy articulated here also means that the task of theology is not easy. If theology must incorporate both science and philosophy, then "doing theology" is, despite understandable desires to the contrary, necessarily difficult. But this should be expected: The world as revealed by modern and postmodern science is exceedingly complex. The complexities are increased, furthermore, when we add the varieties of religious, ethical, and aesthetic experience. If a philosophy or theology is simple, accordingly, we can be fairly certain that it is grossly inadequate. Most of the theologies of the twentieth century, of course, tried to avoid these complexities—by basing themselves on biblical revelation alone or on one of the types of philosophy mentioned in the previous paragraph. But these theologies have proven to be inadequate, not least because they failed to show how what they were saying could be true and important, given the fact that what has passed for "the scientific worldview" seems to rule out meaningful religious and even ethical discourse altogether. And the failures of these theologies and philosophies of religion, I am convinced, are responsible to no small degree for the many failures of the religious communities in modern times. If these communities are to be more adequate in the twenty-first century to the desperate needs of both individuals and the public world, they must have a more adequate theological foundation. This book, with all of its difficulties, is my attempt to make a helpful contribution to this cause.

In dealing with these issues, I have done so with specifically *Christian* faith primarily in mind. Besides the fact that this is my own tradition, it is also the tradition in which most of the discussion of the relation between science and theology has occurred. However, most of the issues discussed in this book belong to what has traditionally been called *natural* or *philosophical* theology (or, in one meaning of the phrase, the *philosophy of religion*), so that, in spite of my Christian biases, I hope that most of my discussion will be regarded as relevant to other religions as well, especially other theistic religions.

Given the centrality of physics in most discussions of science and religion, I should perhaps mention that I had originally planned to include two chapters on the relation of physics to time: one on the relation of panexperientialism to pantemporalism and one on the relation of temporalistic theism to relativity physics. Because of limitations of space, however, both had to be omitted. Although it would have been desirable to include both or at least one of these chapters, their absence is justified by two considerations. First, I have published essays on these topics elsewhere (Griffin HG; PAP; PUST), which the interested reader can use to supplement the picture provided here. Second, this book is not a discussion of the relation of religious thought to the sciences as such—of the sort provided by Holmes Rolston's *Science and Religion*—but of its relation to the philosophical position(s) known as "scientific naturalism." It is not essential, therefore, to have discussions of all the various sciences, even one as important as physics.

2

THE MODERN CONFLICT
BETWEEN RELIGION AND
SCIENTIFIC NATURALISM

Insofar as science from the late seventeenth century through the twentieth century has been informed by the "modern scientific worldview," the relation between science and religion during this period has been characterized by increasing conflict. This has especially been the case since the latter half of the nineteenth century, when the first version of the modern scientific worldview, which combined a *mechanistic* doctrine of nature with a *dualistic* doctrine of the human being and a *supernaturalistic* doctrine of reality as a whole, was replaced by the second version, in which the dualism and supernaturalism were replaced by *materialism* and *atheism* (while the mechanistic doctrine of nature was retained). The relation between science and religion during the modern period, especially the late modern period, has primarily been conflictual.

Given the recent reaction against the "conflict thesis," my overall characterization of the modern period may seem out of date. The new and true view—one might argue by appealing to recent studies, such as David C. Lindberg and Ronald L. Numbers' *God and Nature: Historical Essays on the Encounter between Christianity and Science* (GN, 1–14) and John Hedley Brooke's *Science and Religion: Some Historical Perspectives* (SR, 1–12)—is that the idea that there has been "conflict" or "warfare" between science and religion is an exaggeration, based upon polemical, selective, and distorting readings of the evidence. The actual relation between science and religion has been far more complex, so that no simple thesis, such as the conflict thesis, can be adequate. We can no longer say, accordingly, that during the modern period the relation between science and religion has been characterized by increasing conflict.

That response, however, would involve a confusion of two distinct issues. The "conflict thesis" involves the historical claim that science and religion have *always* been (and always will be) in conflict with each other. This historical claim is usually based upon the philosophical claim, as Brooke indicates (SR, 2), that science and religion are *essentially* in conflict with each other. My thesis about the modern period, by contrast, involves neither of these claims. Philosophically, as indicated in the first chapter, I hold the position that harmony and conflict are both possible, with the actual relation being determined by how both "science" and "religion" are understood. This philosophical belief leads to the historical expectation that, insofar as one can even speak intelligibly of "science" and "religion" as two abstractions, the "relations" between them will have taken a wide variety of shapes—an expectation that is amply borne out by Brooke's discussion, which is aimed largely at showing historically the inadequacy of any of the simple theses, such as the harmony thesis or the conflict thesis (SR, 33–42, 163). Given a proper understanding of the nature of the conflict thesis, therefore, its rejection is fully consistent with my claim that the relation between science and religion in the modern period has been increasingly characterized by conflict. Indeed, Brooke himself suggests (SR, 12) that some "revisionist histories, structured around a critique of the conflict thesis, have . . . gone too far in the opposite direction," implying less conflict than there really has been. In any case, whereas the conflict thesis speaks of an essential and permanent conflict, my claim involves a conflict that is contingent and—we can hope—temporary.

A further feature of my claim should be emphasized. I said, in my opening statement, that the relation between science and religion has been increasingly characterized by conflict "[i]nsofar as science from the late seventeenth century through the twentieth century has been informed by the 'modern scientific worldview'." This qualification allows for the possibility that not all understandings of science have been thus informed throughout this period. Indeed, an essential part of my argument in this book is that a "postmodern science" began emerging in the twentieth century, and that this emergence, in circles in which it has been effective, has resulted in a new harmony, or at least convergence, of science and religion. The result is that the relation between science and religion in the twentieth century must be described in terms of two opposing trajectories: one of increasing conflict and one of convergence. This description, however, must also indicate that the conflict trajectory has thus far remained the dominant one in mainline circles, with the convergence trajectory characterizing only limited, far-from-mainline circles. Accordingly, even after recognizing this alternative trajectory, we can still say that the relation between science and religion in the modern period, understood as the period from the late seventeenth century through the twentieth century, has been characterized by increasing conflict.

The remainder of this chapter will flesh out and illustrate this claim. The first section discusses the transition from the supernaturalistic dualism of the early modern worldview to the atheistic materialism of the late modern world, which was the transition that led to "scientific naturalism" as usually understood. The second section illustrates the conflicts with religion that result from scientific naturalism thus conceived. The third section returns to the distinction between the minimum and maximal construals of scientific naturalism, this time giving a historical account.

From Dualistic Supernaturalism to Materialistic Atheism

One of the reasons there has been confusion as to the extent to which religion is in conflict with the "modern scientific worldview" has been a failure to recognize that there have been two major versions of this worldview, only the latter of which is irreconcilable with theistic religion. Insofar as this essential difference between the two versions is not understood, the radical nature of the conflict between religion and the presently dominant version of the "modern scientific worldview" may not be fully appreciated. After all, one can point out, René Descartes, Robert Boyle, Isaac Newton, and most of the other founders of the modern scientific worldview were devout Christians. And the same is true of many of the figures in the scientific community in the eighteenth century and much of the nineteenth century. There must not, one could conclude, be any essential conflict between this worldview and Christian belief.

Once we see, however, that the presently dominant version of the modern scientific worldview is different in kind from the first version, these historical facts will no longer be confusing. Indeed, the recognition that a decisive shift occurred in the middle of the nineteenth century, due in part to Darwin's influence, explains why, although it was common for elite members of the scientific community to be religious believers prior to that period, this is no longer the case. This shift involved the rejection of beliefs that were central to the first version: the belief in a soul or mind that, besides possessing free will, could survive bodily death, and the belief in a personal deity who, besides creating the world, could act within in. To understand why this shift took place, it is important to understand the precise nature of the early modern beliefs in God and the soul. At the root of the distinctive nature of these beliefs lies what was then called the "mechanical philosophy of nature."

The Mechanical Philosophy of Nature

According to this mechanical philosophy, nature was composed exhaustively of matter. For example, in speaking of the word "nature," Descartes said: "I employ this word to signify matter itself" (Collins DPN, 26). One thing connoted by calling it "matter" is that it had no "inside" in the sense of experience of any sort. Implicit in this insistence was the denial that the units of nature had any "occult" powers, meaning powers hidden from view. One implication of this denial was the denial that matter had any power of self-motion or self-determination, any power to act spontaneously. This doctrine of matter as inert meant, of course, that any movement in it had to be produced by something external to it, just as one part of a machine, such as a clock, moves only if it is moved by another part. Indeed, as Franklin Baumer has emphasized (RRS, Ch. 2), the famous clock at Strasbourg was widely used as a symbol for the universe. In technical language, the mechanistic doctrine meant that all natural causation is *efficient* causation, meaning the causal influence of one thing upon another. There is no *final* causation, in the sense of self-determination, in which something would exert inner influence upon itself. Nothing in nature, in other words, is to be explained by appealing to inner "appetites" or "purposes." The only kind of motion in nature is locomotion, the motion of a bit of matter from one place to another; there is no motion in the sense of internal becoming from a state of potentiality to a state of actuality.

Ontological Dualism

This mechanical doctrine of nature, however, did not imply a deterministic view of the human being. The reason for this is that the human being, according to this early modern view, is not just a body but a combination of body and mind or soul.[1] The mind is different in kind from nature. Whereas nature or matter is devoid of experience, the soul is conscious. Descartes, in fact, considered consciousness to be its defining characteristic. And whereas matter is devoid of spontaneity, the soul is, as Plato had said, a self-moving thing. Human freedom, therefore, could be affirmed. Furthermore, as will be discussed in Chapter 5, the doctrine of the soul as different in kind from the body was also used to reaffirm its immortality.

However, although this dualism between mind and body had these benefits, it also created problems, the most severe of which was the problem of "dualistic

---

1. Although for some purposes the difference between "mind" and "soul" (or "psyche") is important, in this work I use them interchangeably.

interaction," the interaction of wholly different kinds of things. The body, including the brain, was composed of spatially extended, impenetrable, inert bits of matter. How could it either influence or be influenced by a purely spiritual substance, which, far from being impenetrable, is not even spatially extended? That mind and body did interact was, of course, presupposed: The idea that we are responsible for our bodily actions presupposes not only that the mind is a self-determining reality but also that its self-determining activity guides the body's behavior. And the idea that the body influences the mind in return is presupposed by the natural interpretation of various features of our experience, such as the belief that we perceive the outer world by means of our physical senses. And yet the dualists could not explain how mind and body could interact. More precisely, insofar as they could explain how mind and body could interact—or at least *appear* to interact—they could do so only by appealing to God: The divine omnipotence could make possible the otherwise impossible. Mind-body dualism, in other words, required supernaturalistic theism.

Supernaturalistic Theism

As used here, the expression "supernaturalistic theism" is not redundant. "Theism" is belief in a personal deity who created the world and is active in it. The adjective "supernaturalistic" adds the further idea that the divine being's activity in the world is in no way constrained by any principles, except perhaps purely logical ones. There are, in particular, no basic causal principles, simply belonging to the nature of things, that God cannot interrupt. The reason why this "supernatural" intervention can occur is that what we commonly call "nature" does not in fact exist "naturally." It exists, instead, purely contingently, having been created *ex nihilo*. Because our world was created from absolute nothingness, it contains no principles that, being co-eternal with God, cannot be interrupted.

To illustrate this supernaturalistic theism, which was affirmed in a most uncompromising form in the sixteenth century by John Calvin, we can turn to the nineteenth-century Calvinist Charles Hodge, who said in response to the question as to how God relates to the laws of nature:

> The answer to that question . . . is, First, that He is their author. He endowed matter with these forces. . . . Secondly, He is independent of them. He can change, annihilate, or suspend them at pleasure. He can operate with or without them. "The Reign of Law" must not be made to extend over Him who made the laws. (ST, I: 607)

The connection between creation *ex nihilo* and the possibility of miraculous interventions is stated even more clearly by contemporary Calvinist Millard Erickson, who says:

> He [God] has created everything else that is . . . by bringing it all into existence without the use of preexisting materials. . . . Nature . . . is under God's control; and while it ordinarily functions in uniform and predictable ways in obedience to the laws he has structured into it, he can and does also act within it in ways which contravene these normal patterns (miracles). (CT, 54)

Given this understanding of the nature of divine power, all things are possible to God except the logically impossible, such as creating round squares. The mere fact that mind and body have nothing in common could not prevent divine omnipotence from creating the world so that the movements in one would be correlated with the movements in the other. This understanding of divine power also meant, as the statements by Hodge and Erickson show, that extraordinary events could be interpreted as divine interruptions of the usual laws of nature.

Indeed, given another feature of the mechanistic view of nature, with which this supernaturalistic theism was combined, a supernaturalistic interpretation of certain types events was not simply permitted but even required. That is, besides restricting causation to efficient causation, the mechanical philosophy of nature insisted that all efficient causation is *by contact,* just as, again, it is in a machine. The denial implicit in this doctrine—the denial of "action at a distance"—was, in fact, arguably the central point of the mechanistic doctrine (as will be suggested in Chapter 5). In any case, this denial implied that, if the best naturalistic interpretation of a given event would involve action at a distance, such as telepathic influence or psychokinesis, the event must be explained by means of supernatural causation. All events are, of course, supernaturally caused in the sense that God is the "primary cause" of all events. God has chosen, however, to cause most events through "secondary" or "natural" causes. But a few events occur without natural causes, so they must uniquely be explained by reference to God's causation. The supernaturalism of the first version of the modern scientific worldview, accordingly, was required not only by its mind-body dualism but also by its acceptance of miraculous interventions.

From Supernaturalistic Theism to Deism to Atheism

In spite of the fact that this dualistic supernaturalism was established as the framework for the scientific community only through a difficult struggle with two alternative worldviews, and in spite of the fact that there were many reasons for its victory over them (two points that will be developed in Chapter 5), it began to unravel almost as soon as it was established, due to various problems. One was that supernaturalism seemed to be contradicted by the world's evil, especially natural evil, such as the Lisbon earthquake of 1755. The problem of evil had, of course, always created difficulties for supernaturalism, which implies that any given evil in the world could have been prevented by God. By portraying the natural world as wholly passive, however, the mechanistic doctrine of matter had made the problem of natural evil even more obvious. This problem was one of the factors leading many eighteenth-century thinkers, such as Voltaire, to reject supernaturalistic theism in favor of deism, according to which God, after creating the world and establishing its laws, no longer acts in it.

Another factor behind this move to deism was the growing distaste for the idea of supernatural interruptions of the laws of nature. Although some thinkers retained the belief that God, having created the laws of nature, could freely interrupt them, a growing number came to regard such interventions as unseemly. One argument was that God, being all-wise as well as all-powerful, would have ordered the world so well at the creation that no interventions would be needed. Another factor was the growing sense that all events are so intertwined in the universal causal nexus that occasional interruptions of it here and there are inconceivable. As Darwin said in his *Autobiography,* "the more we know of the fixed laws of nature the more incredible do miracles become" (Brooke SR, 271). The perception of all events as governed by immutable laws led, furthermore, to the growing disbelief that there really had been any events, even in biblical times, that required a supernatural explanation. Stories about such events were increasingly interpreted as symbolic myths or pious frauds.

In any case, the deistic view was still a form of supernaturalism, insofar as it assumed the world to have been created *ex nihilo,* which implies that God (metaphysically) *could* intervene in the world. In another sense, however, it was a naturalistic doctrine, affirming that *in fact* no such interventions occur, so that *all* events have natural causes. In this way, a supernaturalistic theology could be joined with an almost purely naturalistic science, the only exception being the assumption that the origin of the universe requires a supernaturalistic explanation. However, the doctrine that the world was designed by an omnipotent creator still created a problem of evil. Indeed, the idea that God had programmed the world in advance to evolve just as it did, with all the pain and destruction involved in "natural selection," served, if anything, to *increase* the problem of evil. Also, the leaders of the scientific community became increasingly reluctant to allow for *any* explanations that posited causal forces differing in kind from the kinds of causal forces now operating, even to explain the origin of our universe. This deistic compromise between naturalism and supernaturalism, accordingly, proved primarily to be a halfway house to complete atheism. (The problems in the deistic compromise will be treated further in Chapter 3.)

From Dualism to Epiphenomenalism to Materialism

Besides this transition from supernaturalism to atheism via deism, the shift from the first to the second version of the modern scientific worldview also involved a shift from dualism to materialism via epiphenomenalism. One reason for the decline of dualism was the fact that it depended upon supernaturalism for its intelligibility. The growing reluctance to appeal to God to explain apparent anomalies led to a growing dissatisfaction with dualism. Another reason was the growing belief in the lawfulness of all events. The idea that all the behavior of matter is governed by laws of nature, combined with the recognition that the human body is comprised of the same material elements as everything else, led to the conclusion that all bodily activities must be as law-governed as the events in the laboratory or the interactions of billiard balls. This conclusion contradicted the position of dualists, according to which the mind, besides being influenced by the body, is also able to initiate activity of its own and then influence the body. For example, believing that the very nature of "science" is to describe all happenings in terms of laws, Darwin said that this doctrine, by allowing the mind to introduce "caprice" into the world, would make science impossible.

Many scientists, therefore, adopted epiphenomenalism, according to which the mind is a nonefficacious by-product of the brain. Thus conceived, the mind is real, in the sense of being a center of experience distinct from the brain. But it has no power, at least no power to exert influence back upon the body. In this way, one could keep one's determinism without losing one's mind. But this compromise was problematic. For one thing, epiphenomenalists still could not explain how mind, with its experience, could have arisen naturally out of nonexperiencing matter. Also, it was completely arbitrary to stipulate that, although matter could affect mind, mind could not affect matter (as some epiphenomenalists admit [Campbell BM, 131]). Like deism, therefore, epiphenomenalism turned out to be primarily a halfway house. The dominant view in scientific and science-based philosophical circles has for some time been materialistic identism, according to which the mind is, in *some* (perhaps unspecifiable) sense, identical with the brain. In either case, the human mind is said to have no autonomy in relation to the body. Accordingly, human behavior, including human experience, is said to be entirely determined by purely physical forces.

THE RESULTING CONFLICT BETWEEN RELIGION AND SCIENTIFIC NATURALISM

The previous section traced, very briefly and schematically, the transition from the worldview with which science was primarily associated at the end of the seventeenth century to the one with which it was primarily associated by the end of the nineteenth century. Although sometimes lumped together as the "mechanistic worldview" or the "modern scientific worldview," these two worldviews have more in conflict than they have in common. They do share the mechanistic view of nature. In the first worldview, however, this mechanistically understood nature is subordinate to both the human soul and the Divine Creator. Indeed, as we will see in Chapter 5, the mechanistic account of nature was adopted largely to protect the reality of both God and the soul. In the second version of the mechanistic worldview, God and the soul have been eliminated. The attempt is made to understand everything, including the origin of the universe and human consciousness and action, mechanistically. In the light of these enormous differences between the first and the second versions of the "modern scientific worldview," it is no wonder that, whereas the first one resulted primarily in a continuation of the perception that science and religion are in harmony, the latter one resulted primarily in the perception of science and religion as in conflict.

To illustrate this perception, we can begin with Bertrand Russell's 1902 essay, "A Free Man's Worship," in which he said that the presuppositions of religious belief and practice are undermined by "the world which Science presents for our belief." In this essay's purple passage, Russell declared:

> That Man is the product of causes which had no prevision of the end they were achieving; that his origin, his growth, his hopes and fears, his loves and beliefs, are but the outcome of accidental collocations of atoms; . . . that all the labours of the ages, all the devotion, all the inspiration, all the noonday brightness of human genius, are destined to extinction in the vast death of the solar system. . .—all these things, if not quite beyond dispute, are yet so nearly certain, that no philosophy which rejects them can hope to stand. (BW, 67)

A science-based philosophy, in other words, cannot affirm divine providence, human freedom, or any form of immortality.

A similar view was expressed by Joseph Wood Krutch in 1929 in a book titled *The Modern Temper.* In a new preface to this book written in 1956, Krutch summarized its thesis (which he no longer accepted) in these terms:

> The universe revealed by science . . . is one in which the human spirit cannot find a comfortable home. That spirit breathes freely only in a universe where what philosophers call Value Judgments are of supreme importance. It needs to believe, for instance, that right and wrong are real, that Love is more than a biological function, that the human mind is capable of reason . . . , and that it has the power to will and to choose instead of being compelled merely to react in the fashion predetermined by its conditioning. [But] science has proved that none of these beliefs is more than a delusion. (MT, xi)

This view, that science rules out a religious interpretation of the universe, was reiterated by Jacques Monod in 1972. Sounding very much like Krutch and Russell, he said that "science outrages values" and that, if "the essential message of science" is accepted, then man must at last wake out of his millenary dream; and in doing so, wake to his total solitude, his fundamental isolation. Now does he at last realize that, like a gypsy, he lives on the boundary of an alien world. A world that is deaf to his music, just as indifferent to his hopes as it is to his suffering or his crimes. (CN, 172–73)

The basis for this bleak view is Monod's contention that science is based on the postulate of "objectivity," understood to be "the *systematic* denial that 'true' knowledge can be got at by interpreting phenomena in terms of final causes— that is to say, of 'purpose'" (CN, 21). Although Monod realizes that "animism," understood broadly to include any doctrine that interprets the world in terms of purposes, is essential if we are to feel at home in the universe, he believes that we must reject the "ancient animist covenant between man and nature" because all animist systems are "fundamentally *hostile* to science" (CN, 170, 171).

A similar view of purposes was held by B. F. Skinner, who spoke of "traditional 'knowledge,'" which must be corrected or displaced by a scientific analysis." The traditional idea Skinner primarily had in mind is the idea that people have freedom. "The hypothesis that man is not free," said Skinner in *Science and Human Behavior*, "is essential to the application of scientific method to the study of human behavior" (SHB, 6, 447). Continuing this theme in a later book, *Beyond Freedom and Dignity*, Skinner argued that to be "natural" is to be completely determined by one's environment. The notion of the "autonomous," which "initiates, originates and creates," he said, is the notion of the "miraculous" (BFD, 12, 191).

This idea that science rules out freedom is also articulated by a contemporary "psychobiologist," William Uttal, who says that a science of the mind requires the assumption that mind is reducible to matter, meaning that the necessary and sufficient conditions for all psychological properties are contained in physiological processes, so that the state of the mind is "totally dependent on the arrangement of the matter" (PM, 10, 82). This reductionistic position, Uttal points out, "conflicts in a fundamental way with the predominant theologies of our society," including "the nearly universal acceptance of the idea that the mind directly affects the body" (PM, 5, 28).

To move from psychobiology to sociobiology: Edward O. Wilson, speaking of "the collision between irresistible scientific materialism and immovable religious faith" (OHN, 179), leaves no doubt about which one is to give:

[M]ake no mistake about the power of scientific materialism. It presents the human mind with an alternative mythology that until now has always, point for point in zones of conflict, defeated traditional religion. . . . Every part of existence is considered to be obedient to physical laws requiring no external control. The scientist's devotion to parsimony in explanation excludes the divine spirit and other extraneous agents. Most importantly, . . .

the final decisive edge enjoyed by scientific naturalism will come from its capacity to explain traditional religion, its chief competitor, as a wholly material phenomenon. Theology is not likely to survive as an independent intellectual discipline. (OHN, 200–01)

Wilson's statement, in using "scientific naturalism" interchangeably with "scientific materialism," brings out explicitly the equation that is presupposed by the other authors.

Taken together, these quotations articulate various dimensions of this form of scientific naturalism, with its atheism and reductionistic materialism, which together imply that values and purposes play no role in the universe in general and human beings in particular, that our moral ideas have no basis in reality, that everything is determined by mechanical causes, that there is no life after death, and that life has no ultimate meaning of any kind. All of these points have been made in a recent statement by William Provine, a historian of science, especially Darwinism. Provine says:

[M]odern evolutionary biology. . . tells us . . . that nature has no detectable purposive forces of any kind. . . . Modern science directly implies that the world is organized strictly in accordance with deterministic principles or chance. . . . There are no purposive principles whatsoever in nature. There are no gods and no designing forces that are rationally detectable. . . .

Second, modern science directly implies that there are no inherent moral or ethical laws. . . .

Third, human beings are marvelously complex machines. The individual human becomes an ethical person by means of only two mechanisms: deterministic heredity interacting with deterministic environmental influences. That is all there is.

Fourth, we must conclude that when we die, we die and that is the end of us. . . . There is no hope of life everlasting. . . .

[F]ree will, as traditionally conceived, the freedom to make uncoerced and unpredictable choices among alternative possible course of action, simply does not exist. . . . [T]he evolutionary process cannot produce a being that is truly free to make choices. . . .

The universe cares nothing for us. . . . Humans are as nothing even in the evolutionary process on earth. . . . There is no ultimate meaning for humans. (PE, 64–66, 70)

If this is how "scientific naturalism" is understood, it would be hard to maintain that it is compatible with a significantly religious view of reality. It certainly is not compatible with historic Christian faith or anything that could credibly claim to be in significant continuity with it. The hope for a harmony between religion and scientific naturalism can only be fulfilled, therefore, if scientific naturalism can be understood in a less restrictive way.

THE HISTORICAL RISE OF MINIMAL SCIENTIFIC NATURALISM

I had introduced such an understanding in Chapter 1, arguing that naturalism should be understood, as the term itself suggests, simply as the denial of supernaturalism, with "supernaturalism" understood as the belief in a supernatural being who can and perhaps does interrupt the basic causal processes of the world.[2] In contrast with naturalism$_{sam}$, our term for the sensationist-atheistic-materialistic doctrine constituting the maximal version of naturalism, naturalism in this minimal sense could be called naturalism$_{ns}$ (for nonsupernaturalist). Why has naturalism in this restricted sense, being limited to the denial of the possibility of supernatural interruptions, *not* been understood to constitute a sufficient as well as a necessary condition for "scientific naturalism"? This section provides a historical explanation of why naturalism$_{ns}$ came wrapped in naturalism$_{sam}$.

Although, as we have seen, the transition from the first to the second versions of the modern scientific worldview involved other issues, the shift to a worldview in which there were no interruptions of the universal causal nexus was the central issue. This shift was presupposed, for example, in David Friedrich Strauss's epochal *Life of Jesus,* the first edition of which appeared in 1835. The presuppositions of the modern world, Strauss said, include the conviction that "all things are linked together by a chain of causes and effects, which suffers no interruption" (LJ, 78). This presupposition had already been expressed by Friedrich Schleiermacher, usually called the "father of modern liberal theology." For example, in *The Christian Faith,* which first appeared in 1821, Schleiermacher said: "It can never be necessary in the interest of religion so to interpret a fact that its dependence on God absolutely excludes its being conditioned by the system of Nature" (CF, 178).

Although Strauss was able to abide by this dictum, the significance and the difficulty of making this shift can be illustrated by the fact that Schleiermacher made an exception with regard to the origin of the life of Jesus (CF, 398–415), for which he was criticized by Strauss (LJ, 771). It can also be illustrated by Charles Lyell, who was famous for his doctrine of uniformitarianism. According to the ontological dimension of this doctrine (as distinct from the *geological*

---

2. Supernaturalism in the fullest sense says that such interruptions actually occur. However, it is also supernaturalism, even if in a restricted sense, to affirm the existence of a divine being that *could* produce such interruptions but chooses not to do so—as does, for example, Peter Forrest in *God Without the Supernatural.*

dimension, to be mentioned in Chapter 8), all things are to be explained in terms of the same set of causal factors. This means, in particular, that causal factors not observable today are not to be used to explain events in previous ages. Lyell was often vilified for this principle by supernaturalists, because it implied the denial of occasional divine interventions. And yet Lyell himself violated this principle with regard to the human mind, saying that to explain its origin we must "assume a primeval creative power which does not act with uniformity." Divine intervention, Lyell maintained, added "the moral and intellectual faculties of the human race, to a system of nature which had gone on for millions of years without the intervention of any analogous cause" (Hooykaas NL, 114). Lyell was representative of many thinkers in the early part of the nineteenth century. As virtual deists, they for the most part assumed that all events could in principle be understood naturalistically. But they did allow for a few exceptions, such as the origin of life, the origin of the human mind, and perhaps the origin and resurrection of Jesus—exceptions that were allowed by their first exception, the origin of the universe.

A crucial figure in the transition from virtual to complete deism was Charles Darwin. In a letter to Lyell, Darwin rejected the idea of any divine intervention to explain the origin of the human mind, saying: "If I were convinced that I required such additions to the theory of natural selection, I would reject it as rubbish. . . . I would give nothing for the theory of Natural selection, if it requires miraculous additions at any one stage of descent" (DLL, II: 6–7).

In fact, the main reason that Darwin's theory was so exciting to some, so threatening to others, was that it promised a fully naturalistic explanation of all features of the world, including those traits traditionally assumed to separate human beings, as created in the image of God, from beasts. Darwin himself remained a deist, believing that the origin of the universe could only be explained in terms of a supernatural creator. Later scientists and science-based philosophers, however, have assumed that even this event could be explained naturalistically (perhaps by applying the Darwinian principle of the "survival of the fit" to spontaneously arising universes). With this shift, the scientific community moved to a thoroughgoing naturalism: Besides the fact that the origin of the universe is no longer explained by reference to a supernatural being, it is also assumed that there *is* no supernatural being who could, in principle, intervene in the world's causal processes.

In our time, this scientific naturalism, according to which the web of natural causation is inviolable, has become virtually axiomatic in science-related circles: It is taken to be both presupposed by the scientific enterprise and ever more confirmed by scientific discoveries, which have, time after time, removed gaps in our knowledge of natural causes that might have earlier seemed to point to occasional supernatural injections. The "God of the gaps" has become otiose in science. Whatever else "naturalism" may be thought to involve, it always is taken to involve the denial of such interventions. For example, philosopher Sterling Lamprecht, in an essay on "Naturalism and Religion," says that naturalism sees the world as "an interrelated whole without intrusions from some other 'realm'" (Krikorian NHS, 20). Likewise, in a recent book titled *Religion, Science and Naturalism,* Willem Drees says that naturalism rejects the belief "that God intervenes occasionally in the natural world" (RSN, 14).

The Widespread Equation of Naturalism with Naturalism$_{sam}$

My suggestion is that, if this rejection of supernaturalism were taken as not only necessary but also sufficient for defining scientific naturalism—which would result in what I have called "naturalism in the minimal sense," or "naturalism$_{ns}$"— scientific naturalism might turn out to compatible with a religious interpretation of the world after all. Historically, however, scientific naturalism has hardly ever been understood in this minimal way. It has almost always been understood in a maximal sense, according to which it entails, besides the rejection of any divine interruption of the world's basic causal processes, also materialism, reductionism, determinism, sensationism, the rejection of objective moral norms, and the rejection of divine influence of *any* sort. In the previous section, I quoted various advocates of scientific naturalism who have understood it in this way.

This maximal understanding of scientific naturalism is also held by many thinkers who, in the name of religion, reject it. For example, in *Darwin on Trial,* attorney Phillip Johnson says: "For present purposes, the following terms may all be considered equivalent: scientific naturalism, evolutionary naturalism, scientific materialism, and scientism" (DT, 116n). For Johnson, therefore, the naturalist's assumption that the world is closed to any supernatural interruptions of its normal causal processes is stated as the assumption that the world is "a closed system of *material* causes" (DT, 116; emphasis added). Although for Johnson the essence of naturalism is the "absence from the cosmos of any Creator" (DT, 117), he also, to show what scientific naturalism is, cites (DT, 126) the much more sweeping statement, quoted earlier, by William Provine.

Historians also typically accept the equation of scientific naturalism with a materialistic, atheistic worldview. For example, Peter Bowler, explaining why theistic evolution was no longer taken seriously in the scientific community by the end of the nineteenth century, says that this community had accepted the Darwinian principle "that evolution was to be explained solely in naturalistic terms, leaving no room for the supernatural" (ED, 15). He accepts the idea, in other words, that to affirm theistic influence in the evolutionary process would necessarily be to affirm supernaturalism. Likewise, Colin Russell, in discussing "Victorian Scientific Naturalism," characterizes it as "the view that nature's activity can be interpreted without recourse to God, spirits, etc., i.e., without bringing in the *super*natural" (SSC, 256). Russell thereby seems to accept the contention that to speak of divine influence in the world would *ipso facto* be to affirm supernaturalism.

The equation of naturalism with atheism, which leads to the equation of theism of any sort with supernaturalism, is commonplace not only in scientific and history-of-science circles but also in the humanities, as illustrated by J. Samuel Preus's 1987 book, *Explaining Religion.* Preus contrasts a "naturalistic" explanation of the existence of religion with "theological" and "religious" interpretations (ER, ix, xi), thereby implying that theological and religious interpretations are necessarily supernaturalistic. Indeed, he explicitly equates the idea that religion has "some objective, transcendent ground" with the idea that it has a "supernaturalistic ground" (ER, xvi); and he equates "the naturalistic paradigm" with "an altogether nonreligious point of view" (ER, xiv). Therefore, he says, his naturalistic approach "takes theological interpretations seriously as part of the religious data but not of their explanation" (ER, 211). For Preus, in other words, naturalism is not simply the rejection of divine intrusions into the world that interrupt the basic causal processes of the universe operating in most events. Naturalism is taken to include the rejection of all explanations of religious experience in terms of a genuine experience of a truly existing Holy Reality, whether or not this Holy Reality be thought capable of supernatural interruptions of the world's basic causal processes. Or, to put the point in other terms, naturalism is understood to include the denial that these basic causal processes include influences from a Holy Reality. This is because Preus, like almost everyone else, identifies naturalism as such with naturalism$_{sam}$, with its sensationism, atheism, and materialism. (The fact that Preus's nickname is "Sam" is, of course, purely coincidental.)

Preus's book illustrates especially well why the linguistic reform I am suggesting is so important. The response of many readers, I assume, will be that the term "naturalism" has for so long been identified with naturalism$_{sam}$ that the attempt to break this identification, so that naturalism would be compatible with theistic religion, is futile. It would be better, they would counsel, to find some other term for my view, thereby simply accepting the fact that the term "naturalism" is beyond redemption. To accept this counsel, however, would be disastrous. If we accept the equation of naturalism with atheism, then a theistic position must, by definition, be called "supernaturalistic." And much more unbreakable than the association between "naturalism" and "atheism," surely, is the association between "supernaturalism" and "supernatural interruptionism." Belief in theism, accordingly, would almost inevitably connote rejection of naturalism$_{ns}$, which is the fundamental ontological belief of the scientific community. The hope for a worldview that could be shared in common by the scientific and the religious communities would be doomed on purely linguistic grounds. The linguistic reform proposed here, which would reidentify "scientific naturalism" with naturalism$_{ns}$ instead of naturalism$_{sam}$, is, for this reason, an integral part of this book's proposal to overcome the apparent conflict between science and religion.

Although the main negative import of naturalism, construed maximally, is the denial of divine influence in the world, it has also typically been taken to exclude even more beliefs generally associated with religion. As the quotation above from Colin Russell shows, naturalistic explanations are thought to exclude reference not only to divine influence but also to "spirits." This exclusion most obviously applies to the disembodied spirits of spiritualists and psychical researchers. But it also applies to an *embodied* spirit, in the sense of a mind or soul sufficiently distinct from the body to be able to act with some autonomy. Given the view that the natural web of causation operates deterministically, "naturalizing the mind" means that it, too, must be conceived as fully determined by the causal forces acting upon it. For Victorian scientific naturalists such as Charles Darwin and John Tyndall, science required the absolute reign of law, thereby the rejection of any "caprice" in nature, whether from divine action, disembodied spirits, or human free will (Gillespie CD, 139, 147–51). This was one of the main reasons for the transition from dualistic interactionism to materialism or at least epiphenomenalism.

Another basis for this transition was the equation of *natural causation* with *causation by material bodies.* As historian Neal Gillespie puts it, this form of naturalism centered around "the belief that all events are part of an inviolable web of natural, even material, causation" (CD, 148). This characterization supports Phillip Johnson's statement, quoted above, that naturalism assumes the world to be "a closed system of *material* causes" (DT, 116). This widespread equation of naturalism with materialism or physicalism continues in our time. For example, philosopher David Papineau, in a 1993 book titled *Philosophical Naturalism,* says that he had originally intended to call the book "Philosophical Physicalism" (PN, 1).

Historical Reasons for the Equation of Naturalism with Naturalismsam

Given the acceptance of the maximal construal of naturalism, according to which it involves materialism and thereby identism and atheism, scientific naturalism cannot be reconciled with religion, especially theistic religion. But why should we continue to accept this maximal construal? One possible answer would be that science requires it. That answer, however, can be countered both historically and systematically. The historical argument, which is now commonplace among historians of science, is that "good science" has been carried out from within other frameworks, such as the Neoplatonic-magical-spiritualist tradition to be discussed in Chapter 5 and the recent holism of the physicist David Bohm. The systematic argument is that, far from presupposing maximal naturalism, scientific activity actually presupposes the falsehood of most of its elements—for example, its determinism and its rejection of nonphysical entities, such as mathematical objects, logical principles, laws, and objective norms. This argument will be made in Part II of this book. A second reason for continuing to equate scientific naturalism with the worldview of materialistic reductionism might be that this worldview is true. However, the arguments to which I have just referred, according to which scientific activity necessarily presupposes the falsity of the distinctive elements of this worldview, is simultaneously an argument for its actual falsity.

To understand the real reason for the equation of scientific naturalism with atheistic, mechanistic materialism, we need to understand the previous positions, including the motives for them, from which the currently dominant scientific worldview derived. In Chapter 5, I will show that science became associated with the worldview of dualistic supernaturalism in the seventeenth century largely because of theological motives. There is, however, a still more fundamental theological notion lying at the root of our current impasse, this being the scheme of primary and secondary causation.

The Scheme of Primary and Secondary Causation

According to this scheme, God is the primary cause of all events. Being omnipotent, God is the *sufficient* cause of all events; no supplementation by other causes is needed. However, God usually does not bring events about unilaterally but through secondary causes, sometimes called "natural causes." These secondary or natural causes are *also* sufficient causes insofar as attention is focused on the *nature* of the events that are brought about, on their *whatness*. If, by contrast, the question of the *thatness* of the events is raised—that is, why a world, with its chains of causal events, exists at all—one must then refer to God, as the creator of the whole universe. However, if one ignores this more radical question, of why there is anything rather than nothing, one need not refer to God to give a complete account of most events. Divine influence, therefore, is usually not constitutive of the nature or whatness of worldly things and occurrences. For the usual course of events, therefore, a science limited to secondary or natural causes can in principle be complete. As long as the question of why something exists at all is not raised, a sufficient explanation of what something is and why something happened can be given, at least in principle, without referring to divine influence.

This scheme has had tremendous appeal to both theologians and scientists. On the one hand, it has allowed theologians to defend the uniqueness of divine causation. Arguing that divine causation occurs on a "different level" from natural causation, they could say that God is not one natural cause among others, that divine causation is not to be treated as simply one more type of finite causation.

On the other hand, this scheme has allowed scientists to ignore theological ideas in developing their detailed accounts of how the world works. This scheme, accordingly, has allowed theologians and scientists to operate autonomously, with each protected from possible conflict with, and thereby interference from, the other.

There was one snag, however. The doctrine was only that God *usually* brings about effects through secondary or natural causes. In a few cases, God was said to bring about the effects directly, without using secondary causes as instruments. The most obvious of such cases were the miracles in the Bible and the lives of the saints. Indeed, a miracle was *defined* as an event brought about directly by God, without the employment of secondary causes. But the list of events produced directly by God's primary causation also included events not usually classified as miracles, such as the creation of the world, of life, of the various species of plants and animals, and of the human soul, the inspiration of the prophets and the biblical writers, and the incarnation in Jesus. Given all these events without natural causes, there could not really be a fully naturalistic science of either nature or history.

This was so because of the most fateful implication of the primary-secondary scheme for thinking about the relation between divine and worldly causation. According to this scheme, to speak of any divine influence that is constitutive of the nature of events in the world was *ipso facto* to speak of a supernatural interruption of the chains of natural causes and effects. Because most events were said to be fully explicable, at least in principle, without reference to divine influence (except for the *invariable* sustaining influence of God), the affirmation that *variable* divine influence played a constitutive role in particular events implied saying that this divine influence interrupted the causal principles involved in most events. For example, to say that the prophet Jeremiah's experience of God's call was constitutive of his experience was to say that God's causal influence on Jeremiah in those moments was different in kind from God's causal influence in most human experience. Likewise, if we say that the creation of life involved divine causation beyond the undifferentiated divine influence that sustains all beings in existence, we have to say that the creation of life involved an interruption of the causal principles involved in most events.

To make this point clearer, we can contrast this way of thinking of the relation between divine and worldly causation with another possible way that might have been adopted. Theologians might have said that worldly causation is never fully sufficient to account for the whatness of any event—that divine influence, besides sustaining the world in existence, also plays a constitutive role in everything. Given that perspective, it would be natural to assume that this constitutive divine influence is variable, differing both in content and effectiveness from event to event. And yet this variability in content and effectiveness could be affirmed while holding that, *formally* speaking, the divine influence is always the same. One would then be able to speak of some events as special acts of God while assuming that divine causation never interrupts the (formal) causal principles that ordinarily prevail. Variable constitutive divine influence would be understood as *part* of the normal pattern of causes and effects, *not* an interruption of this pattern. Such a position could be called "naturalistic theism," or "theistic naturalism." It would be naturalistic, because it would reject the idea of any supernatural interruptions. But it would be theistic, because it would regard variable divine causation as constitutive of natural and historical processes. Although this variable divine influence would be especially significant in some events, making them special acts of God, it would be a causal factor in *all* events, so that these special acts would not involve supernatural interruptions of the normal causal pattern. It is this alternative framework, in its Whiteheadian version, that will be employed in the present book.

Had some such framework been adopted by the seventeenth-century founders of modern science, the history of the relation between science and religion would have been very different. But these founders, such as Galileo, Mersenne, Descartes, Boyle, and Newton, retained the primary-secondary framework, in which divine influence is normally not constitutive of either natural processes or human experience. Indeed, they adopted a set of further assumptions that served to make the exclusion of any constitutive divine influence in the world seem self-evident.

Modern Assumptions Excluding Constitutive Divine Influence

These further assumptions were the mechanical doctrine of nature and the sensationist theory of perception. The *mechanical doctrine of nature*, which said that the physical world is composed of particles that are, in Newton's words, "massy, hard, impenetrable," implied that natural things have no "inside" in which divine influence could work. This mechanical doctrine also entailed that all natural causation involved contact between physical particles. This view, which made *any* influence of mind on body unintelligible, meant that influence of the divine mind on the physical world, if affirmed, would have to be regarded as supernatural. The *sensationist theory of perception* said that the human mind could perceive only those things that could excite its body's physical sense organs, which means that we can perceive only physical things. We cannot naturally, therefore, perceive other minds, including a divine mind.

Thanks to this mechanistic, sensationist viewpoint adopted by modern thought, the primary-secondary framework came to be used much more rigorously than in previous times to deny to divine influence any constitutive role in the ordinary course of events. As we will see in Chapter 5, this mechanistic, sensationistic position was originally adopted to rule out various forms of naturalism, including various forms of theistic naturalism, in favor of a wholly supernaturalistic understanding of the universe in general and of Christianity's miracles and revelation in particular.

However, those imbued with the spirit of the scientific enterprise increasingly refused to accept supernatural interruptions of the natural causal nexus. Given the framework of ideas that had become prevalent, this refusal meant the total exclusion of divine influence from the world. The primary-secondary scheme for thinking of the relation between divine and worldly causation, especially as reinforced by the sensationist theory of perception and the mechanistic understanding of the physical world, implied that divine influence could not be part of the natural processes of the world. If there were any constitutive divine influence in the world, therefore, it would have to be an exception to the normal course of affairs. In this way, the denial of supernaturalistic interruptions led to the complete denial of any variable divine influence in the world whatsoever. The only possible positions, accordingly, were pantheism, atheism, or the deistic idea that, after creating the universe, God did *all* things through secondary or natural causes. As we have seen, this deistic idea soon collapsed into complete atheism, which further entailed the collapse of dualism into materialism, with its reductionism and determinism.

A Way Beyond Conflict

The resulting framework is usually assumed in discussions of the conflict between science and religion. Accepting this framework, those who argue for the continued validity of religious beliefs must either (1) redefine these religious beliefs so radically that they do not conflict with this materialistic version of scientific naturalism, (2) urge a return to the early modern view, with its supernatural incursions into the world, or (3) recommend the deistic compromise, according to which the supernatural creator has so perfectly programmed the system of secondary (or natural) causes to carry out the divine purposes that no interruptions are needed. None of these strategies, I argue in the next chapter, has any hope for success in overcoming the perceived conflicts between science and religion.

My positive argument is that the perceived conflict between science and religion can only be overcome if we reject the maximal construal of scientific naturalism while retaining the minimal construal. Given the historical development I have sketched, it was probably almost inevitable that the nineteenth-century rejection of supernaturalism seemed to be part of a package deal, simply one feature of the transition to a wholly atheistic, sensationistic, mechanistic, reductionistic, deterministic approach. Given the inherited framework, the only way to have minimal naturalism, it seemed, was to adopt maximal naturalism. For example, given the assumption that the behavior of matter is rigidly determined by laws of nature, to deny divine interruption of those laws was to deny divine influence in nature altogether. The equation of these two denials can be seen in the contemporary science writer Isaac Asimov. Describing the historical period in question in his book *In the Beginning,* he says:

> Scientists grew increasingly reluctant to suppose that the workings of the laws of nature were ever interfered with. . . . Certainly, no such interference was ever observed, and the tales of such interferences in the past came to seem increasingly dubious. In short, the scientific view sees the Universe as following its own rules blindly, without either interference or direction. (IB, 11)

As the last sentence illustrates, Asimov assumes that the absence of divine "interference" means that there is no divine "direction" of any sort, so that everything happens "blindly."

The same assumption about matter—that its movement is rigidly determined by laws of nature—also entailed that the movement of matter in the human body could not be diverted by any self-determination, initiative, or "caprice" of the mind. Rather, it was generally assumed, the mind must be totally determined by the motions of the matter in the brain. This assumption meant that, if there was a divine being, this being, besides not being able to influence "nature," also could not influence human experience and thereby "history." The idea of prescriptive, inviolable laws of nature, accordingly, led to a naturalism in which there could be no divine or even distinctively human influence in the world.

However, although the adoption by the scientific community of the atheistic, reductionistic kind of naturalism was perhaps virtually inevitable, another kind of naturalism *could* have developed. Philosophers, theologians, scientists, and historians of science might have realized that the very notion of matter as "obedient" to "laws of nature" presupposes the idea of a cosmic lawmaker. They could have, for example, come to this realization through study of the writings of Boyle, Newton, and the other supernaturalists who founded the early modern worldview (with its "legal-mechanical" doctrine of nature, as it will be called in Chapter 5). In any case, reflection upon this connection could have led those wanting a thoroughgoing naturalism to reject the idea of imposed laws of nature, which would have led to quite a different kind of naturalism. Such a naturalism, in which the so-called laws of nature are really its most long-standing *habits,* was, in fact, articulated near the end of the nineteenth century, most notably by Charles Peirce and William James. It is this kind of naturalism, as developed by Alfred North Whitehead, that lies behind the present book.

This alternative naturalism, however, has thus far remained a minority position. The majority position has retained, in effect, the doctrine of imposed laws of nature, even though this means, in a nontheistic framework, retaining the imposition while rejecting the Imposer. This implicitly self-contradictory position lies at the root of the widespread belief, shared by religious and anti-religious thinkers alike, that religion and scientific naturalism are necessarily in conflict.[3] An essential condition for overcoming this perceived conflict is to adopt a different kind of naturalism, one that can be considered distinctively postmodern, because it rejects the distinctively modern assumptions—the sensationist doctrine of perception, the mechanistic doctrine of nature, and the resulting exclusion of divine presence from the world—that led to naturalism$_{sam}$. This, at least, is my central thesis. Thinkers with different perspectives, of course, have suggested other solutions. Before laying out Whitehead's naturalism, which I propose as the basis for a new harmony between science and religion, I will, in the next chapter, examine three more customary approaches to effecting this harmony.

---

3. Richard Lewontin, one of our most reflective biologists, has provided a particularly candid state-ment of the way in which the equation of religion with supernaturalism has led science-based thinkers to insist on a materialistic version of naturalism ("Billions and Billions of Demons," *The New York Review of Books*, January 9, 1997: 28–32). Referring to "the struggle between science and the supernatural," Lewontin says that he and many fellow scientists—such as Carl Sagan, whose book is being reviewed—accept materialistic explanations "*in spite of* [their] patent absurdity" because of "a commitment to materialism." Admitting that such scientists are compelled to accept a material explanation of the phenomenal world *not* by "the methods and institutions of science" but only by this commitment to materialism, Lewontin adds: "Moreover, that materialism is absolute, for we cannot allow a Divine Foot in the door. . . . To appeal to an omnipotent deity is to allow that at any moment the regularities of nature may be ruptured, that miracles may happen" (31). As the ideological leaders of the scientific community become aware that there is a form of theistic religion that rejects this omnipotent deity and thereby the possibility of such ruptures, they may become free from the perceived need to affirm the materialistic version of naturalism in spite of the "patent absurdity" of many of its doctrines.

3

HARMONIZING SCIENCE AND RELIGION: THREE ALTERNATIVE APPROACHES

Given the analysis of the possible conflicts between scientific and religious communities provided in the previous chapters, four theses follow as to the possibility of harmony.

1. A true harmony between the scientific and religious communities will be possible only insofar as they share a worldview in common.
2. The scientific community is right to insist upon naturalism in the minimal sense, according to which the world's most fundamental causal principles are never interrupted.
3. If there is to be harmony, the religious community must realize that its vital beliefs are compatible with naturalism in this minimal sense.
4. If there is to be harmony, the scientific community must realize that it requires naturalism *only* in this minimal sense.

If these four theses are accepted by the religious and the scientific communities, respectively, science and religion could in principle be harmonized in a naturalistic worldview perceived to be adequate for both scientific and religious interests.

However, all of these theses can be and, indeed, *are* challenged by religious thinkers of different persuasions. Some supernaturalists accept the first thesis while rejecting the second, arguing instead that science should return to a supernaturalistic perspective. This position, as articulated by Alvin Plantinga and Phillip Johnson, will be examined in the first section. Other supernaturalists accept the second thesis while rejecting the first and the third, arguing that harmony can be achieved even though the religious community retains a supernaturalistic framework, because science can be understood as a limited enterprise with its naturalism regarded as purely methodological. I will examine this position, advocated almost a century ago by Rudolf Otto and currently favored by William Hasker, Ernan McMullin, and Howard Van Till, in the second section. A third possibility is to accept the first three theses while rejecting the fourth, arguing that harmony will be possible only if the religious community accepts the maximal as well as the minimal form of naturalism. The third section examines this position as articulated by Willem Drees in *Science, Religion and Naturalism*. Although my conclusion will be that each of these alternatives is seriously inadequate, I will also point out that each position contains important insights that need to be preserved in a more adequate position.

The Proposal for a Theistic Science

One of the main movements on the current intellectual scene is a challenge to scientific naturalism by religious thinkers. From the viewpoint of the present book, the way in which this challenge has been most prominently raised is both promising and problematic. On the one hand, this challenge performs a valuable service in pointing out the extent to which natural science has improperly been identified with scientific naturalism in the maximal sense. On the other hand, by not distinguishing the maximal and minimal meanings of the concept and thereby also rejecting naturalism$_{ns}$, most of those issuing the challenge are intensifying the perceived conflict between science and religion rather than providing a way to overcome it.

This challenge is not entirely new. Scientific naturalism has been under attack by fundamentalist Christians throughout much of the twentieth century. This attack has been intensified in recent decades by the rise of the movement for "creation science" or "scientific creationism," which tries to argue on scientific grounds for a view corresponding closely with that suggested by a literal reading of the book of Genesis. Creationism in this sense, with its argument for a "young Earth" (no more than 10,000 years old), cannot be taken seriously by those who know the evidence. So long as this movement was the only vocal critic of scientific naturalism, the criticism could be easily dismissed.

However, a more sophisticated challenge, based on the view generally known as "progressive creationism," has recently become prominent. This view accepts the evidence that the Earth has been billions of years in the making, but it rejects the idea that later species evolved through natural processes out of earlier ones, holding instead that each species has resulted from a special divine creation. Although this doctrine has been held in some conservative Christian circles for several decades, it recently has served as the basis for an intensified challenge to scientific naturalism in general and neo-Darwinian evolutionism in particular, especially through the writings of philosopher Alvin Plantinga and attorney Phillip E. Johnson, whose writings I will use to explore the contention that the best way to bring about harmony between science and religion is for the scientific community to reverse its rejection of supernaturalism. Their position on science and religion can be summarized in four theses: (1) Religious belief, especially Christian belief, presupposes a theistic worldview. (2) Contemporary mainline science, especially evolutionary science, presupposes a naturalistic (materialistic, atheistic) worldview. (3) Christian faith, therefore, appears to be contradicted by science. (4) This appearance of contradiction can be overcome only by showing that the evidence, rather than supporting the contemporary scientific community's endorsement of a naturalistic worldview, suggests that a "theistic science" would be more adequate.

For this position really to be a proposal to overcome the cultural conflict between the scientific and religious communities, it would have to be understood as a proposal that the scientific community actually adopt the proffered theistic science. Only if this were to occur would the present conflict between the worldviews of the two communities be overcome. The stances of Plantinga and Johnson, however, seem to differ on this score. Plantinga's argument seems to be aimed less at changing the scientific community than at rationally justifying the right of theists to maintain their beliefs and encouraging them, towards this end, to develop an alternative type of science for themselves. Johnson, evidently more hopeful that rational argumentation about evidence can be persuasive regardless of prior prejudices, seems to be trying to bringing about a change in the scientific community itself. I will evaluate this proposal in terms of this latter understanding.

The Case Against Naturalism$_{sam}$

At the heart of the proposal by Plantinga and Johnson is their argument that there is no good reason to continue to assume that science, to be science, must exclude reference to divine agency. This argument is couched in terms of an attack on the necessary association of science with (scientific) naturalism, by which they mean the materialistic, atheistic position that I have called maximal naturalism, or naturalism$_{sam}$, according to which, in Johnson's words, nature is "a closed system of material effects" (RB, 46), which implies that "purposeless natural forces" did "all the work of creating formerly credited to God" (RB, 17).

The question raised by Plantinga and Johnson is why we should accept this perspective.

The strongest argument, says Johnson, is that "science is based on [atheistic] naturalism, and the success of science has proved that naturalism is, if not absolutely true, at least the most reliable way of thinking available to us" (RB, 49). In particular, scientific naturalists "claim to have demonstrated that God as Creator is superfluous, because purely natural forces were capable of doing and actually did do all the work of creation" (RB, 49). However, replies Johnson, rather than proving naturalism, contemporary science simply assumes it (RN, 191). One reason why naturalism is simply assumed in scientific circles, Plantinga says, is the belief that "*natural* science" by definition "involves a *methodological naturalism* or provisional atheism: no hypothesis according to which God has done this or that can qualify as a *scientific* hypothesis" (WFR, 27). Given that stipulation, the *best scientific hypothesis* for how the world's present order came to be will, by necessity, be a nontheistic hypothesis. Plantinga and Johnson are right to dismiss this argument as question-begging. As Plantinga says, "We want to know what the *best* hypothesis is, not which of some limited class is best" (WFR, 28).

Another *a priori* argument for accepting atheistic naturalism, Johnson says, is the assumption by most intellectuals that the theory now dominant in the scientific community must be the best one. Science's successes, such as with electricity and penicillin, are taken to justify confidence in its (neo-Darwinian) account of the origin and development of life. In response, Johnson points out that:

> all statements made in the name of science are not equally reliable. We believe in the efficacy of electricity and penicillin on the basis of experimental verification; many of us disbelieve claims that scientists know how life originated because we know how inadequate the experimental evidence is to justify those claims. (RB, 68)

In other words, the many successes of science as an empirical discipline do not imply the truth of the *metaphysical* theory currently dominant in the scientific community.[1]

With these *a priori* arguments for the truth of nontheistic naturalism dismissed, Johnson turns to the real question: Has the neo-Darwinian account really rendered theistic explanations superfluous? Summarizing the negative argument of his book *Darwin on Trial,* Johnson says that "there is no evidence for, and very much evidence against, the Darwinian assumption that some . . . process of step-by-step gradual change produced the basic body plans of plants and animals"

---

1. I sort out in Chapter 8 the respects in which neo-Darwinism is a metaphysical, as well as an empirical, theory.

(RB, 15). He also argues that the "mind cannot really be explained as a strictly material phenomenon" and that the rise of conscious reasoning minds, in particular, cannot be explained "by a blind watchmaker that cares for nothing but survival and reproduction and therefore ought to have been satisified with cockroaches and weeds" (RB, 66, 92). Johnson's conclusion is that, on purely rational and empirical grounds, the hypothesis that our world came about without any purposive guidance is far less likely than the hypothesis that such guidance was involved.[2]

The Case for Theistic Science vs. Methodological Atheism

Having argued that scientists should want the best view of the universe, not simply the best nontheistic view, and that neo-Darwinism in particular and materialistic atheism in general cannot explain the facts, Johnson has partly prepared the way for the position that Plantinga has called "theistic science." Understanding this proposal requires understanding the view it is rejecting. This view agrees with Plantinga and Johnson that atheistic naturalism does not provide an adequate account of reality, but it does not agree that scientists *qua* scientists can speak of God. Scientists who are theists must respect the convention that science as such is *methodologically* atheistic.

This position is usually supported by the scheme of primary and secondary causation. Science, from this viewpoint, is regarded as the disciplined study of secondary causes. As such, science is a limited enterprise, which can neither affirm nor deny that these secondary causes are themselves products of a higher, divine causation. Therefore, a "theistic science," which would speak of divine causation, is as oxymoronic as an "atheistic science." The explanations provided by a scientist who is a theist should never use the category of divine causation. This is the meaning of "methodological atheism."

"Deism" is the term usually used for the view that God, after creating the world, no longer acts in it. Those who endorse methodological atheism, however, often protest that their view should not be called deistic, because it holds that God constantly sustains the world in being, with every act requiring divine concurrent activity (Van Till BA, 23, 26). To take account of this point, Plantinga uses the term *semideism* for their view. Even the applicability of *this* term has been protested, on the grounds that it implies that God *could* not act immediately in the world, without the use of secondary causes (McMullin PDSC, 76–77). Plantinga, however, uses the term (correctly) to refer to the position of "those who say God *doesn't* do [something different and special in any events, such as the creation of life], whether or not they go on to add that he *couldn't*" (ENAP,

---

2. Although I have here merely summarized Johnson's conclusions, without reviewing his arguments, I provide support for these conclusions in Chapters 6 and 8.

90). In any case, the substantive issue, to use my terminology, is whether there is any *variable* divine causation in the world or whether, after creating the world, the divine influence in the world is limited to an *invariable* sustaining activity. Semideism holds the latter position.

It is its opposition to methodological naturalism, usually based on semideism, that gives the position of Plantinga and Johnson its distinctiveness. In Plantinga's words:

> If after considerable study, we can't see how [some event] could possibly have happened by way of the ordinary workings of matter, the natural thing to think, from [a theistic perspective], is that God did something different and special here. . . . And why couldn't one conclude this precisely as a scientist? Where is it written that such a conclusion can't be part of science? (ENAP, 98)

In advocating a theistic science—which Johnson calls "theistic realism"—Plantinga and Johnson are advocating a form of science that would speak of variable divine causation.

They have several arguments for preferring this theistic realism over methodological atheism based on semideism. One argument is that a position that affirms variable divine causation would be more adequate to biblically based religion, because the Bible certainly seems to affirm that God does not simply sustain the world and concur with its causal processes, but instead does different things in different times and places (Plantinga ENAP, 100). A second argument is that the semideistic view, by agreeing with nontheists that reference to God is not needed to explain anything in the universe (except perhaps that it, with all its potentialities, exists), has contributed to the marginalization of theology in the university and the culture generally. That is, insofar as theologians say that divine causation is on a "different level" from the causation studied by science, being the primary causation supporting the whole chain of secondary causes, the assertion is vacuous with regard to the kinds of explanations scientists and historians try to provide. Johnson says that such a theology, largely acquiescing in its status as a "theory of nothing" (RB, 201), can be tolerated in the intellectual community because it can be ignored, because the divine causation to which it refers is superfluous.

In one sense, to be sure, Johnson's claim is an exaggeration: From the viewpoint of the person who already regards the world as God's creation, to modify neo-Darwinism by saying that the evolutionary process has worked out as it has because it was created to do so by a loving, intelligent creator is far from a vacuous, superfluous addition. It is the addition that makes all the difference in one's outlook on life. In another sense, however, Johnson is right: From the viewpoint of those who do not think in theistic terms, this way of speaking of God, being an accommodation of theistic belief to nontheistic explanatory modes, can simply be ignored, because it leads to no alternative suggestions as to the causal processes involved in the evolutionary process. The adoption of theological realism, Johnson rightly holds, is a necessary condition for theology's regaining its status as an intellectually important enterprise.

The most important argument against the semideistic position is that the acceptance of atheism *methodologically* suggests that it is true *metaphysically*: This argument seems implicit in Plantinga's statement that "if you hew to . . . methodological naturalism . . . , you are very likely to wind up with the very sort of science you would aim at if your naturalism weren't merely *methodological* but . . . the sober metaphysical truth of the matter" (ENAP, 99 n.14). Rendering the argument explicit, Johnson says:

> [I]f employing [methodological atheism] is the only way to reach true conclusions about the history of the universe. . . , and if . . . trying to do science on theistic premises always leads nowhere or into error (the embarrassing "God of the gaps"), then the likely explanation for this state of affairs is that [atheism] is true and theism is false. (RB, 211)

Of course, defenders of methodological atheism may point out, this explanation is not *necessitated,* because one can imagine various hypotheses as to why a supernatural creator would want methodological atheism to work fairly well (for example, the resulting epistemic distance from God may be supposed to be necessary for really robust soul-building). But Johnson is speaking of the *likely* explanation, and insofar as he is asking how the situation would look to intellectuals not already firmly convinced of the truth of theism, his estimation seems correct. After all, we accept this reasoning with regard to most other issues: It works best to act at Christmas as if Santa Claus does not (literally) exist, we assume, because Santa Claus really does not exist. The acceptance of methodological atheism, accordingly, can plausibly be taken as an implicit argument in favor of *metaphysical* atheism. Johnson is not thereby arguing that methodological atheism is *inconsistent* with theism; he recognizes that logical compatibility can be achieved by means of "strained compromises." His claim is only that the acceptance of atheism in science makes theism *unconvincing* (RTH, 490), and this claim seems correct.

Problems in the Plantinga-Johnson Proposal for a Theistic Science

In spite of its many strengths, however, the proposal by Plantinga and Johnson to harmonize science and religion by developing a theistic science contains many problems. At the root of these problems is the equation of theistic realism with supernatural interruptionism.

Many of the points made by both Plantinga and Johnson, to be sure, do not require supernaturalism in this sense. For example, given Johnson's definition of naturalism as "the doctrine that nature is 'all there is' " (RB, 7), the rejection of naturalism in favor of supernaturalism might simply involve the affirmation of the existence of a divine actuality distinct from the totality of finite existents. Likewise, given Johnson's description of naturalism as the doctrine that "creation was by impersonal and unintelligent forces" (RB, 17, 107), the opposite doctrine is "creationism" in the sense of "the very broad proposition that a purposeful supernatural being—God—is responsible for our existence" (RB, 25). Although Johnson refers to God here as a "supernatural being," his point only requires that the word "God" refers to a purposeful being who is a *distinct actuality,* even a *distinct agent,* beyond the totality of finite actualities. This position, accordingly, could be called *supernaturalism_{da}*. It would not necessarily entail the doctrine that this agent can *interrupt* the normal causal processes of the world, which could be called *supernaturalism_{ir}*. The Whiteheadian naturalistic theism that I am proposing involves supernaturalism_{da}, which is what makes it a form of theism. But it rejects supernaturalism_{ir}, which is what makes it a form of *naturalistic* theism. Despite this fact, however, it can affirm much of what Plantinga and Johnson want to affirm, including variable divine influence in the world.

Johnson and Plantinga go beyond this affirmation, however, to specify that the Creator is also supernatural *in the sense of being able to interrupt the world's normal causal processes.* Understanding the Creator as a being who freely established all the laws of nature, Johnson says: "The universal lawmaker has the power to make exceptions" (RB, 92). This understanding of God lies behind his critique of "modernist naturalism," which "excludes consideration of miracles, defined as arbitrary breaks in the chain of material causes and effects" (RB, 46). Against this naturalism, Johnson defends the reality of such breaks, thereby rejecting the perspective of scientists "who see the success of science as inextricably linked to the presumption that no supernatural mind or spirit ever interferes with the orderly . . . course of natural events" (RB, 46). Likewise, Plantinga (WFR, 22) says that "God has often treated what he has made in a way different from the way in which he ordinarily treats it," a point that he illustrates by referring to miracles. And, not one to minimize the implications of divine omnipotence, Plantinga, in discussing God's creation of the world, says that "God could have accomplished this creating in a thousand different ways," such as in the way the semideists believe that God actually did it (WFR, 21). In rejecting naturalism_{sam}, accordingly, Johnson and Plantinga also reject naturalism_{ns}.

Because of its supernaturalism, their theistic realism defends exactly the kind of interventionism that is shunned by those who reject a "God of the gaps." Johnson is right, I hold, to criticize theologians who are "so paralyzed by fear of the 'God of the gaps' fallacy" (RB, 105) that they deny, at least in effect, that there is any variable divine causation in the world whatsoever (which leaves them with little if any basis for speaking of God at all). The divine causation spoken of in the Plantinga-Johnson version of theistic realism, however, is not merely variable divine *influence* within the normal causal processes (which my form of naturalistic theism affirms), but divine *interruption*, bringing about "arbitrary breaks" in the chain of natural causes and effects, whereby natural causes are *replaced* by the divine causation.

It is this feature of the Plantinga-Johnson proposal that prevents it from providing a basis upon which the modern conflict between science and religion might be overcome. They are right, to be sure, to say that the allowance for supernatural interventions would not mean an end to science. For one thing, science throughout the medieval period and even the first half of the modern period operated within a supernaturalistic framework. Also, those who suggest that the only alternative to a consistent naturalism is the view that "all events are the product of an unpredictable divine whimsy," Johnson rightly points out, are guilty of the fallacy of an excluded middle: A third possibility is that although the "universal lawmaker has the power to make exceptions, . . . such exceptional acts do not make the laws unimportant" (RB, 92). However, although the scientific community *could* return to a supernaturalistic$_{ir}$ framework, it is extremely unlikely that it *will* do so. Scientific naturalism$_{ns}$ has been held so long and so deeply that it is now almost universally regarded as belonging to the very spirit of science. For example, in response to William Alston's claim that "the odd miracle would not seem to violate anything of importance to science," because it would probably not conflict with any actual observations, Willem Drees says:

> By undermining scientific reasoning, this argument for the possibility of occasional miracles does undermine something "of importance to science," even though it does not conflict with any observations. The argument undermines the integrity of science . . . , the underlying spirit, the larger web of belief, intentions and procedures of which it is a part. (RSN, 94–95)

In the same vein, Nancey Murphy says that it is true "by definition" that "scientific explanations are to be in terms of natural (not supernatural) entities and processes" (PJOT, 33). Because this attitude is so widespread, the attempt to get the scientific community to reject naturalism$_{ns}$ would surely be quixotic. Indeed, William Hasker says that the theistic science proposed by Johnson and Plantinga "might better be termed 'quixotic science' " (DTR, 483).

Of course, some thinkers, such as Drees, might argue that the *materialist version* of naturalism has now been held so long and so deeply that my effort to convince the scientific community to reject *it* is equally quixotic. There are, however, two crucial differences. First, the conviction that naturalism$_{ns}$ is true is clearly far deeper and more widespread than the acceptance of naturalism$_{sam}$, as shown by the fact that many leading science-based thinkers have rejected the latter without rejecting the former. Second, naturalism$_{sam}$ can be shown to be in tension with many inevitable presuppositions of science itself, as I will argue in the next section. We can, therefore, realistically hope for a paradigm change from naturalism$_{sam}$ to a more open form of naturalism—at least if the case for this change can be completely dissociated from the effort to overcome naturalism$_{ns}$ as well.

A closely related problem with the Plantinga-Johnson position is that it does not reconcile theism with evolution. Johnson rejects any form of "theistic evolutionism" not only because he associates this concept with methodological atheism (DT, 166–69), but also because he rejects the view that later species have evolved out of earlier ones, favoring instead a "progressive creationism"[3] according to which at least many of the later species have been created *ex nihilo*. Plantinga, likewise, expresses the opinion that "the claim that God created mankind as well as many kinds of plants and animals separately and specially" is more probable than the claim that all living beings have a common ancestry (WFR, 22). The doctrine that the various species have been created *ex nihilo* is the theory that Darwin most strongly opposed. The possibility that the evolutionary community will return to this view is extremely remote.

A third problem with the position of Plantinga and Johnson is that it involves a tension between the mode and tempo of creation. Plantinga, as we saw, affirms that God could have created the world "in a thousand different ways" (WFR, 21). And Johnson (RB, 209) says that belief in a divine designer "does *not* mean that organisms were necessarily created by instantaneous fiat as opposed to gradual development (although they might have been)." The notion that they *might have been* created by instantaneous fiat raises the question why, if God *could* have created the world all at once, such a long, slow process would have been used. One can, to be sure, propose possible, nonfalsifiable answers, but there is certainly *no natural fit* between the tempo of the creative process and the hypothesis that our present world was created *ex nihilo* (rather than out of things that, besides existing prior to our particular world, contained power of their own that could not be fully controlled by divine power). This lack of fit becomes all the more evident when one assumes, as do most supernaturalistic theists, that God's sole or at least primary reason for creating the universe was for it to be a suitable environment for the divine-human drama: Why would an omnipotent being take ten to twenty billion years simply to set the stage for the main act?

---

3. Johnson, to be sure, does not explicitly affirm this view and has even expressed reluctance to accept the label (RH, 299; RTH, 492). But, given his disassociation from Genesis literalism, nontheistic evolution, and theistic evolution, his rejection of saltationism as "a meaningless middle ground somewhere between evolution and special creation" [DT, 61], his defense of miraculous interruptions, and his view that "the common ancestry thesis is false as a general theory" (RTH, 491), progressive creationism seems the only remaining alternative.

A fourth problem for the Plantinga-Johnson form of theistic realism, if truly *all* things are considered, is the traditional problem of evil: If the world is the product of a personal creator who is both perfectly good and omnipotent in the traditional sense, having the power unilaterally to bring about and prevent events in the world, why is the world so filled with evil? Why has God not intervened to prevent especially horrible forms of evil, such as the Holocaust? Again, many *possible,* nonfalsifiable answers can be given, and Plantinga, for one, has been involved in providing such possible answers, but they are not *plausible* answers, at least for those not already strongly committed to the truth of supernaturalistic theism (a fact that Plantinga admits). Although I cannot here engage in a discussion of the problem of evil faced by supernaturalistic theists, I have extensively discussed this problem, with some attention to Plantinga's position, in other books (ER; GPE).

Although the problem of evil is not properly part of the discussion of the relation between science and religion, the way in which this problem prevents many scientists and philosophers from considering the kind of theism advocated by Johnson is illustrated by Steven Weinberg's response to him. Referring to some especially beautiful birds and trees, Weinberg (DFT, 250) says:

> I have to admit that sometimes nature seems more beautiful than strictly necessary. . . . But the God of birds and trees would have to be also the God of birth defects and cancer. . . . Remembrance of the Holocaust leaves me unsympathetic to attempts to justify the ways of God to man.

As long as the idea of a Divine Creator is identified with the God of supernaturalistic theism, who *could* but *does not* prevent evils such as birth defects, cancer, and the Holocaust, the arguments *for* God, such as the arguments from order and beauty, are canceled out by the argument *against* God posed by the world's disorder and evil.

I am examining the possibility that the Plantinga-Johnson proposal could provide a way to overcome the conflict between the worldviews of the scientific and religious communities, which would require that the scientific community adopt their proposed theistic science. I have thus far given four reasons, all rooted in the ontological supernaturalism$_{ir}$ of their theism, for considering this possibility very unlikely. A fifth and final reason is the fact that their ontological supernaturalism is connected, as is usually the case, with epistemic supernaturalism, the view that some doctrines are to be believed simply because of their alleged origin in a supernatural revelation. I have shown elsewhere (CFSN) that Johnson's epistemic supernaturalism undermines his admirable attempt to provide a basis for a common reason, through which the polarizations of our time might be overcome. Plantinga, even more clearly an epistemic supernaturalist, describes the Bible as "a special revelation from God himself, demanding our absolute trust and allegiance" (WFR, 8), and he indicates that clear statements in the Bible could override extremely strong scientific evidence to the contrary, such as the view that the Earth is very old (WFR, 20). That, of course, is the position that leads to what I have called conflicts of the second kind, in which a religious community remains committed to beliefs about particular facts after science has demonstrated them to be almost certainly false. One reason why the scientific community would resist returning to ontological supernaturalism is precisely because it would reopen the possibility of conflict of this kind within the scientific community itself.

In sum: Plantinga and Johnson have provided an impressive case against scientific naturalism$_{sam}$. But, as Nancey Murphy (PJOT, 32) and William Hasker (MJP, 184–86) point out, the scientific community will give up its present worldview only when it sees a more attractive alternative. Johnson recognizes the truth of this point (RH, 303; RTH, 490–91). However, because he and Plantinga have challenged naturalism$_{sam}$ with the opposite extreme of supernaturalism$_{ir}$,[4] their proposal depends on the entirely unlikely prospect that the scientific community will return to the supernaturalistic framework from which it has only recently freed itself. Their position does not, therefore, present a persuasive case against the thesis that, if science and religion are to be harmonized, naturalism$_{ns}$ must be accepted by both the religious and the scientific communities. Indeed, far from undermining this thesis, the Plantinga-Johnson position serves to underline its importance.

The Proposal to Place Scientific Naturalism Within a Supernaturalistic Framework

Although the proposal for a theistic science rejects the last three of the four theses with which I began this chapter, it accepts the first thesis, which says that a true harmony between the scientific and religious communities will be possible only insofar as they share a worldview in common. The second alternative to my proposal rejects this first thesis, saying instead that there can be harmony even though the scientific community employs only naturalistic explanations while the religious community presupposes a supernaturalistic framework. This harmony requires two stipulations: The religious community must give up the idea of supernatural interventions into the world, at least for the most part, and the naturalism of the scientific community must be understood as purely method-

---

4. Although I have, for the purposes of this section, employed the distinction between supernaturalism$_{da}$ and supernaturalism$_{ir}$, the term "supernaturalism" (without qualifying subscript) should be used only for the latter doctrine, according to which God does, or at least could, interrupt the world's most basic processes, because the term almost inevitably carries this connotation. Accordingly, the term "supernaturalism," when used in the ontological sense, will henceforth, as in earlier chapters, always be used for the doctrine here called supernaturalism$_{ir}$.

ological, not metaphysical. This methodological naturalism was briefly discussed in the previous section in order to clarify, by contrast, the proposal for a theistic science. In the present section, it will be examined as a serious proposal in its own right.

Proponents of this view accept the idea that, in William Hasker's words, "scientific explanation cannot appeal to supernatural intervention" (DTR, 483). In line with this position, Hasker cites with approval Steven Weinberg's statement that "the only way that any sort of science can proceed is to assume that there is no divine intervention and to see how far one can get with this assumption." Likewise, responding to Plantinga's suggestion that the best explanation of the origin of many kinds of plants and animals is an interruption of the ordinary course of natural processes, Ernan McMullin says that to count as "scientific," an explanation must refrain from invoking any special action of God. "Scientists *have* to proceed in this way," McMullin contends (PDSC, 56–57). In the same vein, Howard Van Till says that science methodologically excludes a "God of the gaps" (WFRC, 42–43).

What makes this position compatible with supernaturalism, say its advocates, is the stipulation that the naturalism of science is purely methodological. McMullin, for example, says that science's naturalism "is a way of characterizing a particular *methodology,* no more. In particular, it is not an ontological claim about what sort of agency is or is not possible" (PDSC, 57). As a way of clarifying the position, Van Till distinguishes between a narrow and a broad sense of "naturalism." Naturalism in the narrow sense "simply refers to the idea that the physical behavior of some particular material system can be described in terms of the 'natural' capacities of its interacting components and the interaction of the system with its physical components." Naturalism in this sense, Van Till stresses, says nothing about "the ontological origin of [the material system's] existence" or about "the ultimate source of its capacities for behaving as it does, or its purpose in the larger context of all reality" (SC, 127). Naturalism in the broad sense, by contrast, "superimposes the strong metaphysical stipulations that neither the existence nor the behavioral capacities of material systems derived from any divine source." It is only naturalism in this broad sense that would be irreconcilable with Christian theism.

As this discussion implies, the reconciliation of science with Christian supernaturalism is to be effected by placing the former, regarded as a strictly limited enterprise, within the latter. For example, Van Till, speaking of science as "an incomplete picture of reality because of its inability to probe beyond the creaturely realm," says that it "needs to be placed within the framework of an all-encompassing, biblically informed, theistic worldview." Rejecting Plantinga's vision of a theistic science, Van Till says that his own religiously informed science would

> differ from ordinary science not by the occasional insertion of immediate
> divine acts into a world whose internal economy is either deficient or

defective, but rather by its recognition that every aspect of the world's functionally complete economy is radically dependent on the Creator's ceaseless activity as the world's Originator, Sustainer, Governor and Provider. (WFRC, 45)

In terms of our earlier discussion of the scheme of primary and secondary causation, this proposal involves not only the doctrine that science is strictly the study of secondary causes, whereas theology deals with God as primary cause, but also the doctrine that, at least in the domain studied by the natural sciences, God works *entirely* through secondary causes. There can be no competition between theological and scientific explanations, accordingly, because divine and creaturely causation operate on "different levels." Van Till says that his position involves

a theologically-based recognition that God is not one component among many others in the economy of the created world—God's creative action, operating at a level different from creaturely action, undergirds *all* that occurs, not only that which eludes our first efforts toward scientific description. (WFRC, 44–45)

Rejecting the position of "progressive creationism," according to which new species are directly created by God, Hasker says that "divine purpose enters into the process at another level—not in the origins of specific creatures, but in establishing the process as a whole" (DTR, 485).

In "establishing the process as a whole," God ordained not only that there would be a world with an evolutionary process, but also that this process would go in the direction it did, because God endowed the world with all the necessary capacities and tendencies. The world, says Van Till, was "gifted by God from the outset with all of the form-producing capacities necessary for the actualization of the multitude of physical structures and life forms that have appeared in the course of Creation's formative history" (BA, 23).

The advocates of this view regard it as superior to the view that God needs to intervene now and then to create new species. That view, Van Till says, would imply that God deliberately chose "to withhold certain form-producing capacities," leaving the world deficient (BA, 25–26). Against "the effort to find gaps in the developmental economy of the creation," Van Till asks, rhetorically: "Is not a gapless economy far more awesome?" McMullin, reflecting on the creator suggested in Genesis, Job, Isaiah, and the Psalms, ask rhetorically: "Would such a Being be likely to 'intervene' in His creation in the way that Plantinga describes?" Rather, says McMullin, God "must be thought of as creating in that very first moment the potencies for all the kinds of living things that would come later, including the human body itself" (PDSC, 75). Hasker likewise suggests that a deity lacking in either power or wisdom is implied by Plantinga's progressive creationism, because it "seems to show us a God who prefers, whenever possible, to carry out his work of creating through the evolutionary process but who just couldn't bring this off in the case of biological life. The symphony of creation is interrupted" (DTR, 487).

Although this position favors a science that is naturalistic (methodologically), its overall worldview is fully supernaturalistic. God is omnipotent, the sole source of the universe, which was brought into being out of nothing (Van Till BA, 36; SC, 124). Furthermore, McMullin stresses, although God has chosen to work through secondary causes, "God *could* also, if He so chose, relate to His creation in a different way, in the dramatic mode of a grace that overcomes nature" (PDSC, 76–77). Furthermore, these authors say not only that God *could* act supernaturally in the world, but also that God actually *does so*. The rejection of the God of the gaps, say both Van Till (BA, 37; SC, 124), and McMullin (PDSC, 74), in no way rules out belief in miracles, understood as extraordinary acts of God in the world. And Hasker, after having said that "scientific explanation cannot appeal to supernatural intervention," adds that the Christian should "refuse to make the assumption that every actual event has a scientific explanation." It may be, for example, that "there just *is no* scientific (i.e. naturalistic) explanation of the origin of life" (DTR, 483).

This allowance may seem surprising, given all they have said about the superiority, from both a scientific and a theological perspective, of a world without supernatural interventions. They claim, however, that no contradiction is involved. McMullin justifies the exceptions in terms of different "domains," appealing to the "old and valuable distinctions between nature and supernature, between the order of nature and the order of grace, between cosmic history and salvation history" (PDSC, 74). Van Till overcomes the apparent contradiction by pointing out that his position only opposes the idea of interventions *necessary* to make up for *deficiencies* in God-given capacities. "Miracles," by contrast, "are acts freely performed by God for their timely revelatory and redemptive value, not obligatory acts needed to compensate for earlier omissions" (SC, 124).

Evaluation of the Proposal

Now that we have the position before us, we need to ask whether it really, as its proponents suggest, provides a basis for overcoming the perceived conflicts between science and religion. This position, in the first place, clearly provides some help to individual scientists and theologians. It allows scientists who are theists to practice science in the same way as do their nontheistic colleagues. And it allows theologians to believe in God without challenging currently orthodox scientific theories about the causal categories needed to account for how things work within the world. In the second place, this proposal, if accepted by the religious communities, would lessen the conflict between them and the scientific community, because it would mean that the religious communities would revise their doctrines, including their interpretation of Scripture, so as not to conflict, at least as much as in the past, with the methodological naturalism involved in the scientific community's interpretation of the world.

Although these benefits are real, however, this proposal is plagued with problems. Three of these—that it is inadequate religiously insofar as it does not allow for variable divine influence in the world, that it has contributed to the marginalization of theology, and that its methodological atheism suggests the truth of *metaphysical* atheism—were already mentioned in the previous section But it has even more problems. In the first place, closely related to the issue of the superfluity of divine causation is the issue of its very intelligibility. This position holds that every nonmiraculous event has two sufficient causes: the divine, primary causation, and the natural, secondary causation. A sufficient cause of an event, however, is by definition a cause sufficient to bring it about. If one cause provides a sufficient explanation for any event, any other causal explanation is ruled out, unless that second explanation is simply an aspect of the first or reducible to it. This principle, called the principle of "causal-explanatory exclusion" by Jaegwon Kim (SM, 291), implies that if the divine causation is sufficient to bring an event about, it makes no sense also to speak of natural causation upon it. The reverse implication is that, if the natural causation is sufficient to bring it about, it makes no sense to speak of divine concurring causation.

A second problem is that this proposal, in spite of what it may do for individuals, does not provide a basis for overcoming the contradiction between the worldviews of the religious and scientific *communities*, which is the primary problem. Although this proposal would forge an agreement that science is to be naturalistic *methodologically*, it leaves the two communities as far apart as ever on the *metaphysical* issue: The worldview of the religious community remains supernaturalistic, while the maximal naturalism currently dominant in the scientific community is unchallenged.

To be sure, this maximal naturalism *is* challenged insofar as this proposal says that naturalism$_{sam}$ cannot provide an adequate, self-consistent cosmology and that the cosmic evolutionary process is far more intelligible if regarded as the product of a supernatural creator. This challenge would constitute a proposal to overcome the conflict between the worldviews of the two communities, however, only if the proposal to place methodological naturalism within a supernaturalistic framework were presented as a position for the scientific community as such to adopt. But this has generally not been the case. Rather, the position has been presented primarily as a defensive move on behalf of Christians and other theists, to allow them to retain their religious beliefs while accepting methodological naturalism. There is no reason in principle, however, why the proposal could not be presented more ambitiously, as a proposal to the scientific community of a better worldview for itself. Because it is only as such that this position could become a proposal to overcome the cultural problem constituted by the clash between the worldviews of the religious and the scientific communities, it is in this light that I will evaluate it. Doing so means, of course, asking whether it could become the general worldview for global civilization, at least Western civilization, in the 21st century.

The obvious problem with this proposal might seem to be that it would mean proposing that the scientific community return to a supernaturalistic framework, which, I have suggested, is extremely unlikely. In that discussion, however, we were considering the idea that science as such could affirm supernatural interruptions of the normal causal processes. It is precisely this idea that the present proposal, with its insistence that science must be methodologically naturalistic, would exclude.

Nevertheless, the position *is* supernaturalistic. It holds, in the first place, that the world was created by God *ex nihilo,* which implies that, because all the causal principles of the world have been freely created, they *could* be freely interrupted. Even if it is stipulated that these principles will never *in fact* be interrupted, it remains the case, by hypothesis, that there is a causal agent who *could* interrupt them. This point by itself is already a significant violation of the naturalism of the scientific community, for which an interruption of the web of natural causes and effects is inconceivable. It might seriously be proposed, nevertheless, that the scientific community could return to this earlier understanding of the universe, in which the inviolability of the causal nexus is due to divine decision, not metaphysical necessity. It could be pointed out that Darwin, for example, held this position.

One problem with this solution, however, is that when it is believed that there *could* be such interruptions, it is rather natural to assume that there *have been* some. If God could interrupt the normal processes, would we not expect that God would do so now and then, to accomplish an especially important purpose? And this, as we have seen, is just what Hasker, McMullin, and Van Till hold. They seek to minimize the importance of this fact for the science-religion discussion by distinguishing between two domains, the religious and the scientific, then suggesting that the interruptions occur only in the former, so that scientific explanations would never need to speak of such interruptions.

No such division of domains, however, can hold up. For example, Van Till, having pointed to the importance of "the distinction between the scientific and religious domains of concern," says that "the object of study by the natural sciences is the *physical universe,* no more, no less. . . . Nonphysical things are not the object of study by these sciences" (SHH, 10, 11). This statement reiterates the early modern division, according to which science deals with the physical world, while God and the soul belong to the religious domain. The notorious problem with this division is that soul and body interact. If the soul influences its body on the basis of a free decision, then the physical world, which includes human bodies, is not an autonomous realm that can be adequately described in terms of a deterministic science. And if the soul's activity was supernaturally influenced by God, then a scientist, to describe accurately the behavior of the person's body, would have to bring supernatural intervention into the account.

Hasker and McMullin also affirm interventions that would conflict with a purely naturalistic science. Hasker (DTR, 484) and McMullin (PDSC, 79) both argue that neo-Darwinism, which excludes any appeal to supernatural interventions, is fully compatible with Christian theism. And yet Hasker does not believe "that the kinds of processes described by neo-Darwinism are capable of producing some of the unique attributes of human beings" (DTR, 486); and McMullin, after having distinguished the order of nature from that of salvation, says that "the story of salvation *does* bear on the origin of the first humans," meaning that "God somehow 'leant' into cosmic history at the advent of the human" (PDSC, 74–75). McMullin admits, indeed, that one cannot simply rule out conflicts "by laying down that Christian doctrine can have *no* implications for matters that fall under scientific jurisdiction." This is impossible because it is not the case that "the two domains are, in principle, so safely walled off from one another that no conflict can possibly arise." He believes, however, that "the domain of such potential conflict is quite limited" (PDSC, 62).

From the viewpoint of those who have accepted scientific naturalism$_{ns}$, however, the discussion of *how many* interventions would be affirmed is beside the point: The very idea of such interventions is inconceivable. The otherwise helpful suggestion made by methodological naturalists, that the evolutionary process will be more intelligible if regarded as initiated and guided by a purposive creator, will surely fall on deaf ears if it carries with it the idea of *any* supernatural interruptions of the normal causal processes.

A Completely Consistent Methodological Naturalism: Rudolf Otto's Proposal

Would the proposal to place methodological naturalism within a supernaturalistic framework be viable if the proposal were that all events *without exception* are produced by means of secondary or natural causes, so that there are *no* interruptions? Such a proposal was made early in the twentieth century by Rudolf Otto in a book entitled *Naturalism and Religion,* with further elaboration in an essay on "Darwinism and Religion." Otto's position was based on the idea that "with the awakening of criticism and reflection, and the deepening of investigation into things. . . . , there soon awakens a profound conviction that a similar mode of causal connection binds all things together" (NR, 19). As this statement suggests, Otto accepted naturalism$_{ns}$, at least in the sense that no supernatural interventions actually occur.

Although Otto said that "there can be no religion without the supernatural," he distinguished between a lower and a higher supernaturalism. The lower supernaturalism, which Otto rejected, is "the notion that by individual special acts, the supernatural intrudes upon the normal course of things, intervening in a friendly or an inimical matter" (DR, 129). Citing Luther's axiom that "God works through means," Otto said that "we should not look for Him in that which is exceptional or in the vulgar sense a 'miracle' " (DR, 129). This lower supernaturalism, he argued, is doubly problematic:

> On the one hand it breaks the order of nature; on the other it inserts the eternal divine causation into the temporal stream, and makes it to function as one cause side by side with others, both of which things definitely conflict with the religious sense. (DR, 132)

We can, of course, read such things in the writings of authors discussed above, only to find that they allow exceptions. Otto, however, endorsed the view "that no exception should be made over the origin of man, and that even here the lower supernaturalism should not be reintroduced" (DR, 134–35). And life itself, he argued, must have also come about through purely natural factors and laws (DR, 137). More generally, he held, the theist has no reason to oppose Darwin's attempt "to explain the whole development of the living world by his methods in a rigid unbroken chain of cause and effect" (DR, 138).

All of this is fully consistent, Otto held, with the higher supernaturalism, which simply maintains that the whole world, with all its causal processes, is the product of purposive design. For religion, Otto said, "all interest lies in the fact that everything has come about in such a way that it reveals intention, wisdom, providence, and eternal meaning" (NR, 37). And this theological, teleological portrayal of reality, he maintained, is actually aided, rather than hindered, by the scientific view of the world as involving an unbroken causal nexus.

> To the belief that eternal wisdom has purposely determined the existence of life and human intelligence in this world, the scheme of the world, as resting on a strictly causal basis, is not inimical but helpful if not necessary. All the processes of evolution working themselves out strictly in accordance with natural law ... are then nothing but so many different means for making certain of this end. (DR, 137–38)

Otto evidently did not present this view simply as a means for religious people to retain their theism, but as a better, more intelligible view of the world, which would also be better for the scientific community. Consistently with this interpretation, Otto provided reasons to think the atheistic, materialistic, reductionistic form of naturalism to be unintelligible. He refers, for example, to Kant's antinomies as showing that "if we were to take this world as it lies beyond us for the true reality, we should end in inextricable contradictions" (NR, 69). He points out, furthermore, that a reductionistic scientific naturalism, which reduces mind to deterministic matter, is self-refuting, because science is "only possible if mind and thought are free and active and creative" (NR, 295). In particular, science presupposes that the mind has the freedom to search for truth and to follow the laws of logic instead of the mechanical laws of nature (NR, 321–22). Furthermore, the fact that the mind freely moves its body is "so obvious and so unquestionable that no naturalism can possibly prevail against it" (NR, 308). The reductionistic form of naturalism needs to be replaced, accordingly, not simply for the sake of overcoming the conflict between naturalism and religion, but because this form of naturalism is inadequate for the scientific community itself. In so arguing, Otto partly anticipated the position of the present book.

Otto, nevertheless, did not present a position that the scientific community could regard as acceptable. For one thing, his position still affirms one supernatural act—the very creation of our universe. Although uniformitarianism reigns once the universe is created, the universe as such came into existence through a mode of causation different in kind from that which prevails now. This exception means that naturalism$_{ns}$, which has become the deepest philosophical conviction of the scientific community, is not fully embraced.

Furthermore, although Otto said that the self-contradictions and inadequacies of the reductionistic form of naturalism provide good reasons to reject it, his own position could not be defended in terms of its own self-consistency and adequacy to the facts, as he himself admitted. Pointing out that his position contained some insoluble "riddles" (NR, 14–16, 361), he focused on two: the mind-body problem and the problem of evil.

The mind-body problem was severe for Otto because, in rejecting materialism, he accepted an absolute, Cartesian dualism between mind and matter, thereby "an absolute difference . . . between physical and mental causality" (NR, 311). Because we cannot understand how the body could have given rise to mind, he said, the mind can only be regarded as an "inexplicable guest" in the material world (NR, 153, 300). Likewise, how the mind, "which is in itself not energetic," can "determine processes and directions of energy is undoubtedly an absolute riddle" (NR, 349). Otto did provide an implicit answer, however, by referring to God as "unrestrained omnipotence" (NR, 360), thereby suggesting, like many dualists before him, that divine omnipotence, not being constrained by anything except logical impossibilities, could make mind and matter interact in spite of their heterogeneity.

However, even if it helped to lessen the mind-body problem, the idea of God as unrestrained omnipotence only served to increase Otto's other major riddle, the problem of evil. The extent to which Otto recognized the seriousness of this problem is suggested by his statement, with this problem in view, that faith means "going against appearances," requiring "a courageous will to believe," because "many gaps and a thousand riddles will remain" (NR, 15). Otto does suggest that his position, according to which all things are accomplished through secondary causes, provides a partial answer:

The riddle of theodicy thus becomes easier, for what surrounds us in nature and history has not come direct from the hand of eternal wisdom, but is in the first place the product of the developing, striving world. (NR, 366)

However, although Otto's position is deistic in one sense, it certainly does not say that God left the world to work itself out in one way or another. He said, rather, that although the lower supernaturalism may regard only a few events as due to divine providence, the higher supernaturalism thinks otherwise: "Everything, even the most trivial matter, comes to be regarded as part of the divine ordering, as due to divine providence and therefore as of divine causation" (DR, 129). This would seem to mean that divine providence must be held responsible not only for all of the horrible evolutionary contrivances, which bothered Darwin so much (see Chapter 8), but also for every evil in human history. Furthermore, as suggested earlier, the idea that God *could* intervene into the world inevitably creates a strong presumption that God *should* do so occasionally, to accomplish worthwhile purposes. The riddle of theodicy, therefore, does not seem to be any less problematic for Otto's position than it is for less deistic positions.

Otto claimed that it is not necessary for faith to find a solution to all such riddles, that it can rest content with demonstrating the insufficiency of the reductionistic view, thereby showing that nature points "to something outside of and beyond itself" (NR, 361, 372). However, if Otto could rightly dismiss reductionistic naturalism by pointing out its incoherencies and inadequacies, why should not others dismiss dualistic supernaturalism by pointing to the equally serious incoherencies and inadequacies in *it?* That, indeed, was what happened, as this dualistic deism collapsed into materialistic atheism.

In sum: Otto's position, like the less consistent version of methodological naturalism examined earlier, does not provide a hopeful basis for overcoming the conflict between the worldviews of the religious and scientific communities. Neither version, accordingly, provides a persuasive case against the thesis that, if science and religion are to be harmonized, the religious and the scientific communities must share an acceptance of naturalism$_{ns}$.

THE PROPOSAL TO ACCOMMODATE RELIGION
TO MAXIMAL NATURALISM

My claim that a harmony between science and religion can be brought about only if the religious community renounces supernaturalism is, of course, widely accepted in liberal religious circles. In some of these circles, the main resistance would be to my thesis that the scientific community should limit its naturalism to this rejection of supernaturalism, thereby renouncing its allegiance to naturalism in the maximal sense. Because this materialistic version of naturalism has widely been equated with science itself, or at least the "modern scientific worldview," it is not surprising that some religious thinkers have claimed that this kind of naturalism is not incompatible with a religious, even a specifically Christian, outlook. I will explore this claim in terms of one of the most recent and thorough articulations of it, Willem Drees's *Religion, Science and Naturalism.*

Maximal Naturalism and Minimal Religion

I have claimed that the materialistic version of naturalism is not compatible with any *significantly* religious interpretation of reality, certainly not with a Christian interpretation recognizable as such. Although Drees sometimes seems to suggest otherwise (RSN, 43n., 159), he points out that there is no conflict in his own mind between himself as a scientist and as a religious believer—which he calls himself (RSN, 28)—because his approach is "minimalist with respect to religion" (RSN, 4).

This judgment of minimalism is certainly correct. In Drees's outlook, there is no divine action in the world, which, as Drees himself points out (RSN, 93), eliminates most historic Christian beliefs. Drees's materialism also entails identism—"in some way, I am my brain" (RSN, 184)—which implies denials of both freedom and life after death. His materialism also entails that there is no realm of objective moral values or norms, to which moral judgments might correspond (RSN, 216, 218, 221). As this list indicates, Drees's religious beliefs are so minimal as to be virtually nonexistent. He does seemingly bring in a belief by referring to "a sense of transcendence" (RSN, 237). Believing that one question that cannot be answered naturalistically is why anything exists at all, he uses the word "God" to point to the ground of the world (RSN, 266–68). However, his ideas about God are not "affirmed as realist claims, but rather accepted as speculations and regulative ideals" (RSN, 237). The fact that his reference to God involves no real belief is brought out clearly by his rejection of the idea that the Christian tradition is "true to the way things 'really are' or to the way reality ultimately is" (RSN, 279). He says, instead, that we should want to keep Christianity alive because it is "useful and powerful" (RSN, 278).

Drees feels compelled to accept such a minimal religion because he has accepted a maximal naturalism. Building on Peter Strawson's distinction between soft (non-reductive) and hard (reductive) naturalism, Drees says: "I use the label 'naturalism' for 'hard naturalism'; 'materialism', 'physicalism', and 'physical monism' may be construed as near synonyms" (RSN, 11). Indeed, he characteristically speaks of his position as "materialist naturalism" (RSN, 53, 258).

Articulating this naturalism in terms of a number of claims, he begins with "ontological naturalism," which says:

> The natural world is the whole of reality that we know of and interact
> with; no supernatural or spiritual realm distinct from the natural world
> shows up *within* our natural world, not even in the mental life of humans.
> (RSN, 12)

The main implication of this ontological naturalism is the affirmation "that natural
processes are not occasionally interrupted or suspended" (RSN, 248). If this
were all that ontological naturalism denies, it would be identical with what I
have called naturalism minimally construed. For Drees, however, the only religious views of transcendence that are consistent with ontological naturalism are
those that "do not assume that a transcendent realm shows up *within* the natural
world" at all (RSN, 18). Ontological naturalism, in other words, rules out *any*
kind of divine influence *in* the world (RSN, xi, 140, 222).

As the last clause in the indented quotation above makes clear, Drees's
ontological naturalism excludes divine influence not only in nature, in the sense
of the physical world, but also in human experience. One implication of this
point is that religion can in no way be assumed to arise out of divine influence
or, stated otherwise, out of a human experience of God. His ontological naturalism, Drees says (RSN, 26), implies *methodological* naturalism, which is taken
to entail "a non-religionist approach in the study of religions" (RSN, 27). If
there were a "divine causal role" in the origin of religious belief, Drees says,
"the naturalist account [would be] incomplete, and therefore wrong" (RSN, 222).
This naturalist account of the origin of religion has implications, Drees points
out, for our ideas about the divine: "Some causal contribution of God in the
temporal processes that brought someone to faith is essential to the likelihood
that claims concerning God's existence may be true" (RSN, 222). If divine
activity plays no role in the creation of such ideas, therefore, "it is extremely
unlikely that our ideas about gods would conform to their reality" (RSN, 251).
That is certainly the case.

In explaining why all divine activity in the world is ruled out by ontological naturalism, Drees refers to "scientific insights about the lawful behaviour of
natural processes" (RSN, 94). Scientific discoveries have led, he says, to "our
understanding of the world as a tightly knit web of processes described by laws"
(RSN, 92). Given that view of the world, divine influence could occur only as
an interruption of this "tightly knit web of processes," but Drees sees no "religiously relevant gaps in the natural world, where the divine could somehow
interfere with natural reality" (RSN, xi). Like our previous authors, Drees assumes the analysis inherited from the scheme of primary and secondary causation, according to which most events are fully explicable without reference to
constitutive divine influence, so that any such influence would involve an interruption of the way in which events are normally brought about.

The basis for Drees's denial of any such influence is fleshed out by the
claims he calls "constitutive reductionism" and the "physics postulate." Accord-

ing to *constitutive reductionism,* "Our natural world is a unity in the sense that all entities are made up of the same constituents" (RSN, 14). Thus stated, this claim might mean only that all the actual entities of which the world is composed are of the same basic type, which Whiteheadian naturalism says, adding that, besides the simplest entities of this type, which physics studies, there are higher-level, more complex ones, such as cells and human beings, which are compounded out of the simpler ones. The word "reductionism," however, indicates that this is not Drees's meaning. Rather, he means that "different entities are constituted from the same basic stuff, say atoms and forces," in such a way that the behavior of the larger entities, whether rocks or human beings, is entirely a function of their most fundamental constituents, which are studied by physics (RSN, 14).

This point leads to the *physics postulate,* which says: "Physics offers us the best available description of these constituents, and thus of our natural world at its finest level of analysis" (RSN, 14). This claim, upon which the previous one depends, lies at the root of the difference between his naturalism and Whitehead's. For Whitehead, this claim exemplifies the "fallacy of misplaced concreteness," in which the abstractions employed by one of the particular sciences for analyzing something are mistaken for the concrete actuality itself. It is assumed, for example, that there is nothing more to electrons than what physicists, in terms of their limited concerns, need to say about them. If this *is* a fallacy, Drees commits it wholeheartedly, saying that "physics is the science of the fundamental aspects of natural reality" (RSN, 188), even that "physics is fundamental as inquiry about the fundamental ontology of the world" (RSN, 17). Whitehead, by contrast, assigns this inquiry to philosophy, in the sense of metaphysics, which seeks to develop a coherent, adequate view of the fundamental aspects of all actual entities by drawing upon the insights of all the special sciences, including human psychology. In fact, as we saw in Chapter 1, Whitehead believes that philosophy, in carrying out this task, is to draw upon the moral, aesthetic, and religious experiences of human beings. Drees, by contrast, endorses "the primacy of science in the realm of knowledge" (RSN, 3), with "physics as the most fundamental science" (RSN, 14). Coherence is sought by means of regarding all other phenomena as consequences of the entities and forces studied by physics (RSN, 3). Coherence is sought, in other words, by reductive naturalism (RSN, 3).

Drees explains this reductionism by using Wilfrid Sellars' distinction between our *manifest images* of things and the *scientific images* provided by scientific theories (RSN, 9–10). An example is afforded by the difference between our manifest image of a table, as an inert, solid substance, and the scientific image of it as comprised of billions of tiny, buzzing particles in vast amounts of empty space. The scientific image is superior because it can account for the manifest image and much more besides. The scientific image of anything, Sellars said, "is in principle the adequate image" (SPR, 36). Having invoked this distinction,

Drees then applies it (as did Sellars) to the human being, saying that "our concept of a person (with an inner life, emotions, responsibilities, etc.), as it is central to most religious views, is rooted in our manifest images of the world" (RSN, 10). The implication is that our inner life, with its emotions and its apparent freedom, is reducible, at least in principle, to the body's most elementary constituents. These conclusions about the mind-body relation in general and freedom in particular are the most important, and most debatable, implications of Drees's constitutive reductionism combined with his physics postulate. I will return to this issue later.

For now, we need to look at his "evolutionary explanations postulate," which says:

> Evolutionary biology offers the best available explanations for the emergence of various traits in organisms and ecosystems; such explanations focus on the contribution these traits have made to the inclusive fitness of organisms in which they were present. Thus, the major pattern of evolutionary explanation is functional. (RSN, 19–20)

As the reference to "inclusive fitness" indicates, Drees, in speaking of "evolutionary" explanations, means *Darwinian* explanations. Applying this type of explanation to religions, he says: "The primary pattern of evolutionary explanation is functional: religions arose, and therefore probably contributed to the inclusive fitness of the individuals or communities in which they arose," or perhaps "they arose as a side-effect with the emergence of some other trait" that contributed to inclusive fitness (RSN, 250). This kind of explanation is assumed to be not merely necessary but also sufficient, so that it is not to be supplemented with the idea that religions arose partly out of a response to divine activity.

Why Accept Materialistic Naturalism?

Now that we have Drees's position before us, the question is, Why does he accept this maximal construal of naturalism, with its reductive materialism? In the previous chapter, I suggested that, given the history of science-related thinking from the mid-seventeenth century, it was probably impossible for most thinkers in the latter part of the nineteenth century to imagine another kind of naturalism: Because the mechanistic view of nature was virtually self-evident, naturalism seemed to require materialism. Drees, however, knows that there are other possibilities. He contrasts his version of naturalism with a "richer naturalism," by which he primarily means the "religious naturalism" of Whitehead and theologians influenced by him (RSN, 252–58). Indeed, saying that he regards process theology's reinterpretations to be "possible" (RSN, 95), he calls it one of the two most challenging alternatives to his own position (RSN, 236). The question, accordingly, is why Drees believes materialistic naturalism to be superior to this richer, religious naturalism.

Drees seems to believe that one reason in favor of materialistic naturalism is that it is less metaphysical. Indeed, by speaking of Whiteheadian naturalism as a *metaphysical* interpretation of science (RSN, 2, 148, 254), he seems to imply that his own view is *not* metaphysical. However, in line with his recognition that theories are never strictly implied by the data (RSN, 9), he says:

> My naturalism is a metaphysical position. It goes beyond the details of insights offered by the various sciences as an attempt to present a general view of the reality in which we live and of which we are a part. (RSN, 11)

His contention, accordingly, seems to be only that his naturalism is *less* metaphysical, which is suggested by his next sentence:

> However, it is a rather "low-level" metaphysics in that it stays close to the insights offered and concepts developed in the sciences, rather than that it imposes certain metaphysical categories on the sciences or requires a modification of science so that it may fit a metaphysical position taken a priori. (RSN, 11)

However, any view is metaphysical if it attempts to state how things really are, rather than resting content with a purely phenomenological or instrumentalist account. Drees's naturalism is, in this sense, no less metaphysical than Whitehead's. To declare that the "concepts developed in the sciences" are in principle fully adequate for describing what electrons, atoms, molecules, and living cells are really like is no less metaphysical than Whitehead's claim that they are not. Indeed, one could well argue that, in claiming that all events in the world—including the teachings of Gotama and Jesus, the discoveries of Newton and Einstein, and the music of Mozart and Mahler—are simply products of the subatomic particles constituting their brains, reductive materialism is the most audaciously speculative metaphysical position of all time.

Drees says that his project is to work "within the consensus view of the natural sciences" (RSN, 242). In claiming that his metaphysics does not require "a modification of science," Drees is really only stating his agreement with the currently dominant consensus as to the best framework for interpreting scientific data. But consensus, as Drees knows, is no guarantee of truth. He points out, for example, that today's "creationists advance positions which were part of the scientific consensus in geology and palaeontology some 200 years ago" (RSN, 243).

Drees does, however, finally state the only terms in which his form of naturalism could be justified: It must prove itself to be "comprehensive, coherent, and fruitful" and "without an equally satisfactory alternative" (RSN, 23). The "romantic" interpretation of science provided by process theology is not equally satisfactory, Drees claims, because it postulates "more fundamental entities or relations than one needs to account for all our experiences" (RSN, 2). As this statement reveals, Drees's argument finally hinges upon "Occam's razor," understood as the principle that one should not postulate more entities or relations than are absolutely necessary to account for all the phenomena. This is a good principle. The question, of course, is whether Whiteheadian naturalism is guilty of violating it, or whether Drees's less rich naturalism turns out to be too poor to account for "all our experiences." Drees's conclusion is that, although Whitehead's alternative cannot be excluded *a priori*, "currently there seems to be no compelling reason to abandon . . . materialistic naturalism" (RSN, 53). I will argue, by contrast, there *are* compelling reasons to abandon it, even apart from its inability to provide a basis for reconciling science and religion.

The Assumed Adequacy of Materialistic Naturalism

In general, Drees's attempt to demonstrate the adequacy of materialism to all our experiences could be described as casual. His assumption seems to be that, given the twofold fact that materialism is less metaphysical than the other forms of naturalism and is, in any case, the consensus view, it occupies a privileged position. Materialism can be assumed to be adequate, in fact, unless proven otherwise beyond a shadow of doubt. When there is debate within the scientific community about phenomena that, if genuine, would threaten materialism, we can assume, without examination, that those phenomena are not genuine. Even if materialists have not yet come up with an adequate theory to explain this or that indubitable fact, we can remain satisfied with materialism, trusting that an adequate theory will eventually be developed. Or, in some cases, we can simply dismiss the problem, saying that the demand for an adequate theory in those areas is unrealistic. My characterization of Drees's apparent assumption constitutes a serious charge, but a careful look at his book bears it out.

One problem is that the effort to understand the coherence of the sciences in terms of materialistic naturalism is, arguably, possible only by excluding various kinds of relevant data from the outset. Drees admits that "the coherence of our knowledge . . . might be seen as an artifact: we might have restricted ourselves to phenomena which could be dealt with in a coherent way" (RSN, 13). Drees, however, argues against this perception, saying instead that "there is a coherence across the variety of sciences which is not an artefact [*sic*] due to the way we organise science, but which tells us something about the natural world" (RSN, 13). Having already suggested that the coherence of the sciences achievable in terms of reductionistic materialism *is* in part an artifact of the fallacy of misplaced concreteness, I will later suggest that it is also an artifact of the omission of human subjectivity as such. For now, I will focus on the omission of the kind of human experience studied by parapsychology.

Drees seems to be aware of the fact that the phenomena of parapsychology, if genuine—meaning that the apparent influence at a distance cannot be explained away—would be threatening to the materialistic version of naturalism. And he is aware that the alleged phenomena cannot be rejected *a priori*. Speaking of "the claims in parapsychology regarding telepathy across spatial or temporal distances, apparently without a mediating physical process," he says that the rejection of such claims, which are "at odds with the scientific consensus," is "not beyond dispute" (RSN, 242). Also, although he classifies parapsychology with creation science, astrology, homeopathy, "and the like," thereby suggesting that it is "not worth the effort needed" to study it (RSN, 243), he concedes that "it is legitimate for some individuals to study claims about parapsychological phenomena" (RSN, 243). Nevertheless, he says:

> I personally do not consider it sufficiently promising to spend much time exploring parapsychology, since I consider the likelihood of positive results very slim and the possibilities of developing my work within the consensus view of the natural sciences more important. (RSN, 242–43)

The circularity involved in this reply is obvious: On the one hand, the main reason for sticking with the consensus view of materialistic naturalism is that it is said to be adequate to account for all human experiences. On the other hand, this materialistic view is used as the criterion for deciding *a priori* which alleged human experiences actually occur. Drees has not, accordingly, dispelled the suspicion that the account of the coherence of the sciences offered by materialistic naturalism is an artifact of selection.

Although I will provide further examples of selectivity below, the case of parapsychology is more important than it may appear at first sight. Drees's whole account of religion, as we have seen, is based on the assumption that it in no way arises from a direct, nonsensory perception of a Holy Reality transcendent to the totality of finite, physical things. His approach to morality, aesthetics, and mathematics is likewise based on the assumption that these human practices are in no way based upon a nonsensory perception (intuition) of a realm of abstract, ideal objects. The main basis for holding that no such perceptions occur is the sensationism of naturalism$_{sam}$, which says that we have no capacity for nonsensory perception. Parapsychological evidence for extrasensory perception, if it is genuine, shows that this assumption is false, thereby undermining one of the bases for trying to understand all cultural phenomena nontheistically.

This point provides a transition to a set of problems revolving around Drees's "evolutionary explanations postulate," according to which all natural and cultural phenomena can be explained in purely functional terms (inclusive fitness for survival), with no influence from God or a realm of abstract, eternal forms (such as truth, beauty, and goodness). One challenge to this postulate to which Drees responds is Plantinga's claim that an ontological naturalism, in Drees's sense, cannot give a satisfactory definition of "proper functioning." That is, the idea that bodily organs or limbs are functioning properly is not logically equivalent to any of the generally proffered criteria, such as those couched in terms of frequency or contribution to reproduction. We need not debate the cogency of Plantinga's argument but only to look at Drees's way of responding. He begins by suggesting that "an evolutionary understanding of proper function as advocated by Ruth G. Millikan (1984, 1993) is able to deal with the alleged counterexamples and objections." He offers, however, no hint as to what her approach is, and does not even indicate the pages in Millikan's books in which the crucial material is to be found. He then says:

> If I grant for the sake of the argument that there is currently no completely satisfactory definition of proper function. . . , Plantinga's conclusion to the falsehood of naturalism does not follow. The absence of a satisfactory general naturalist definition of proper functioning is not evidence that there can be no such a definition. The definition may have eluded us so far. However, it is more likely that the request for a strict definition is too demanding. (RSN, 153–54)

The problem with these responses is that they make the position unfalsifiable in principle. Drees says, rightly, that his form of naturalism is to be judged in terms of whether it can do justice to all our experiences. But when confronted by a concrete challenge to this adequacy, Drees defers the response to the indefinite future, then adds that the challenger may be asking too much. Drees seems to say, in effect, that nontheistic naturalism is itself to be the standard for deciding where it is appropriate to hold this form of naturalism to strict standards of intelligibility.

Drees also responds to the claim that Darwinian explanations cannot be provided for morality, religion, the appreciation of beauty, and the ability to do higher mathematics. With regard to the latter, Drees says: "The ability to do advanced mathematics can be understood evolutionarily as the use of cognitive capacities which evolved for other purposes (plasticity)" (RSN, 155). The problem with this appeal to plasticity is that it makes Darwinism itself so plastic as to be unfalsifiable. The original claim, to recall, is that "[e]volutionary biology offers the best available explanation for the emergence of various traits in organisms and ecosystems" and that "such explanations focus on the contribution these traits have made to the inclusive fitness of organisms in which they were present" (RSN, 19–20). But then, when it is pointed out that the mathematical ability of a Pythagoras, a Newton, or an Einstein would hardly contribute to the survivability of early humans in a hunting-gathering society, the answer is that this mathematical ability must have come about as a necessary concomitant of some other trait that did contribute to the chances of survival in that context. Given the ability to resort to such a dodge with regard to every trait that cannot be explained in terms of inclusive fitness for survival, Darwinism becomes as unfalsifiable as the systems its proponents dismiss as "metaphysical."

The importance of this dodge should not be underestimated. The claim that "Darwinian histories" are adequate in principle to account for all evolutionary emergents, including distinctively human forms of experience, involves the claim that no forms of human experience are to be explained in terms of nonsensory intuitions of some timeless, Platonic realm transcending "nature" understood as the totality of finite, physical entities. By contrast, many theists—both supernaturalistic theists, such as Plantinga, and naturalistic theists, such as Whitehead (who had been a mathematician)—believe that mathematical experience does imply contact with such a realm, so that a strictly Darwinian account, taken to deny such contact, is inadequate. Indeed, in Whitehead's case, one factor that moved him toward theism was the inability to understand how mathematical objects—being ideal, nonactual entities—could be efficacious in the world in general and in human experience in particular[5]—in fact, how they could even exist—unless they existed in a primordial actuality. This is a serious challenge to Drees's materialistic version of naturalism that comes not from religion and morality but from the attempt to render physics itself intelligible, insofar as it presupposes mathematics. And yet Drees offers no substantial answer to this challenge. Having rejected the existence of a timeless, Platonic realm (RSN, 216, 218), he asserts that mathematics, as a "second-order" activity, "may be construed without reference to a realm of abstract objects apart from the natural realm with all its particulars" (RSN, 221). This, however, is a mere assertion. Drees provides no way of understanding the objectivity of mathematics apart from the presumed intuition of such a realm.[6]

Drees's response to the problem raised by religion is no more satisfactory. As we saw earlier, he applies the functional, evolutionary explanation, saying that "religions arose, and therefore probably contributed to the inclusive fitness

---

5. This problem has played a big role in the philosophy of mathematics. Part of the problem is due to the sensationist assumption that we could not perceive ideal entities even if they existed. Although Kurt Gödel famously suggested that we should have as much confidence in mathematical "intuition" as in sensory perception (WC, 268), many philosophers of mathematics have rejected this suggestion (Chihara AGT, 217; Putnam WL, 503). The problem became even more vexed when it was argued by Paul Benacerraf (MT) that we can perceive things only if they can exert causal efficacy on us. This notion has been widely accepted. For example, Penelope Maddy says that "to be abstract is also to be causally inert" (RM, 37). Reuben Hersh has pointed out how this problem was created by modern nonbelief in a divine being who could make such entities causally efficacious (WIM, 12).

6. Most philosophers of mathematics, including those who reject the idea of a "Platonic" mathematical realm for the reasons discussed in the previous note, agree that most mathematicians in practice presuppose the objective existence of such a realm (Hersh WIM, 7; Maddy MR, 2–3).

of the individuals or communities in which they arose" (RSN, 250). But Drees does not defend the adequacy of any actual view as to how they might have done this. After very briefly mentioning several views as to what the functional role of religion might be, and without responding to criticisms of exclusively functional explanations, he simply says: "Here I will not defend one particular view of the function of religions, but rather reflect on some general implications of such naturalist views of religions as functional cultural practices" (RSN, 250). In short, far from showing the adequacy of a nontheistic naturalism to account for the existence of religion, he simply assumes it.

The fact that Drees evidently feels no need to defend the adequacy of his position seems to be based upon his assumption that it occupies a privileged position, so that it can be assumed to be innocent of inadequacy unless absolutely proved guilty. For example, having pointed out that theologians might argue the need for theism by considering "the incompleteness of any naturalist explanation," he says:

> In my view, limitations in our knowledge are not to be seized upon for religious apologetics; the absence of evidence does not count as evidence of absence. If we do not know which actual Darwinian history explains a certain feature, it does not follow that there is no actual Darwinian history. It would only be evidence of absence if we were quite sure that we had explored all the possibilities in such a way that decisive pieces of evidence could not have eluded us. (RSN, 247)

The privileged position of Darwinian naturalism in Drees's mind is shown by the fact that he does not counsel critics of theism, before rejecting the idea of divine influence in the world, to make sure that all the possibilities for conceiving such influence have been considered. Rather, all such possibilities can be ignored *a priori* as long as there is a chance that a Darwinian explanation may exist, even if it has eluded Darwinian thinkers for 150 years.

His defense of the adequacy of a materialistic, Darwinian approach to morality is, if anything, even more problematic. Having reviewed the attempts of sociobiologists, such as E. O. Wilson, to provide a Darwinian explanation of morality, Drees points out their inadequacy:

> The view that all moral judgements are forged upon us by our past . . . seems to me to be insufficient for morality; it still identifies the moral justification with an explanation of how we came to have preferences which we do turn out to have; there is no room for a contrast between "what is" and "what ought to be." (RSN, 218)

Given this inadequacy of sociobiological explanations of morality, he states forthrightly the resulting problem for his version of naturalism:

However, upon a naturalist view as developed here, there seem to be no other sources for substantial moral judgements than the heritage of our biological and cultural past. There is no room for the justification of ethical decisions in relation to entities in some Platonic realm, as if we come to hold moral principles by intuiting an absolute moral order. (RSN, 218)

Nevertheless, he suggests, the situation may not be hopeless. "A procedural view of moral justification such as offered by [John] Rawls (1971)," he suggests, "may be compatible with an evolutionary view" (RSN, 219). However, other than citing *The Biology of Moral Systems* by R. D. Alexander, he gives no defense of this possibility except the following statement: "Ethical objectivity need not be linked to a realm of ethereal entities, such as abstract values. Rather [he argues, quoting Philip Kitcher], it 'involves the existence of a standard beyond personal wishes, a standard in which the wishes of others are given their place'" (RSN, 219). He does not explain, however, where this "standard" exists or how it differs from "ethereal entities, such as abstract values," which he had declared to be nonexistent.

His discussion then becomes even more problematic. Suggesting that this procedural form of ethical justification could be complemented by sociological dimensions, he makes the following statement, which suggests that he does not understand the implications of his own worldview:

> [W]e reflect upon our moral intuitions, and thus consider whether they have certain general features which we consider desirable. . . . In our reflection, we may test our moral judgements by criteria such as generality and disinterestedness. . . . We owe our intuitions to the evolutionary past, but they can be considered and corrected, since we have the ability to evaluate our primary responses. . . . [G]enuine ethical behaviour does not come to us "by nature," but rather requires moral effort. . . . Formal analysis, the application of criteria such as disinterestedness and coherence, and the moral deliberation of many people together are important for the credibility of morality, precisely because they surpass and may correct the conclusions of our ordinary biological and psychological mechanisms. . . . [O]ne could say that our moral intuitions are explained by sociobiology, but that these intuitions need not be our best ethical conclusions, since we can reconsider them. (RSN, 219–21)

This statement contains at least three points that are inconsistent with Drees's reductive materialism. First, having denied the existence of abstract values, pejoratively calling them "ethereal entities," Drees presupposes them, speaking of "criteria" such as "disinterestedness." Second, after saying that all of our moral intuitions come from our evolutionary past, being explained in terms of biological and psychological mechanisms, he then assumes the existence of

other, higher intuitions, in terms of which we can "evaluate our primary responses," thereby arriving at "our best ethical conclusions." This notion of
higher intuitions (through which we make the transition from the "desired" to
the "desirable") implies the nonsensory perception of a realm of moral norms,
perhaps even the existence of a divine mind making these norms efficacious—
all the ideas from which his account was supposed to prescind. A third problem is that the idea of "moral effort" presupposes a notion ruled out by Drees's
version of naturalism: freedom, in the sense of self-determination in terms of
a goal.

The Problems of Subjectivity and Freedom

This tension with regard to freedom is, in fact, arguably the major problem in
Drees's position. I will approach this problem, which Drees scarcely acknowledges, by beginning with the problem of subjectivity, which he does clearly
acknowledge.

     As we saw earlier, Drees's view that scientific explanation is always reduc-
tionistic entails that our "manifest image" of ourselves, as persons having an
inner life with emotions and apparent freedom, must be ontologically reducible
to (meaning explainable in principle in terms of) the entities and forces of
physics. Drees is aware, however, that many thinkers have considered human
subjectivity to be the other topic, along with the very existence of the world,
most likely to "escape the omnicompetence of the natural sciences." However,
calling this claim "disputable" (RSN, 114), Drees disputes it simply by register-
ing his "impression" that a reductionistic approach might work so that consciousness "can perhaps be understood naturalistically" (RSN, 102, 183). Nevertheless,
realizing that the problem is serious, he adds:

> There is, of course, a difference between the experience from within and
> a description from the outside. While I experience love, hate, or boredom,
> the scan shows electrical and chemical processes. Can scientific insights
> and philosophical analysis explicate how the experience from within has
> come into existence. . . ? (RSN, 114)

Saying, furthermore, that "subjectivity seems to be a major challenge to a naturalist view of reality," he "point[s] to philosophical literature which attempts to
answer that challenge" (RSN, 183). However, as with other issues, he considers
it unnecessary "to make a choice among various competing approaches," thinking it sufficient to "indicate some of the ways in which a naturalist view of the
mind might be developed" (RSN, 184). Materialism can be assumed to be true,
he seems to assume, even if no satisfactory materialistic solution to the mind-
body problem has yet been found.

This attitude is especially disturbing in the light of Drees's recognition of the fact that Thomas Nagel, one of the most astute analytic philosophers to have dealt with the mind-body problem, has concluded that the kind of reductionism required by Drees's general position is impossible. Drees can accept Nagel's analysis insofar as it involves a rejection of eliminative materialism, according to which first-person language, such as "consciousness," "emotions," "feelings," and "decisions," can be eliminated in favor of physicalist, third-person language, such as language about the firing of neurons (RSN, 183, 187, 189). But Drees cannot agree with Nagel's conclusion that "the subjectivity of consciousness is an irreducible feature of reality" so that "it must occupy as fundamental a place in any credible world view as matter, energy, space, time, and numbers" (VN, 7–8). This view, Drees sees, is incompatible with "constitutive reductionism" combined with "the belief that physics is the science of the fundamental aspects of natural reality" (RSN, 188).

Faced with the challenge of Nagel's position, Drees simply says: "I consider the position unlikely" (RSN, 189). Although he recognizes that Nagel might be right, he says that

> it is too early to give up. I consider it likely that somewhere in the realm indicated by Churchland, Dennett, and Searle . . . , there is a possibility for a future theory of mental life which is naturalist in my sense. (RSN, 188)

Again, he has argued the adequacy of his version of materialism by appealing to the future. This appeal, furthermore, flies in the face of the fact that, as we will see in Chapter 6, a growing number of materialist philosophers are contending, even more emphatically than Nagel, that the problem is insoluble *in principle*, so that further time is not relevant.

Equally serious is the inconsistency in Drees's position with regard to freedom. Drees clearly presupposes the reality of freedom. He refers to "moral effort" and "the capacity to make moral deliberations" (RSN, 219, 220). He speaks of "mak[ing] a choice" (RSN, 184) and even refers to freedom as "self-determination" involving "rational reflection on my past actions and potential consequences of various options" (RSN, 216). How is this all possible if our behavior, like that of chairs and tables, is explainable in principle in terms of the particles and forces studied by physics?

The qualifying phrase "in principle" is important: The one major claim of Drees's naturalism not yet mentioned is "conceptual and explanatory non-reductionism," which says:

> The description and explanation of phenomena may require concepts which do not belong to the vocabulary of fundamental physics, especially if such phenomena involve complex arrangements of constituent particles or extensive interactions with a specific environment. (RSN, 16)

This principle, Drees believes, distinguishes his position from the other alternative position that, along with a richer naturalism, he considers most challenging, namely, the dismissal of all forms of religion on the basis of "a more radical naturalism" (RSN, 236). It is only by insisting on conceptual and explanatory non-reductionism, he says, that he can have even a minimalist religion. Unless one can talk about *consciousness, values, rationality,* and *choices*, religion would be nonsensical. The question, however, is whether Drees's ontological reductionism allows him to use such language legitimately.

The basic problem is that, although he speaks of explanatory (as well as conceptual) nonreductionism, he does believe that all complex phenomena, including human behavior, are *in principle* explainable in terms of their constituent particles. He is an explanatory nonreductionist only in the sense that, because of our limited cognitive capacities, we will never *in fact* be able to explain all things in terms of the principles of physics. We have to use concepts appropriate to the various levels, such as those of chemistry, biology, and psychology (RSN, 16–17). But the resulting relative independence of the various sciences does *not* imply "ontological non-reductionism" (RSN, 16). Although we cannot conceptualize how this is possible, the truth, we are to assume, is that all human experiences and behaviors, including all human decisions, are ontologically reducible to the causal forces working at the subatomic level. What Drees calls *explanatory* non-reductionism, accordingly, is really only *conceptual* non-reductionism. Indeed, in an apparently approving summary of Daniel Dennett's position, Drees says that "even though this view is eliminative at the level of understanding, it is not (as Churchland's proposal is) eliminative at the level of language." As a consequence, Drees says, "we will continue to say that we drank water 'because we were thirsty'" (RSN, 186). Yes, we will continue to *say* this, but what we are supposed to *understand,* according to Drees, is that we really drank the water because our subatomic particles made us do so. The same would be true, of course, for our ethical acts.

The difficulty in Drees's position here is shown by his discussion of whether concepts applying to the levels above physics are superfluous. He at first says they are not: "Naturalism need not exclude the meaningfulness and non-superfluous character of concepts which are involved in explanations in sciences other than physics" (RSN, 15). Later, however, he indicates otherwise. Discussing the idea that higher theories, such as psychology, are in principle (even if not practically) dispensable, he says:

> If a theory is superfluous, it is not thereby wrong. Rather, if one could derive the superfluous theory T_1 from the more fundamental theory T_2 the first theory would not be autonomous, but it would still be a good theory for the domain with which it deals. (RSN, 192)

It cannot be a "good theory," however, if the causal concepts it employs are misleading because all the causality really occurs at the subatomic level. Drees sees this point clearly when discussing the traditional scheme of primary and secondary causation, according to which God's complete determination of all events allegedly does not deprive the creatures of their own causal agency. Speaking of the problem of "double agency," Drees says that "once one allows for two different sufficient causes causing a single event, one of them seems superfluous" (RSN, 261). He is right. But he evidently fails to see that, by analogy, if our actions are fully caused by subatomic particles, language attributing agency to the mind is superfluous. Far from contributing to a "good theory," concepts referring to the mind's "deliberations," "decisions," and "moral efforts" would obfuscate the true causal relations.

That is the conviction behind Paul Churchland's eliminative view, according to which all such mentalistic language is to be eliminated in favor of an account provided by the neurosciences. Against this view, Drees argues that the commonsense ontology is compatible with the ontology implicit in the scientific theory. Drawing on the analogy of the two tables, he says:

> Just as quantum physics does not eliminate solid tables, but leads us to a different conception of them, so too would a different conception of mental states in some future psychological theory, for instance in terms of neurology, not thereby eliminate the states. (RSN, 194)

The issue, however, is not whether the conscious states would be eliminated, but whether any efficacy could be assigned to them. Drees's position entails a negative answer. Indeed, he himself contrasts his reductionist approach with that of "authors who appeal to top-down causation in order to understand the mind-brain relationship" (RSN, 102). His reductionism implies, by contrast, that all vertical causation goes *upward*, from the subatomic particles to the person as a whole, so that the person as a whole has no autonomous power to exert influence back upon these particles. Drees insists, in fact, that unpredictability at the quantum level does not imply "indeterminacy or openness to non-natural influences, either from humans or from God" (RSN, 247). With that point, which implies that an influence based upon a partially free decision would be "non-natural," we return full circle to Drees's assumption that naturalism entails that all events in the world are controlled by deterministic laws of nature, so that nothing, including human behavior, can be validly explained by reference to the partially autonomous choice of a human mind.

The conclusion of all this is that Drees, besides not showing that a significantly religious outlook can be harmonized with the materialistic version of scientific naturalism, also has not supported his claim that this version of naturalism can "account for all our experiences." He has not, therefore, shown that the "richer naturalism" based on Whitehead's philosophy is unnecessary either for reconciling science and religion or even for science itself.

Summary

None of the proposals examined in this chapter provide an adequate basis for overcoming the present conflict between the worldviews of the scientific and the religious communities. Each proposal, however, does contain important points. The Plantinga-Johnson proposal for a theistic science recognizes not only that a true harmony will require that the scientific and religious communities share a common worldview, but also that this worldview cannot be that of naturalism$_{sam}$. More distinctively, this proposal recognizes that there is no reason why, if the evidence seems to require it, science could not speak of divine activity, even *variable* divine activity, that is constitutive of events in the world. Science, accordingly, need not be methodologically atheistic.

The problematic part of the first proposal, that theistic science would speak of supernatural interruptions of the world's normal causal processes, is rightly rejected by the second proposal, which holds that any realistic proposal must recognize that science as such will remain methodologically naturalistic. The distinctiveness of this second proposal—when carried out consistently (as by Otto) and extended to become a proposal to overcome the conflict of worldviews—is the recognition that, for harmony to be achieved, adjustments are needed in *both* communities: The religious community must fully reject the idea of supernatural interruptions of the world's normal causal processes, while the scientific community must return to an overall theistic framework. This proposal, however, shares the weakness of the first proposal, which is the equation of theism with its supernaturalistic version, so that, if harmony is to be achieved, science with its methodological naturalism would have to be placed within a supernaturalistic worldview.

The strength of the third proposal is its recognition that, if there is to be harmony, it will have to be on the basis of a worldview that is naturalistic$_{ns}$. This third proposal also recognizes that the acceptance of naturalism in this sense will provide the basis for a better religion, although this insight could not be demonstrated by Drees because of his equation of naturalism$_{ns}$ with naturalism$_{sam}$. The proposal of the present book, which is that Whiteheadian theistic naturalism provides a framework adequate for both science and religion, intends to integrate all of these insights into a coherent position.

Crucial to this proposal is its suggested linguistic reform, which, if accepted, would mean an end to the practice, now rampant, of contrasting "naturalistic" and "theistic" views of the universe. By accepting this contrast, we not only imply that a naturalistic view is necessarily atheistic, but also that a theistic view is necessarily supernaturalistic. We thereby rule out the compatibility of a theistic science with naturalism. The convention that "science cannot speak of God" can be challenged, but this challenge will ignored as long as the proposal for a theistic science seems to be a proposal to return to a supernaturalistic science. We should, accordingly, cease speaking as if a view by virtue of being "naturalistic" were thereby opposed to a theistic view of the universe. When we want to refer to atheism, there is a perfectly good word available: "atheism."[7] The proposed linguistic reform, therefore, is that "naturalism" be used only to denote what the term itself connotes, which is the rejection of supernaturalism. Given this usage, we can talk about a "theistic naturalism" and even a "theistic science" without implying anything contrary to the scientific community's most fundamental ontological commitment, which is to naturalism$_{ns}$.

This point is crucial to the prospects for a worldview held in common by the scientific and the religious communities. Unless science itself again becomes theistic, the worldview of civilization will not be theistic. The proposal to think of science as simply a limited way of knowing, valid only for a limited portion of reality, simply will not work. The fact that the word "science" *means* "knowledge" is not simply a linguistic truth with merely historical import. For most people now, to say that some idea is "nonscientific" is little different from calling it "anti-scientific," which is in turn little different from calling it false. Unless it becomes the consensus of the scientific community that it is necessary to refer to divine influence to explain the world (including human experience), the perceived conflict between science and theistic religion will not be transcended.

If a worldview involving divine influence is to have a chance at becoming the worldview of the scientific community, it is essential to remember that this community requires only a naturalism that excludes the possibility of supernatural interruptions of the world's normal cause-effect relations. What often happens, however, is that this definition becomes equated, more or less unreflectively, with some more restrictive conception of what "naturalism" means, such as the idea that "nature is all there is," with "nature" understood to mean "the physical world" or "the world as knowable through sensory perception." With *that* conception of naturalism, a worldview such as Whitehead's appears *not* to be a form of naturalism, so that it is excluded from the search for a naturalistic worldview that can do justice to all of our experience. By keeping in mind that the naturalism required by the scientific community is *only* naturalism$_{ns}$, we can see that a worldview involving divine influence would not necessarily be excluded.

---

7. The same point can be made against the widespread practice of using "naturalism" for materialism or sensationism. The terms "materialism" and "sensationism" are available to designate these doctrines without ambiguity.

4

WHITEHEAD'S SCIENTIFIC AND RELIGIOUS NATURALISM

The conflicts between science and religion, I am arguing, can be effectively overcome only through the adoption of a form of naturalism that is satisfactory from both a scientific and a religious viewpoint. In the previous chapter, I examined various ways in which this conclusion might be avoided—by overcoming the association of science with naturalism$_{ns}$, by combining a semideistic form of supernaturalism with a methodological naturalism, and by accommodating religion to naturalism$_{sam}$. Having argued that none of these approaches succeed, I turn in the present chapter to Whitehead's philosophy, which, I am proposing, provides the kind of naturalism that is needed to bring about harmony between scientific and religious convictions.

The chapter develops in the following way. In the first section, I put Whitehead's philosophy in the context of the twentieth-century quest for a more open, non-reductionistic naturalism. In the second section, I lay out Whitehead's naturalistic theism, thereby supporting one of the theses of this book, which is that genuine theism need not involve supernaturalism. The third section is devoted to an introduction to Whitehead's more open naturalism (which not only Drees but also J. B. Pratt, to be discussed below, called his "richer naturalism"), indicating in a preliminary fashion some ways in which it permits experiences and beliefs that are ruled out by the materialistic version of naturalism. In the fourth section, I point to some ways in which recent developments in science support this richer, more open naturalism.

Toward a More Open Naturalism

Deweyan Naturalism

The cultural conflict between religion and naturalism, I have argued, could be overcome only through a worldview that combines two elements: a nonsupernaturalistic religion, on the one hand, and a larger, nonreductive naturalism, on the other. In 1944, a movement with this twofold focus published a volume entitled *Naturalism and the Human Spirit*. Although the various contributors to this volume had been shaped by diverse influences, the paramount influence was that of John Dewey, himself a contributor to the volume. We can, therefore, refer to the vision expressed in this volume as "Deweyan naturalism." I will employ primarily the contributions by Yervant Krikorian, who, besides serving as editor also contributed "A Naturalistic View of Mind," Sterling Lamprecht, who contributed the essay on "Naturalism and Religion," and John Herman Randall, Jr., who wrote the epilogue, "The Nature of Naturalism."

As Lamprecht put it, the warfare between science and religion has been due to the fact that "a premature naturalism or an antiquated religion or both have been current" (NHS, 17). The Deweyan naturalists would overcome both of these sources of conflict. Before looking at their updated religion, which involved the rejection of all forms of supernaturalism, we need to look at their more mature naturalism, which, as Randall put it, involved the transition from a modern to a "post-modern" naturalism. (Appearing in 1944, this must be one of the earliest uses of the term "post-modern," especially in application to a philosophical position.)

Modern naturalism, being mechanistic, materialistic, and reductionistic (NHS, 17, 244, 360), had ruled ideals out of nature, thereby denying reality to the religious, moral, and artistic life of humanity (NHS, 22, 369). Modern naturalism's hostility to religion was based not upon anything inherent to a naturalistic approach, but upon a too restrictive view of nature, which encouraged a reductionistic method (NHS, 28, 371). The result was the distinctive problem of "modern philosophy," namely, "the conflict of the moral and religious tradition with newer scientific concepts" (NHS, 368). But now the maturing of science has made possible the emergence of "'post-modern' naturalistic philosophies" (NHS, 369). This newer naturalism opposes the reductionist position—which hitherto "often arrogated to itself the adjective 'naturalistic'"—as strongly as it opposes all forms of supernaturalism (NHS, 58). This new naturalism, which includes a natural teleology, "can find a natural and intelligible place for all human interests and aims, and can embrace in one natural world, amenable to a single intellectual method, all the realities to which human experience points," including religious experience (NHS, 379, 369). These are heady claims; we need to look more closely to see whether, or at least in what sense, they are true.

A major emphasis of this movement is that the term "naturalism" refers primarily to a method. The emphasis is on "continuity of analysis," which means applying one method—the scientific, empirical method—to all problems (NHS, vi, 18, 242, 357). This emphasis, which involves the rejection of epistemic supernaturalism, thereby implies the rejection of ontological supernaturalism: Lamprecht's statement that the world is regarded as "an interrelated whole without intrusions from some other 'realm'" (NHS, 20) was quoted in Chapter 2. Lamprecht's more complete definition shows that this rejection of "intrusions" into the natural realm is said to be less restrictive than it was in the materialistic, reductionistic forms of naturalism.

> [N]aturalism means a philosophical position, empirical in method, that regards everything that exists or occurs to be conditioned in its existence or occurrence by causal facts within one all-encompassing system of nature, *however "spiritual" or purposeful or rational some of these things and events may in their functions and values prove to be.* (NHS, 18; emphasis added)

On this basis, which allows the natural to include purposeful, rational, and even spiritual causes, the reconciliation of scientific naturalism with "a religious view of the world" is said to be possible (NHS, 17).

Of course, this reconciliation presupposes, as already indicated, a religious view that is not supernaturalistic. For Lamprecht, this means understanding the religious life as "a life in which multiple interests and diverse values are brought into effective and organic unity through central allegiance to some integrating ideal" (NHS, 20). More briefly, the religious life at its best is a "fulfillment of life's most urgent moral needs" (NHS, 21). These statements suggest that a religious view of the world needs only to provide the presuppositions of the moral life. For the Deweyan naturalists, furthermore, these presuppositions are even more minimal than they were for Immanuel Kant. The rejection of supernaturalism means, we learn, that there is no "transcendental God" (NHS, 358). Actually, we are told that naturalism "would remain intact if God were discovered to exist" (NHS, 36). But we are also told that all the arguments for God fail, and that, in any case, naturalism must "so interpret the world as to make it . . . seem 'alien' to those whose beliefs center in a theistic interpretation of the world" (NHS, 36, 30). Naturalism is said to rebel "against the claim that life ought to be centered in and guided by divine will or providence" (NHS, 36).

As these statements suggest, the naturalism of this movement excluded any significant belief in God. This fact is further shown by Randall's statement of its attitude toward the "rather solitary figure of Whitehead," who, as "one of the pioneers of contemporary naturalism, . . . has argued brilliantly and powerfully against dualisms and in favor of a continuity of analysis" (NHS, 367). In spite of the fact that Whitehead had thereby exemplified the movement's criteria for naturalism, there were doubts about the "consistency of his somewhat dubious naturalism." The only reason given (NHS, 367) for these doubts was "a rather widespread suspicion of the principle of concretion"—which was the term Whitehead had given to the first version of his doctrine of God, which was not yet even a personal deity but a mere principle. This principle was, however, transcendent to the world of flux, and the allowance of this transcendence was sufficient to render his naturalism "dubious."

This response to Whitehead reflects a further doctrine of Deweyan naturalism, which is that "nature is the whole of reality," meaning that there is no other realm, even a realm consisting entirely of forms or ideals, that transcends the natural processes constituting nature (NHS, 243). Naturalism, in other words, "means the refusal to take 'nature'. . . as a term of distinction. . . . 'Nature' serves rather as the all-inclusive category" (NHS, 357). The restrictiveness of this conception becomes even more evident when we read from other authors in this volume (Abraham Edel and Ernest Nagel) that naturalism holds to "the primacy of matter" and "seeks to understand the flux of events in terms of the behaviors of identifiable bodies" (NHS, 65, 211).

This aspect of Deweyan naturalism raises the question as to whether it can do justice, as it claims, to all kinds of human experience, including religious experience. Given the denial of the existence of God, how can it do justice, in particular, to the widely reported experience of something holy or divine? In line with Dewey's view (expressed in *A Common Faith*) that the term "God" should be taken to refer to a cluster of ideals created by the human mind, the members of this movement hold that "divinity belongs, not to what is existent, but to what man discerns in imagination" (NHS, 358). This view would certainly not be regarded by most mystics as doing justice to their experience.

Beyond that point, furthermore, Deweyan naturalism cannot even do justice to the moral experience to which they reduce religious experience. On the one hand, ethics is said to be based upon a purely empirical account of human preferences (NHS, 262). On the other hand, however, the Deweyan naturalists did recognize (as does Drees) that ethics deals not simply with what happens to be *desired* or *preferred* by particular human beings, but also with what is *desirable* or *preferable*, and statements about what is preferable are said to be capable of being true or false (NHS, 262–63). The problem is that an analysis that is purely natural or empirical, in their sense of those terms, cannot move from the preferred to the preferable. That move would presuppose ideal standards that transcend the flux of natural processes, including the human processes of preferring this or that, and the Deweyan naturalists deny the existence of such ideals. Although they claim that their enlarged naturalism has room for human ideals, this is true of ideals only in the descriptive or factual sense: There is no room for *normative* ideals. As William Shea has argued in *The Naturalists and the Supernatural,* the difference between a religious and an nonreligious view turns on "whether the values immanent in experience find a resonance in the cosmos" (NS, 25–26). The worldview of Deweyan naturalism, by denying this, does not do justice even to its own minimalist conception of religion.

The inability of these naturalists to do justice to moral experience, let alone religious experience in richer senses of the term, is due also to another feature of their position that is stressed by Shea: their equation of perception with sensory perception (NS, 79–80). Given that equation, their empiricism must be the traditional *sensory* empiricism, as distinct from the *radical* empiricism of William James. And if they are to consider this sensory empiricism to be capable of dealing with all of reality, they must deny the reality of anything—a divine being or even transcendent ideals—that could be experienced only through a *nonsensory* perception.

Besides failing with respect to normative ideals, Deweyan naturalism also fails with respect to freedom, another precondition of the moral life. Otto, as we saw, could defend freedom only by resorting to a dualism between nature and the human soul, whereas Drees, eschewing any distinction between mind and brain whatsoever, could not affirm genuine freedom. These Deweyan naturalists, however, claimed that, given their enlarged naturalism, they could affirm freedom without dualism, even without thinking of the mind as an entity distinct from the brain (NHS, 257–58). But this claim, made by Krikorian, dissipates upon examination. Like Otto and Drees, he rejected any panpsychist or panexperientialist conception, according to which experience and spontaneity characterize the most fundamental units or processes of nature. Krikorian held, instead, that experience, with its purposive freedom, is an emergent property, depending upon complex organization (NHS, 247–49). However, as more recent discussions of the mind-body problem have shown (see Chapter 6), this idea that freedom can somehow emerge out of complex organization of nonpurposive entities is incoherent. As we saw in Chapter 2, William Provine, supporting this conclusion from the perspective of neo-Darwinism, says: "There is no way that the evolutionary process as currently conceived can produce a being that is truly free to make choices." Krikorian's own analysis, in fact, supports this conclusion. The difference between his position and that of the older naturalism, it turns out, is only that between a "strict" and a "mild" mechanism. Even in his so-called mild mechanism, furthermore, entities are related in such a way that, "given their present condition, one can predict their future" (NHS, 244). His defense of free, purposive action depends, accordingly, upon regarding it as compatible with a completely predictive mechanism (NHS, 48, 248). To affirm freedom only in this Pickwickian, compatibilist sense, however, is not to affirm freedom in the sense presupposed by morality, which is the freedom to have done otherwise, as even some materialists say (again, see Chapter 6).

If this form of naturalism cannot do justice to the religious belief in God, normative ideals, and freedom, the same is true for that other belief Kant took to be basic: life after death. The impossibility of life after death is implied by Krikorian's analysis of the mind-body relation summarized above, which leads to the conclusion that the mind's unity is not an entity distinct from the body (NHS, 269). The point is explicitly made by Randall, who says that the rejection of supernaturalism means that there is "no personal survival after death" (NHS, 358). In contrast with James's radical empiricism, which led him to deal with the question empirically (by means of psychical research), the Deweyan naturalists, in spite of their claim to be fully empirical, settled the question *a priori.*

The Naturalism of James Bissett Pratt

The Deweyan movement did not succeed in showing the compatibility of "a religious view of the world" with scientific naturalism, because its "post-modern naturalism" was, in reality, little different from the modern, reductionistic naturalism that it claimed to leave behind. Overcoming the conflict would require a significantly larger, richer naturalism. Such was suggested by James Bissett Pratt in a 1939 book simply titled *Naturalism.*

Like others, Pratt pointed out that naturalism means the rejection of a supernaturalistic worldview: It "appeals to nothing outside of Nature but looks for its explanations to nothing but Nature itself" (N, 5, 93). Naturalism entails, he emphasized, the rejection of an authoritarian method in favor of an empirical method for seeking truth (N, ix–x, 10, 140).

Like the Deweyan naturalists, furthermore, Pratt contrasted two kinds of naturalism. The earlier one, which had until recently been in possession of the field, was crude and dogmatic (N, 19, 43). Besides affirming materialism and atheism, it was unempirical: Turning its tools into idols, it attacked any data that did not fit its preconception (N, 9, 135–38). The more recent naturalism is critical and empirical, realizing that the empirical method means that naturalism cannot be equated with any given system (N, ix–x, 16, 43, 140). The result is a "richer naturalism," involving an integration of new facts with old (N, 9). To indicate what he meant by this richer naturalism, Pratt referred to "Whitehead and his followers" (N, 42).

Unlike the allegedly new naturalism offered by the Deweyan movement, the richer, more liberal naturalism suggested by Pratt does indeed contain several elements that go significantly beyond the materialistic, reductionistic versions. First, naturalism need not hold that all causation is mechanical, but can with perfect consistency recognize the reality of teleological, purposive causation (N, 135–36). In this regard, one of Whitehead's sardonic quips—"Scientists animated by the purpose of proving that they are purposeless constitute an interesting subject for study" (FR, 16)—is cited by Pratt (N, 137–38). In the second place, naturalism can, according to Pratt, hold that some kind of immanent, dynamic teleology pervades not only the whole living world but also the nonliving world (N, 87). Third, naturalism need not equate the mind with brain processes, but can affirm that mind and brain interact (N, 118, 134). In the fourth place, Pratt said: "It is perfectly consistent with a very real Naturalism to take into serious consideration the hypothesis that the Cosmos as a whole is permeated with immanent purpose, that it is a teleological and, therefore, a spiritual organism" (N, 142).

This fourth point would mean thinking in terms of a indwelling mind of the cosmos. With regard to how to conceive the immanent cosmic purpose as efficient, Pratt suggested that "its relation to the total physical universe would, in outline, be somewhat like the relation between the human will and the human body which it inhabits." The idea of an organic universe, accordingly, leads to "the conception of an indwelling mind which expresses itself in all the activities, great and small, of the Cosmos" (N, 165–66). This type of cosmic teleology, unlike the "concept of a transcendent God acting from without upon a dead material world," Pratt said, is "consistent with a liberal Naturalism" (N, 174–75).

Thanks to his richer naturalism plus his belief that religion does not require supernaturalism, Pratt was able to affirm a convergence of religion and naturalism. This convergence, in fact, was Pratt's main thesis. On the one hand, the new naturalism, which recognizes within the cosmos a teleological character and thereby a cosmic mind, "is not far from Religion" (N, 172). On the other hand, religion has, at least in some circles, undergone a complementary movement: "Beginning . . . with a view essentially supernatural, it has tended to a steadily more and more orderly picture of Reality" (N, 167–68). Whereas God was earlier employed "to stop gaps in natural events not otherwise explicable," God is now coming closer and closer "to that Nature toward which Naturalism also seems to be tending" (N, 168). Pratt, in this connection, uses Spinoza's phrase *Deus sive Natura,* "God or Nature." It is clear, however, that Pratt meant this phrase not in Spinoza's sense, which is either atheistic or pantheistic, but in the *pan-en-theistic* sense that God is the mind or soul of the universe, interacting with it analogously to the way in which the human mind interacts with its body.

Pratt's *Naturalism* had, in 1939, finally suggested a basis upon which a religious view of the world and the naturalism required by science could be compatible—by providing a worldview that is fully naturalistic and yet genuinely religious. But Pratt's little book (which was a series of lectures) did little more than point to such a basis. It did suggest how mind-body interactionism and human freedom can be conceived without dualism and, on that basis, how a real theism can be conceived without supernaturalism. Beyond its suggestiveness, however, Pratt's book does not go very far in helping us develop clear, self-consistent conceptions of the mind-body relation and the God-world relation.

And it does not even address the issues of religious experience, life after death, and the status of normative ideals. To fill out the sketch that Pratt has provided, I now turn to the naturalistic worldview by which Pratt's book was primarily inspired, that of Alfred North Whitehead.

WHITEHEAD'S NATURALISTIC THEISM

The conflicts between science and religion, we have seen, have two primary sources: a commitment by much of the religious community to a supernaturalism that conflicts with the naturalism presupposed by science, and a commitment by the dominant portion of the scientific community to a formulation of this naturalism that conflicts with beliefs presupposed by religion. Whitehead's philosophy can be read as one long effort to overcome both of these sources of conflict by developing a naturalistic worldview that is adequate for both scientific and religious interests. As we saw in Chapter 1, Whitehead said that philosophy "attains its chief importance" by fusing religion and science into one rational scheme of thought (PR, 15). The way to do this, he suggested, is to overcome previous exaggerations on both sides, thereby effecting a reconciliation involving "a deeper religion and a more subtle science" (SMW, 185). In the next section, I will introduce Whitehead's way of overcoming the primary exaggeration from the side of science, which is the mechanistic, materialistic view of the ultimate units of nature. In the remainder of the present section, I discuss his way of overcoming the primary exaggeration from the side of religious belief, which is the supernaturalistic version of theism.

Although Whitehead had long been an atheist or least an agnostic, soon after beginning to think metaphysically he came to agree with Aristotle that a dispassionate analysis of "the general character of things requires that there be . . . an entity at the base of all actual things" (SMW, 173–74). In particular, he came to believe, the order, the novelty, and the ideals in the world point to the existence of an all-pervasive actuality worthy of the name God (PR, 40, 46, 164, 247; AI, 115; RM, 96; MT, 103). Whitehead's metaphysical reflections led him to theism. But it was a *naturalistic* theism, consistent with the conviction of scientific naturalism that the basic causal principles of the world are never interrupted.

Reinterpretations of the doctrine of God to make it consistent with scientific beliefs are, of course, nothing new. This kind of reinterpretation has been occurring in modern liberal theologies from almost the onset of modernity, as we have seen, beginning with various forms of deism in the eighteenth century. But these modern liberal reinterpretations have, by denying divine activity in the world, resulted not in a deeper, but in a more superficial, religion. The religious character of Christianity and Judaism have, in fact, been virtually lost. Whitehead's approach, in formulating a "naturalistic theism," may at first glance seem to be one more example of this kind of reductionistic reinterpretation, in which the result is a "theism" not worthy of the name. Whitehead's reinterpretation, however, does not deny divine presence but reconceives divine power.

The Generic Idea of God

The nature of this reinterpretation can best be seen in terms of what I call the "generic idea of God" in our tradition. According to this generic idea, the word "God" refers to (1) a personal, purposive being who is, (2) supreme in power, and (3) perfect in goodness, who (4) created the world, and (5) acts providentially in it, who (6) is experienced by human beings, especially as the source of moral norms and religious experiences, and is (7) the ultimate guarantee for the meaningfulness of human life, (8) the ultimate ground of hope for the victory of good over evil, thereby (9) alone worthy of worship. Modern liberal theology in its early (deistic) stage typically gave up the fifth and sixth features, thereby divine providence and authentic religious experience. In its later stages it usually gave up the first, second, fourth, seventh, and eighth features as well. One form of modern theology has retained some of these other features by giving up the third one, thereby denying the perfect goodness of God (see the essay by John Roth and Sontag in Davis EE). A Whiteheadian theism can retain all nine features by simply modifying the traditional understanding of some of them. The central modification involves the second feature, according to which God is supreme in power. This change in meaning, from coercive to persuasive power, entails a modification of the traditional meanings of the fourth, fifth, and eighth features as well. These modifications constitute the change from supernaturalistic theism to the kind of naturalistic theism affirmed in this book.

The Rejection of Creatio Ex Nihilo

Whitehead's naturalistic theism was formulated in explicit opposition to the supernaturalistic theism of Augustine and Newton, with its "theology of a wholly transcendent God creating out of nothing an accidental universe" (PR, 95). Whitehead's strategy for reconciling the beliefs of the religious and scientific communities, we have seen, involves overcoming exaggerations from both camps. He expresses his conviction that this supernaturalistic version of theism is an exaggeration of the truth by saying that it involves the unfortunate habit of paying "metaphysical compliments" to God (SMW, 179). The target of that criticism was the conception of God "as the foundation of the metaphysical situation with its ultimate activity" (SMW, 179).

In rejecting that idea, Whitehead was rejecting the doctrine of *creatio ex nihilo,* in the sense of creation out of absolute nothingness. As we saw in the quotations from Charles Hodge and Millard Erickson in the first section of Chapter 2, that doctrine of creation implies that the world's basic units have no inherent power and that there are no metaphysical principles to which God's exercise of power has to conform. God can violate even those causal principles that seem so fundamental that they may be called "metaphysical." Because every aspect of the world was due solely to the will of God, God has complete control over everything that happens. Erickson brings out clearly these implications of the doctrine of *creatio ex nihilo:*

> God did not work with something which was in existence. He brought into existence the very raw material which he employed. If this were not the case, God would not really be infinite. There would be something else which also was, and presumably had always been. Consequently, God would have been limited by having to work with the intrinsic characteristics of the raw material which he employed. (CT, 374)

In the worldview of the supernaturalists such as Erickson, there is no "raw material" with "intrinsic characteristics" that could limit God's control of the world. God, therefore, can unilaterally bring about anything that God wishes to bring about, as long as no logical contradiction is involved.

It is this idea of God that lies behind the traditional problem of evil. As Whitehead pointed out, this idea of God implies that God is responsible for "the origin of all evil as well as of all good" (SMW, 179). It is also this idea of God, of course, that places much of the religious community in opposition to the naturalism presupposed by the scientific community. As Erickson says, his religious community "operates with a definite supernaturalism—God resides outside the world and intervenes periodically within the natural processes through miracles" (CT, 304). It is this idea of God that underlay the expectation that the scriptures of our religious tradition would be infallibly inspired, which is the belief that led to epistemic supernaturalism and the resulting religious authoritarianism. It is this idea of God that led, furthermore, to the idea that the various species were created *ex nihilo,* the main idea against which Darwin's theory of evolution, with its "descent with modification," was directed. It is this idea of God, accordingly, that has been behind most of the reasons for the late modern rejection of theism.

The Rejection of Extreme Voluntarism

Whitehead explicitly rejected this idea of God, according to which there is "one supreme reality, omnipotently disposing a wholly derivative world" (AI, 166). Rather than portraying a God who is wholly transcendent, in the sense of being able to exist apart from a world, Whitehead's philosophy portrays not only "the

World as requiring its union with God" but also "God as requiring his union with the World" (AI, 168). The essence of supernaturalism, as reflected in Hodge and Erickson, is an extreme voluntarism, according to which God's relationship to the world is based entirely on the divine will. Whitehead rejects this view, saying that "the relationships of God to the World should lie beyond the accidents of will," being instead "founded upon the necessities of the nature of God and the nature of the World" (AI, 168). As this reference to the "necessities of . . . the nature of the World" indicates, the world's existence is not "wholly derivative," not entirely contingent. This view does not rule out a moderate voluntarism: It still allows one to speak of God as willing, as deciding among options. What it denies is *extreme* voluntarism, according to which the very existence of a realm of finite actualities is contingent upon a divine decision.

This denial, to be sure, does not mean that *our* world, with its electrons, protons, and inverse square law of gravitational attraction, exists necessarily. *This* world came into existence at some point in the past (evidently, it now seems, some twelve to eighteen billion years ago). Its coming into existence, however, was "not the beginning of [finite] matter of fact, but the incoming of a certain type of order" (PR, 96). That is, the creation of our particular world, which Whitehead calls our "cosmic epoch," involved bringing order out of chaos, or at least inducing the rise of a new type of order out of a previous type. When Whitehead speaks of "the World" (capitalized) as necessary, accordingly, he means only that *some* world of finite actualities must exist, and that whatever particular world or cosmic epoch does exist will exemplify certain metaphysical principles. Calling these principles "metaphysical" means that they obtain necessarily, not through a contingent act of will. And, not having been freely instituted, they cannot be freely violated.

The Rejection of (Supernatural) Miracles

Implicit in this distinction between the two meanings of "world" is a distinction between (necessary) metaphysical principles and (contingent) laws of nature. In supernaturalistic theism, no such distinction is made: *All* the principles by which the world normally runs were freely created by God, so they are all equally contingent. This is why the God of Hodge and Erickson can interrupt not only the law of gravity but also the very principle of causation, according to which finite events are usually causally influenced by prior finite events. In Whitehead's naturalistic theism, by contrast, there are, beneath the contingent laws of our particular cosmic epoch, some metaphysical principles, which obtain necessarily and, therefore, can never be violated.

This distinction can be used to clarify the difference between the miraculous, in the sense of the supernatural, and the merely paranormal. A paranormal event, such as levitation, involves merely an exception, perhaps due to an over-riding force, of some contingent law of nature, which simply describes the way things in question usually behave. Our "laws" of nature, in other words, are merely descriptions of the most long-standing *habits* of nature. A supernatural miracle, by contrast, would be the violation of one or more metaphysical principles. Given Whitehead's worldview, the paranormal can happen (see Chapter 7), the supernatural cannot.

God and Creativity

At the heart of the metaphysical principles is creativity, which Whitehead calls the "category of the ultimate." Creativity is the twofold power of every actual entity to exert both final and efficient causation. This idea is embodied in Whitehead's doctrine that every actuality is a momentary event, or *actual occasion*, which creates itself out of the causal influences received from prior actual occasions, then exerts influence upon subsequent occasions. In supernaturalistic theism, this twofold creative power belonged essentially to God alone; any creative power possessed by finite events was a wholly contingent gift of God, which could thereby be overridden or canceled out at will, so that God could completely determine what occurs. This is the doctrine that Whitehead rejects as a false metaphysical compliment. Although he does refer to God as "creator," Whitehead warns against this term's misleading suggestion that "the ultimate creativity of the universe is to be ascribed to God's volition" (PR, 225).

According to his naturalistic theism, the ultimate creativity of the universe is necessarily embodied in finite actualities as well as in the divine actuality. This means that power is inherent in the world as well as in God. It means, more precisely, that every one of the world's units is inherently influenced by all prior units, that every unit inherently has some power of self-determination, and that every unit inherently has the power to inflict itself upon others, for good or for ill. The resulting universal web of finite causation cannot be interrupted, even by God. The implication is that the divine causation in the world is always persuasive, never coercive in the sense of wholly determining.

A God-Shaped Hole Without a God of the Gaps

The idea that God exerts a persuasive influence means that Whitehead's position is crucially different from other most versions of scientific naturalism, according to which a divine being cannot influence the world in an ongoing way. But Whitehead agrees with other naturalists in denying supernatural intervention. He does in one place use the dangerous term "intervention," saying: "Apart from the intervention of God, there could be nothing new in the world, and no order in the world" (PR, 247). His meaning, however, is not that God occasionally intervenes, thereby interrupting the otherwise uniform system of cause and effect. Rather, his position is that God is one of the causal influences upon *every* event. It is this *uniform* divine "intervention" that makes both the order and the novelty in the world possible. Divine causation, therefore, is never an *interruption of* the normal cause-and-effect pattern; it is an essential *factor in* that pattern. Whitehead's naturalism contains a God-shaped hole without a God of the gaps. Theism has been reconciled with ontological uniformitarianism.

Historically, as we have seen, the notion behind the idea of the God of the gaps was that the "whatness" of most events can in principle be fully explained without reference to variable divine causation. The adjective "variable" is essential: Most theologians, even most of those who can be called "deistic," do say that God is present in the world, even actively present, because the divine sustaining causality is necessary to support the very existence (the "thatness") of the creatures (which Plantinga acknowledges by calling the position merely *semi*deistic). This divine sustaining causality, however, is the same for all creatures or events. Within this framework, *variable* divine causation would be constitutive of the whatness only of events that do not have finite causes sufficient to account for all their characteristics. In relation to such events, God would directly be at least partly responsible for their whatness as well as their thatness. Given this framework, the idea of variable divine causation became connected with the idea of the "God of the gaps," in which reference to divine influence was used to fill some gap in the current knowledge of finite causes. When further scientific knowledge led to that gap's being filled, theologians then retreated to another gap, which in turn became filled by advancing scientific knowledge, and so on. Due to its association with this discredited idea of the God of the gaps, the idea of variable divine causation faded from scientific discourse, from most philosophical discourse, and even from much theological discourse (except for the retention of metaphorical God-language).

Theologians seem to be faced with a dilemma. On the one hand, if they refer to some feature of the world as inexplicable apart from divine influence, they are accused of having a God of the gaps. On the other hand, if they, perhaps through nervousness about being thus accused, do not point to any feature of the world that seems to require reference to divine influence, the reason for their continuing to speak of God becomes puzzling to those accustomed to basing affirmations on evidence.[1] Talk of God is, accordingly, either dismissed as groundless and superfluous or indicted for referring to a God of the gaps.

---

1. Drees, in fact, refers to this problem (RSN, 272), citing Ernan McMullin's statement (NSBC, 74) that "since there are no real 'gaps' to fill, we may be left without an argument for God's existence of the kind that would convince a science-minded generation."

The distinction between a God of the gaps and a God-shaped hole provides a way out of this apparent dilemma. As I intend the distinction, "God-shaped hole" is an inclusive category, referring to any feature of a cosmology that requires divine influence to fill it. Only some appeals to a God-shaped hole, however, would be appeals to a God of the gaps. One difference would lie in whether God is appealed to in order to explain some feature of the world that could *in principle* be explained by a finite cause or set of causes. If this is not the case, then the appeal is to a God-shaped hole that is *not* filled by a God of the gaps.

Whitehead's reason for speaking of divine influence in the world fits this description. The eternal nature of God is the ground of all the metaphysical principles. God is also the primary source of the most fundamental contingent laws of nature of our particular cosmic epoch. And God is, at any time in the temporal process of the world, the source of novel possibilities, meaning possibilities that have not been previously actualized in the world. These are functions that must be performed if there is to be any novelty in the world and, indeed, if there is to be a world at all. They are functions, however, that could not in principle be performed by any finite being or combination of beings. Whitehead's worldview, therefore, has a God-shaped hole that is not filled by a God of the gaps.[2]

This distinction also involves a second issue. When the pejorative language of a "God of the gaps" is used, what is usually involved is the idea, which arose from the framework of the primary-secondary scheme, of divine causation as *supplanting* or *replacing* the finite causes that would have normally occurred. In Whitehead's view, by contrast, the divine causation never *replaces* the normal causes but is always *part of* the normal causal pattern. Some theologians might object to this idea by employing another pejorative expression, saying that this view makes God merely "one cause among others." We should be wary of criticisms based on this expression, however, given its use by semideistic theologians to make a virtue out of a necessity—to make it seem a good thing that they did not affirm any variable divine causation in the world. That phrase, in any case, would not accurately characterize Whitehead's doctrine, for at least two reasons. First, this phrase would apply properly only to views that have God performing some function that could, in principle, be performed by finite causes. Second, in Whitehead's position, although divine influence *is* one of the many constitutive influences on every event, it is also unique in many respects. God is, for example, the only causal power that is eternal and universal, being involved in *every* event in the universe. Also, in agreement with most theological views, Whitehead's position entails that the divine causation is necessary for sustaining the very existence of the world. What seems to be unique about Whitehead's position, at least in modern times, is that God's sustaining causa-

---

2. I have developed this distinction more fully elsewhere (IBI, 75-77). This same basic point has been developed from a different perspective by Robert Russell (SP, 193–95, 216–17).

tion, necessary for the thatness of all events, involves God's variable causation, which is partially constitutive the whatness of each event.

Divine Influence and the Metaphysical Principles

One more feature of Whitehead's naturalistic theism is its assimilation of the God-world relation to the metaphysical principles involved in all causal relations. Although many naturalists have assumed that *any* appeal to God to explain something about the world would necessarily be a violation of naturalistic rationality, for Whitehead what is irrational is only "the easy assumption that there is an ultimate reality which, *in some unexplained way,* is to be appealed to for the removal of perplexity" (SMW, 92; emphasis added). His insistence is only that any explanation involving a supreme reality should involve "the same general principles of reality" that are otherwise employed (SMW, 93). This insistence is expressed most famously in his dictum that "God is not to be treated as an exception to all metaphysical principles, invoked to save their collapse," but as "their chief exemplification" (PR, 343).

This dictum, coming right after Whitehead's indication that his position on God and the world represents an alternative to the three positions in Hume's *Dialogues Concerning Natural Religion,* is relevant to the Humean charge that references to God as creator of the world always involve an equivocation on the idea of causation. Whitehead expresses his agreement with this criticism in relation to the supernaturalistic form of the cosmological argument. That argument is invalid, he says, "because our notion of causation concerns the relations of states of things *within* the actual world, and can only be illegitimately extended to a transcendent derivation" (PR, 93; emphasis added). His own notion of God, he adds, is "that of an actual entity immanent in the actual world, but transcending any finite cosmic epoch" (PR, 93). That is, although God in one sense transcends our particular world, it is also the case that God is an actuality *within* our world. Also, God belongs to the same genus as all other actual entities, thereby exemplifying the same metaphysical categories.[3]

---

3. The fact that God exemplifies the metaphysical categories means that just as time or temporality, in the sense of the reality of the distinction between past, present, and future, is real for all other actualities, it is also real for God. Whiteheadian-Hartshornean process theism, therefore, is a version of temporalistic theism. One criticism of temporalistic theism has been that, by implying that there is an unambiguous cosmic "now," it is contradicted by special relativity physics. This criticism has been lifted up by one author, in fact, as the principal objection to Whiteheadian-Hartshornean process theism (Gruenler IG). I have argued elsewhere (Griffin HG) that this alleged contradiction does not exist. (The reason for calling it "Whiteheadian-Hartshornean theism" is explained in the next note.)

Accordingly, the causal relations between God and finite actual entities are not different in kind from the causal relations between finite actual entities.[4] Therefore, the Humean charge of equivocation, valid against supernaturalistic theism, does not apply to Whitehead's theism. And, because the very creation of our cosmic epoch involved the bringing of a new kind of order into existence, not the creation of finite existence out of nothing, the same causal relations were involved then as well. No supernaturalist exception to uniformitarianism, therefore, was involved even in the very creation of our world, a fact that distinguishes Whitehead's theism from deistic positions— such as that of Rudolf Otto and of Cleanthes in Hume's *Dialogues*. This feature of Whitehead's "answer to Hume," which is an essential aspect of his theistic naturalism, has been insufficiently appreciated, even among Whitehead scholars.

In any case, having explained how Whitehead's theism is naturalistic and thereby why it does not create the kinds of expectations that have led to the widespread rejection of theism in the late modern world, I will now summarize the ways in which it is, nevertheless, fully religious, fulfilling the various features of the generic idea of God summarized earlier: God is personal and purposive. The supreme power of the universe is pure goodness, pure unbounded love. This being of pure love created our world and, in fact, continues to create it in each moment. Far from being a remote, inaccessible creator, this God is intimately involved in the origination of each event in the universe. Each experiential event in the world receives from God its "ideal aim," which amounts to prevenient grace, being an urge toward the best possibility open to it. This ideal aim evokes the event into existence. Also, it is through this ideal aim that the event becomes exposed to novel possibilities, through which mere repetition of the past can be transcended.

This is not traditional theism, deism, or pantheism, but panentheism, according to which God is in all things and all things are in God. The presence of God in human beings in particular means that we can directly experience God. Indeed, we directly experience God all the time at the unconscious level. The only unusual thing about "religious experience" as ordinarily understood is that this generally unconscious experience of the divine presence rises to consciousness. This experience of God's providential activity in us is also the ultimate source of our moral and aesthetic ideals. The reverse side of panentheism, our existence in the "consequent nature of God," provides the

---

4. This is the case, at least, in a Whiteheadian position in which the doctrine of God has been revised in accordance with Charles Hartshorne's view that God is not a single actual entity but a "living person," meaning an everlasting series of divine occasions of experience. This revision, I have argued elsewhere (RWS Ch. 4), is necessary to conform to Whitehead's dictum that God should be regarded not as an exception to the metaphysical principles but as their chief exemplification.

basis for the ultimate meaningfulness of our lives, because no matter what happens to us and our world, reality will never be as if we had not existed: We will be retained everlastingly in the divine life. This panentheism, by portraying God as the truly supreme power of the universe, also provides a basis for hope in the ultimate victory of good over evil—an issue that I discuss briefly in Chapter 7 and more fully elsewhere (PPS and DC). By possessing all these features of the generic idea of God, the Divine Reality portrayed in this naturalistic theism evokes worship and otherwise provides the basis for a rich religious life.

WHITEHEAD'S MORE OPEN SCIENTIFIC NATURALISM

Having discussed the primary modification that needs to be made from the side of religion if it and science are to be integrated into a single outlook, I turn now to the primary modification needed from the side of scientific naturalism, which involves the reconception of the ultimate units of nature. At the heart of this modification is Whitehead's rejection of the notion of "vacuous actuality" in favor of the view that the ultimate units are to be understood as occasions of experience—the view that I call "panexperientialism." In this section, I will indicate only very briefly how this panexperientialism opens the way for a naturalism that allows for the possibility of freedom, religious and moral experience, divine influence in nature, and life after death, saving for later chapters a fuller exposition.

Hard-core Common Sense

Before beginning that substantive exposition, however, it is necessary to point to the ultimate criterion that, according to Whitehead, should be used to evaluate any philosophical or theological position, the criterion that I call "hard-core common sense."

The idea that common sense should be a criterion for evaluating positions evokes conflicting reactions. On the one hand, we often reject ideas because they are, we say, "contrary to common sense." On the other hand, it has become customary to think of science as having disproved commonsense ideas time and time again: Common sense thought the Earth was flat, that it was the center of the universe, and that matter was solid, but science showed all of these and many other commonsense ideas to be false. Science is sometimes characterized, in fact, as a "systematic assault on common sense." From this perspective, it would seem retrograde to affirm common sense as a serious criterion, let alone the *ultimate* criterion, in terms of which to judge other ideas, especially those deriving from scientists.

It is to take account of these conflicting attitudes toward common sense that I have introduced the distinction between "hard-core" and "soft-core" commonsense notions. Common sense of the soft-core variety involves beliefs that are widespread at a particular time and place but that, given new knowledge, can be given up. The beliefs in a flat Earth, a geocentric universe, and solid matter were all soft-core commonsense beliefs. Given new knowledge, these beliefs were rightly given up. Indeed, Whitehead himself sometimes uses the term "common sense" to refer to such beliefs, as when he says, "It is part of the special sciences to modify common sense" (PR, 17). He even says, in fact, that his own philosophy is an attempt to modify "the general common-sense notion of the universe" (MT, 129).

Hard-core commonsense beliefs, however, are different in kind. Rather than being parochial beliefs, limited to a particular time and place, they are universal, being common to all humanity. This does not mean, to be sure, that they are consciously and verbally affirmed by all human beings, but only that they are *inevitably presupposed in practice*, even if denied in theory. The fact that they are inevitably presupposed means that, if we do try to deny them verbally, we will inevitably contradict ourselves, in the sense that our behavior (our "practice") will contradict the content of our verbal statement. For example, one of our hard-core commonsense beliefs is the belief in an "external world," meaning a world of actual things (including other people) beyond our own immediate experience. Whitehead uses the term "common sense" for this belief in referring to the direct experience of actualities beyond ourselves as "a presupposition of all common sense" (PR, 18).[5] Skepticism about an external world is called "solipsism." If I try to profess solipsism, however, my behavior will contradict this profession: If I announce to an audience that I am a solipsist, I show that I do not at all doubt the existence of the people constituting the audience; if I restrict myself to writing down my beliefs, I show that I presuppose the real existence of not only my computer but also my body, whose fingers operate the computer. This type of contradiction between theory and practice is called a "performative self-contradiction" by Jürgen Habermas and Karl-Otto Apel.[6]

---

5. There are also some other passages in which Whitehead uses the term "common sense" in relation to what he considers the inevitable presuppositions of practice. For example, he speaks of several notions that cannot "be dropped without doing violence to common sense" (PR, 128). He speaks of "the obvious deliverances of common sense" (PR, 51). And he refers to his own "endeavor to interpret experience in accordance with the overpowering deliverances of common sense" (PR, 50).

6. Performative self-contradictions arise, explains Martin Jay (DPC, 29), "when whatever is being claimed is at odds with the presuppositions or implications of the act of claiming it." See Jay's article for an illuminating discussion of the role of this notion in the "critical theory" of Habermas and Apel.

It is important to see that the two types of ideas, which I call *hard-core* and *soft-core* commonsense ideas, have nothing in common except the word "common sense." Soft-core commonsense ideas are *not* common to all peoples in all times and places; they are *not* inevitably presupposed in practice; and we *can*, therefore, deny them verbally without thereby contradicting any ideas being presupposed in this act of denying them. The fact that soft-core commonsense ideas should not be used as the ultimate criterion for judging a philosophical or theological hypothesis, therefore, says nothing about the propriety of assigning this role to hard-core commonsense ideas. The fact that the expression "common sense" is used for two such radically different types of ideas, even by Whitehead himself, does, of course, lead to confusion if qualifying adjectives are not used. But once the distinction is clearly seen, this problem, being purely verbal, provides no reason not to consider *hard-core* commonsense ideas to be the ultimate criterion by which to evaluate the adequacy of any position. The positive reason to thus consider them is that violating them inevitably leads to self-contradiction, and the "law of noncontradiction" is usually and rightly considered the first rule of reason. To deny any ideas that we inevitably presuppose, even in the very act of denying them, is to be irrational in the most fundamental sense of the term.

I have thus far illustrated hard-core commonsense ideas in terms of our belief in an actual world beyond our own experience. Another idea that we inevitably presuppose in practice is the idea of efficient causation, meaning the belief that some things exert causal influence on other things. Hume's treatment of this "natural belief" is used by Whitehead to introduce his idea that such beliefs should be fundamental for our philosophical theories, not dismissed as *mere* presuppositions of practice. Hume's conclusion that this notion of efficient causation could have no place in theory was based on his sensationist empiricism. Limiting our experience of the world to our sensory perception of it, he argued that, because we do not *see* one thing bring about another thing, we have no basis for including efficient causation, in the sense of the real influence of one event or thing upon another, in our theories. The only meaning for "causation" provided by (sensory) experience, he argued, is the "constant conjunction" of two kinds of occurrences. For some reason, he pointed out, we cannot help presupposing the reality of causation as real influence. But this idea cannot play a role in our scientific, philosophical, or theological theories.

Whitehead, while agreeing with what Hume said about sense-perception, disagreed with Hume's decision not to include the inevitable presuppositions of practice among the theoretical data: "Whatever is found in 'practice' must lie within the scope of the metaphysical description. When the description fails to include the 'practice,' the metaphysics is inadequate and requires revision" (PR, 13). Calling this principle "the metaphysical rule of evidence," Whitehead said that "we must bow to those presumptions, which, in despite of criticism, we still employ for the regulation of our lives" (PR, 151). "Rationalism," he added, "is the search for the coherence of such presumptions."

The central task of philosophy, in other words, is to develop a position that shows how the various notions that we inevitably presuppose in practice can all be true. Part and parcel of this task, however, is showing how these ideas can also be harmonized with ideas deriving from particular types of experience, most especially scientific and religious experiences. No idea, be it called "scientific" or "religious," should be accepted if it contradicts one of our hard-core commonsense notions. This criterion was implicitly employed in the previous section, insofar as part of the reason for affirming a naturalistic, rather than a supernaturalistic, theism is to avoid conflict with the reality of both human freedom and genuine evil, both of which are inevitable presuppositions of human thought and practice. In the present section and in Chapter 6, the fact that human freedom is a hard-core commonsense notion is used as one of the bases for shifting from a materialistic to a panexperientialist version of naturalism.

Panexperientialism and Freedom

Most forms of scientific naturalism have not been able to affirm freedom, because they assumed the ultimate units of nature, and therefore the ultimate units of the human body, to be devoid of experience and spontaneity. This made it impossible to think of the mind as a full-fledged, unified actuality, distinct from the brain, because the emergence of such a mind, different in kind from the constituents of the body, would be unintelligible apart from a supernatural creator. Naturalists, accordingly, have assumed that what we have called the mind must somehow be identical with brain processes, rather than being a distinct entity. The human being is thereby assumed to be structurally analogous to rocks, toasters, computers, and other things to which we would not attribute either a unified experience or freedom. In such things, even if some spontaneity is attributed to the ultimate units of nature (perhaps on the basis of the indeterminacy principle in quantum physics), these individual spontaneities are canceled out by the "law of large numbers" in the visible object. Thinking of the human being as analogous to such things implies that there is no more freedom in a person as a whole than there is in a toaster or a computer.

However, given Whitehead's panexperientialism, which says both that (1) all actualities have experience and that (2) actualities at one level can give rise to higher-level actualities, the mind can be understood to be *numerically distinct* from the brain without being an *ontologically different* type of thing from the brain cells. Whitehead gives us, in other words, a *nondualistic interactionism*. By affirming interaction between mind and brain, his position regards the mind as distinct from the brain and thereby as a genuine individual, to which freedom can be affirmed. But by denying dualism, according to which mind and brain cells would be different in kind, Whitehead's position allows us to understand, in a naturalistic framework, how body and mind can influence each other. The human mind, understood as a series of very high-level occasions of experience, can, accordingly, be attributed that high degree of freedom that we all presuppose it to have. And we can intelligibly attribute responsible freedom to our bodily actions as well. We can have naturalism, therefore, without determinism. These ideas will be developed more fully in Chapter 6.

Panexperientialism and Moral and Religious Experience

At the root of Whitehead's solution to the mind-body problem is his doctrine that perception is not to be simply identified with sensory perception. Rather, sensory perception is a high-level form of perception, derivative from a more fundamental, nonsensory perception, which he calls "physical prehension." We share this more fundamental form of perception, or prehension, with all other actualities, whether they have sensory organs or not. This is why mind and brain cells can interact: They prehend each other's experiences.

This same idea explains how various kinds of experience that are ruled out by sensationist philosophies can occur. One such kind of experience is telepathy. Given the capacity of all individuals for nonsensory prehension, we can directly prehend each other's minds, and occasionally some aspect of this prehension may rise to the level of consciousness. Whitehead, indeed, explicitly points out that his philosophy allows for influence at a distance in general and telepathy in particular (SMW, 150; PR, 308). This same idea also explains how we can experience God, even though God is not a possible object of sensory perception: We directly prehend God, just as our brain cells prehend our minds. Whitehead thereby shows how religious experience, in the theistic sense of a direct experience of God, can occur. Also, by regarding the primordial mind of God as the home of all ideal forms, Whitehead explains both how normative ideals exist in the nature of things and how we can experience them: We experience them by prehending God's appetitive envisagement of them (PR, 31, 33–34, 46).[7]

Panexperientialism and Divine Influence in Nature

Whiteheadian panexperientialism also explains how God can influence not only human experience, but all of nature: All actual entities, having the capacity for

---

7. For some reason, Drees has the impression that process theology is "too much a 'mystical' religiosity, accepting reality as it is, lacking opportunities to articulate a 'prophetic' sense of contrast between the way the world is and the way the world should be" (RSN, 253). However, the fact that Whiteheadian process theologians can refer to our nonsensory perception of normative ideals in the mind of God is one of the most important ways in which our position differs from that of other naturalists, such as John Dewey and, for that matter, Drees himself.

nonsensory prehension, can directly experience God. As Pratt suggested, God can influence the whole world analogously to the way in which the mind can influence its whole body. Whiteheadian panentheism, with its panexperientialism, allows us to develop a theistic evolutionism, according to which the evolutionary process has been significantly directed by divine influence. This idea will be developed at length in Chapter 8.

Besides providing the basis for reaffirming the biblically rooted idea that the world is God's creation, the doctrine of panexperientialism also provides the basis for reaffirming the biblically rooted idea that reality is ultimately historical or temporal. This idea has been undermined by the idea that time does not exist for the particles studied by physics combined with the idea, exemplified by Drees, that physics tells us the ultimate nature of the world. Drees, for example, draws the conclusion that temporal portrayals of the universe "may be of limited validity, and not fundamental" (RSN, 266; cf. 262, 264). By contrast, the idea that experience goes all the way down, with each event prehending prior events, provides a basis for saying that temporal relations are fundamental after all, as I have shown elsewhere (PUST; PAP).[8] This idea is presupposed in the discussion of "Creation and Evolution" in Chapter 8.

Panexperientialism and Life after Death

Of all the ideas usually thought to be essential to a religious interpretation of the world, it is surely life after death that has been thought to be most antithetical to a naturalistic worldview. Whitehead's form of naturalism, however, allows even for it. Whitehead himself did not devote much attention to this issue in his writings. In relation to the question of the ultimate meaning of life, he focused, as mentioned earlier, on our "objective immortality" in the consequent nature of God, which his naturalistic theism allowed and even entailed. That is, if God exemplifies the general metaphysical principles, then, besides influencing the world, God also prehends its activities. Our being prehended and retained everlastingly in the divine experience is our objective immortal-

---

8. One of the main points of those two essays is that Whitehead's idea that enduring objects, such as photons and electons, are really temporally ordered societies of momentary occasions of experience, each of which prehends previous members of that society, provides the basis for saying that time as we know it, with its irreversibility, is real for the entities studied by physics. This doctrine opposes the hitherto orthodox view, according to which time emerges at some point in the evolution of the universe, perhaps with the beginning of entropic processes, perhaps even later with the rise of life, and that in any case time is not *irreversible in principle*, but only usually, at least for the processes studied by physics. Since I wrote those essays, the Whiteheadian view has received empirical support, as experiments have led physicists to conclude that, in the words of the title of a recent article, "Time Proves Not Reversible at Deepest Level" (Weiss TP).

ity. Whitehead (rightly) regarded this as the ultimate religious concern (PR, 340, 350–51). But he did mention in passing a couple of times that his philosophy allows for the possibility of what most people mean by "immortality," which involves continued life after bodily death (RM, 107; AI, 208). Because the mind or soul is numerically distinct from the brain, and because experience, even conscious experience, does not require sensory organs, it is conceivable that the mind could exist in another environment than that of the physical body. Because his philosophy is neutral on this question, he added, it should be decided on the basis of empirical evidence, by which he meant the evidence of psychical research, which his philosophy allows to be taken seriously. Whitehead's philosophy, accordingly, encourages a naturalistic, in the sense of empirical, approach to this question, which will be explored more fully in Chapter 7.

As this brief discussion indicates, Whitehead's naturalism is clearly far more open to a wide range of religiously important experiences and beliefs than is the reductionistic materialism that has thus far provided the dominant basis for scientific naturalism. A scientific naturalism based on Whiteheadian panexperientialism, accordingly, would provide a far more hopeful basis for a reconciliation of the beliefs of the religious and the scientific communities. The scientific community, however, will replace the materialist version of scientific naturalism with a more open version only if the latter seems to be more illuminative of the empirical facts. In the next section, I will briefly discuss some developments that undermine the hitherto dominant view, thereby suggesting the need for a more open naturalism.

SCIENTIFIC DEVELOPMENTS SUPPORTIVE
OF A MORE OPEN NATURALISM

One of the best-known developments of science in the twentieth century, and probably the one that has been responsible for most of the discussion of a new harmony between science and religion, is the principle of indeterminacy in quantum physics. This principle potentially puts an end to science's commitment to determinism (even if most physicists and philosophers of science have thus far tried to interpret indeterminacy in such a way as to minimize the damage to determinism). Quantum indeterminacy could betoken an element of self-determination, and even the slightest iota of self-determination in the most primitive units of nature opens the way for thinking in terms of a hierarchy of self-determination produced by the evolutionary process, in which more complex organisms have increasingly greater degrees of freedom. A more subtle science, therefore, has undermined the determinism that early modern science seemed to require (see Robert Russell SP: 200–04, 212–13).

The reductionism of modern science has also been challenged by several developments as science has become more subtle. Reasons to affirm downward causation from the mind to the brain have been provided by otherwise respectable brain scientists, including Nobel prizewinners Roger Sperry (SMP) and John Eccles (HS). Psychosomatic studies support this view: Mental imagery, for example, changes body chemistry (Murphy FB). The DNA molecule is no longer seen as a passive holder and by-product of its genetic material, but as an organism that actively organizes its parts (Keller FO). As downward causation becomes re-established as a general fact, pervasive of nature at all levels, the notion of downward causation from God, or the universe as an inclusive individual to its parts, will no longer seem so aberrant.

The materialism of the modern worldview and the resulting behaviorism of modern science is challenged not only by the notion that science can speak of the human mind as distinct from the brain, but also by the judgment of some ethologists that we must posit internal experience to explain the behavior of other animals, such as bats and bees (Donald Griffin QAA, 14, 23). It is also challenged by evidence from what are perhaps the most subtle experiments yet, those for "nonlocal interactions" between physical particles, meaning some kind of communication that is not transmitted through energetic waves or particles (Stapp ET; Bohm and Hiley UU, 134–59).

Materialism and behaviorism are undermined, furthermore, by developments that have challenged the sensationism upon which modern science was based. This challenge has been made most explicitly by parapsychology, with its evidence for extrasensory perception (which will be discussed in Chapter 7), and by humanistic and especially transpersonal psychologies, which accept the reality of nonsensory perception. But this challenge is also implicit in many other developments. Studies have shown that bacteria must have a primitive form of memory and perception (Adler and Tse DMB; Goldbeter and Koshland SMM).[9] And the evidence for nonlocal communication between elementary particles suggests that even they enjoy a nonsensory form of perception and, indeed, the views of David Bohm (UU) and Henry Stapp (ET) amount to saying just that. Nobel prizewinner Ilya Prigogine, writing with philosopher of science Isabelle Stengers, also says that we must attribute a kind of perception to matter (OOC, 33, 145, 163, 171, 180, 181). The evidence seems to support Whitehead's view that all individuals have a nonsensory form of perception, which he called "prehension," and that sensory perception, by contrast, is a sophisticated, derivative form of perception enjoyed by animals complex enough to have specialized sensory organs.

---

9. Lynn Margulis (see note 13 in Ch. 8, below) and Dorion Sagan agree, saying (WIL, 184): "No organic being is a billiard ball, acted on only by external forces. All are sentient."

Summary and Conclusion

By modifying both religion and scientific naturalism in the light of each other, Whitehead has provided us with a worldview that is at once religious and scientific. Whether this worldview actually results in a reconciliation of science and religion will depend, of course, upon whether the religious and the scientific communities accept the proffered terms of peace. Whether the religious community does so will depend, to a great extent, on evidence that a sufficiently robust theology, adequate to vital religion, can be articulated on the basis of this form of theistic naturalism. Whitehead himself did not articulate such a theology, but his philosophy arguably provides the requisite basis. Whether the scientific community will accept the proffered terms depends mainly on whether this type of naturalism can be seen to be more adequate than the reductionist type for science itself. Whitehead certainly believed this to be the case, saying repeatedly that his criticisms of "scientific materialism" and his articulation of a more inclusive scheme of thought were offered not merely in the interest of reconciling science with our moral, aesthetic, and religious intuitions, but also in the interest of "science itself" (SMW, 68, 84). Part II of this book constitutes a twofold argument that this Whiteheadian naturalism provides not only a better basis for religious belief and practice than does supernaturalism, but also a better basis for scientific belief and practice than does materialistic naturalism.[10] First, however, it is important to see why the early modern worldview, which eventually led to naturalism$_{sam}$, was accepted in the first place.

10. As I have suggested elsewhere (RWS), Whitehead's position can be called "naturalism$_{ppp}$," because his prehensive doctrine of perception replaces the sensationism of naturalism$_{sam}$, while his panentheism and panexperientialism replace its atheism and materialism, respectively. The thesis of the present book is that the adoption of naturalism$_{ppp}$, which provides a postmodern embodiment of naturalism$_{ns}$, would not only allow for a reconciliation of the worldviews of the scientific and religious communities but would also provide a better basis for scientific and religious thought more generally. In that other book (RWS), I have explained more fully the contention that naturalism$_{sam}$ is inadequate for science itself, and have also developed naturalism$_{ppp}$ in such a way that its religious adequacy can be more fully judged. The adequacy of this worldview for Christian faith in particular is argued most fully in a work in progress (DC).

5

RELIGION AND THE RISE OF THE MODERN SCIENTIFIC WORLDVIEW

I am proposing a postmodern liberal theology as the basis for harmony between science and theology. It is a *liberal* theology, in distinction from conservative-to-fundamentalist theologies, for two reasons. On the one hand, it rejects *epistemic* supernaturalism, and thereby the method of authority, in favor of determining questions of truth on the basis of experience and reason. On the other hand, it rejects *ontological* supernaturalism, which is presupposed by the method of authority, in favor of a naturalistic theism. It is a *postmodern* liberal theology, in distinction from modern liberal theologies, in that it rejects the basic assumptions of the modern worldviews, namely, (1) the sensationist theory of perception, according to which all perception is by means of the physical sensory organs, and (2) the mechanistic theory of nature, according to which the basic units of nature (*a*) are "vacuous actualities," being wholly devoid of experience, (*b*) can exercise only efficient causation, being wholly devoid of any capacity for spontaneity, self-determination, or final causation, and (*c*) can only receive causation from and exert causation on *contiguous* things, meaning that they are incapable or receiving or exerting influence at a distance. In being postmodern as well as liberal, this theology's resolution of the conflicts between science and theology involves, as the above points indicate, a rejection of not only some basic assumptions of traditional theologies but also some basic assumptions of modern science.

Modern liberal theology, by contrast, has posed a challenge only to the basic assumptions of the traditional theologies. That is, it rejects the method of authority and the idea of a deity who intervenes supernaturally in the world, but it does not challenge the above-mentioned assumptions of the modern worldview shared by modern philosophy and science alike. Modern liberal theology has thereby avoided being in conflict with modern science. The result, however, is that modern liberal theology has become largely vacuous. It has avoided affirming anything that would conflict with the worldview of modern scientists by hardly affirming anything whatsoever of a distinctively religious nature. It has avoided conflict with the scientific community by losing contact with the religious community whose faith it was supposed to articulate. Modern liberal theology can, therefore, hardly be said to have overcome the conflict between science and religion.

As indicated in Chapter 1, this book's approach to overcoming the conflicts that have obtained between science and religion involves a two-way interaction between them, mediated by philosophy, resulting in a mutual modification of hitherto orthodox positions in both the religious and the scientific communities. In modern liberal theology, by contrast, the influence has gone all one way: Traditional theology has been modified in the light of science, with "science" understood to be not only empirical facts but also the currently dominant worldview (naturalism$_{sam}$) by which those facts are interpreted. The result is that, whereas religious orthodoxy has been rejected or at least greatly modified, scientific orthodoxy has been accepted wholesale or at least not significantly challenged. Insofar as modern liberal theology has sought to bring its affirmations into harmony with science, this deference to scientific orthodoxy has resulted in an extremely thin theology because this orthodoxy does not allow for any significantly religious interpretation of reality.

This deference to scientific orthodoxy on the part of modern liberal theology has been due in large part to the acceptance of the assumption, long promulgated by the scientific community, including historians and philosophers of science, that the "modern scientific worldview" is both the result of, and the prerequisite for, good science. According to this assumption, in other words, the worldview adopted by the scientific community was originally formulated on the basis of rational reflection upon empirical evidence. And, once hit upon, this scientific worldview has remained, and forevermore will remain, the basis of good science. Science, of course, will continue to advance, in the sense that its methods will lead to ever-new discoveries. But the modern scientific worldview is not itself a subject for debate, as its truth has been amply shown. If we are to have a position that is at once religious and scientific, accordingly, it will have to be in terms of the modern scientific worldview. If this means that our religion must be very thin, this is simply a fact that has to be accepted. Willem Drees's position is only a particularly extreme exemplification of this widespread attitude.

For much of the modern period, theologians had some justification for accepting this view. Few historians or philosophers of science dissented from this story of science as autonomous, based solely on the rational-empirical method, free from ideological distortions. According to this story, although individual scientists, being human, will err, science as such is a self-correcting enterprise. If there had been fundamental problems in the worldview developed by the seventeenth-century geniuses, they would have been weeded out. This, indeed, happened, insofar as those seventeenth-century thinkers, due to their deference to ideas inherited from the theological worldview of the Middle Ages, retained dualistic and supernaturalistic ideas. Through the further development of science in the eighteenth and nineteenth centuries, however, these atavistic ideas were left behind, so that by the beginning of the twentieth century the scientific worldview had become fully rational, rejecting belief in God and souls.

This triumphalist account of scientific materialism was challenged, to be sure, in the late nineteenth and early twentieth century by some philosophers who also had good scientific credentials, such as Henri Bergson, William James, Charles Sanders Peirce, and Alfred North Whitehead (see Griffin et al., FCPP). All of these philosophers rejected scientific materialism as an error based on what Whitehead called "the fallacy of misplaced concreteness." But the ideas of these philosophers were generally rejected or ignored by mainstream thinkers. Their writings, especially those of Peirce and Whitehead, were difficult. Probably more important, they were out of fashion, as most philosophers, including philosophers of religion, turned to anti-metaphysical modes of philosophy, such as phenomenology, existentialism, and linguistic analysis. Those who kept their philosophizing connected to science largely endorsed scientism: The "philosophy of science" consisted largely in explaining scientific concepts and explaining why science is the one true path to knowledge. In this climate, it is not surprising that few theologians were in position to challenge the worldview of the scientific community in its own terms. To most liberal theologians, the only options seemed to be to declare science and theology mutually independent (perhaps on the basis provided by Kant, according to which science describes merely the world of appearance while theology deals with reality) or to accommodate theology, more or less completely, to the worldview of modern science.

In our time, however, all this has changed. Philosophers of science have shown that all theories, including scientific theories, are "underdetermined" by the empirical evidence—that is, that theory is not proved by the data and therefore could not be inferred from the data alone. Other factors determine which theory will be chosen. If this is true for particular theories dealing with some limited set of data, it is all the more true for the overarching worldview for interpreting all data. Thomas Kuhn, in spite of the fact that he later had misgivings about having used the term "paradigm" for an all-inclusive worldview, popularized the notion that "paradigm-shifts" at the level of overarching worldview are made partly on the basis of factors other than rational reflection upon empirical data. At least as important has been the work of those historians of science who have sought to show what some of these other factors have been. One result of their work that has perhaps been especially surprising—given the assumption that the relation between science and theology must be a one-way affair, because, at least since the rise of modern science, it has always been— is the revelation that the rise of modern science itself was heavily conditioned by theological ideas and motives.

The remainder of the present chapter[1] consists of a summary of some of the findings of these revisionary historians of science, to show with some concreteness the ways in which theological motives shaped the philosophy that came to be called "the modern scientific worldview." This historical summary contains implications of relevance for the current discussion of the relation between science and theology. The most general implication is simply that, if we recognize that what has been called "the modern scientific worldview" was in its origins significantly shaped by theological motives, the main basis is removed for assuming that this worldview should today be immune to philosophical-theological critique. But an even stronger implication can be drawn: If we find that the reigning orthodoxy within the scientific community is destructive, and if we find that this worldview has resulted largely from the influence of the Christian theology of an earlier century, then it is especially incumbent upon the theological community today to challenge this worldview.

In any case, having indicated my stance, I turn to the task at hand, which is to summarize some of the ways in which theological ideas, in conjunction with social forces, did indeed shape the movement that has been called the "rise of modern science." The first section discusses the three-way battle of worldviews that occurred in the seventeenth century. The second section explores some of the reasons why the legal-mechanical view emerged victorious in this three-way battle, especially over the other view seeking to supplant the hitherto dominant Aristotelianism.

THE THREE-WAY BATTLE OF
WORLDVIEWS IN THE SEVENTEENTH CENTURY

Until recently, the rise of the "modern scientific worldview" has been described in terms of the replacement of the Aristotelian worldview, with its organic and

---

[1]The body of this chapter is a revised and updated version of an essay, "Theology and the Rise of Modern Science," completed in 1983 as a background paper for a book by Joseph C. Hough, Jr., and John B. Cobb, Jr., *Christian Identity and Theological Education* (Chico: Calif.: Scholars Press, 1985), which was supported by a grant from the Association of Theological Schools.

teleological concepts, by the "new mechanical philosophy," with its exclusion of all such ideas from natural philosophy in favor of mathematical and mechanistic notions. The more recent literature, however, has pointed out that this mechanistic view of nature had been battling at least as vigorously against a third view, or set of views, belonging to what can be called the Neoplatonic-magical-spiritualist tradition. For example, overagainst the old idea that the Aristotelian cosmology crumbled and "modern science" simply took its place, Brian Easlea says:

> It was a case of scholastic Aristotelianism crumbling and protagonists of very different and rival cosmologies engaging in a bitter and protracted struggle for supremacy, both with each other and against the entrenched proponents of Aristotelian-Thomistic cosmology. "Modern science" emerged, at least in part, out of a three-cornered contest between proponents of the established view and adherents of newly prospering magical cosmologies, both to be opposed in the seventeenth century by advocates of revived mechanical world views. Scholastic Aristotelianism versus magic versus mechanical philosophies. (WH, 89)

In this chapter, I describe the protagonists in this "three-cornered contest." However, because Aristotelianism's battle with the mechanical philosophy is sufficiently well known from the standard accounts and its battle with the Neoplatonic-magic-spiritualist tradition is now only of historical interest, I for the most part ignore Aristotelianism, concentrating on the less well-known, and now more relevant, battle between its two challengers.

The Neoplatonic-Magical-Spiritualist Tradition

The Neoplatonic-magical-spiritualist tradition began in the Greek Renaissance in Italy in the latter part of the fifteenth century, which superseded the Latin Renaissance of the fourteenth and early fifteenth centuries. Whereas that earlier Renaissance was humanist, the later one dealt with philosophy, theology, mathematics, and science, with some of the central figures making what they called "magic" central. "Magic" was a broad term, referring to a variety of views, studies, and practices. One of the main distinctions was between "natural" and "spiritual" magic. Whereas *natural* magic involved attracting the natural powers (or "virtues") inherent in things and taking advantage of their natural "sympathies" and "antipathies," *spiritual* magic involved the attempt to tap the spiritual powers of the universe, an attempt that could include what is now called "mediumship" or "channeling" (Yates RE, 79–84, 126–27). The magical dimension of this movement was rooted in the writings attributed to Hermes Trismegistus, supposedly an ancient Egyptian author, which were made prominent by Marsilio Ficino, one of the founders of this second Renaissance. However, Pico della Mirandola, the towering figure of this movement, considered this Hermetic natural magic to be insufficiently powerful. It needed to be supplemented, he argued, by Cabalistic magic, through which one could command the powers of angelic and demonic forces (Yates RE, 79; Easlea WH, 97–98).

Although those involved in this practice emphasized that they invoked only good spirits ("white magic"), their enemies often accused them of invoking demons, or evil spirits ("black magic"). These accusations became especially commonplace as the witch craze, to be discussed later, increased in the sixteenth and seventeenth centuries. Many "magicians," consequently, emphasized that they advocated and practiced only natural magic, which included medicine, natural philosophy (physics), mathematics, astrology, and alchemy. The alchemical side of Hermeticism was especially emphasized in the early sixteenth century by Paracelsus, one of the most important figures in overcoming the hegemony of the Aristotelian-Galenic tradition in medicine (Yates RE, 194). Paracelsus was also one of the members of this tradition who connected its ideas with an egalitarian social philosophy supporting the cause of the peasants against the privileged (Easlea WH, 100–03), one of the aspects of this tradition that made it threatening to the establishment, as will be discussed below.

I will henceforth often refer to this Neoplatonic-magical-spiritualist movement as the "third tradition." It was, of course, not the third to arise historically, only the third to be recognized by historians. In any case, referring to this somewhat motley mixture of philosophies as a "tradition" could minimize the many differences of opinion among the various thinkers and groups thus lumped together. However, even after those differences are honored, there were enough ideas held by all or at least most of those who are generally considered to be members of this tradition to justify this judgment. Among these ideas are (1) the need to experiment, rather than relying on tradition (*vs.* the Aristotelians); (2) the human being as a microcosm, which implies an animistic view of nature, replete with aims, powers, sympathies, and antipathies; (3) deity as present in the world and the world as present in deity (with this idea sometimes understood pantheistically, sometimes more pan*en*theistically); (4) the behavior of things as rooted in their divine or divinely implanted powers (rather than in externally imposed laws); (5) the possibility of effects produced at a distance (*vs.* the Aristotelians and, later, the mechanists); (6) the reality of a form of perception more fundamental than sensory perception; and (7) formally, the desirability of interpreting God's "two books" (Scripture and Nature) in terms of each other, so as to produce one unified, "pansophist" view (*vs.* the Aristotelian theologians, who distinguished absolutely between natural theology, based solely on [universal] experience and reason, and revealed theology, based solely on Scripture).

Some historians of the early modern period have argued that this third tradition was extremely influential upon the attitudes, methods, and actual dis-

coveries of what is now considered "modern science." For example, Frances
Yates, one of the main proponents of this view, attributes to this movement, with
its Platonic-Pythagorean orientation, one of the central features of modern science, the impulse to look for mathematical regularities. "Renaissance magic,"
says Yates, "was turning towards number as a possible key to operations" (GB,
146). The term "operations" in this statement points to a second feature of modern
science that Yates attributes to this tradition, the shift from a *scientia contemplativa*
to a *scientia activa.* Attributing this shift primarily to Pico, she says:

> The profound significance of Pico della Mirandola in the history of
> humanity can hardly be overestimated. He it was who first boldly formu-
> lated a new position for European man, man as Magus using both Magic
> and Cabala to act upon the world, to control his destiny by science.
> (Yates GB, 116)

Pico's Magus, she argues, made it "dignified and important for man to operate"
by making it a religious act, *not* contrary to the will of God (GB, 155–56).

This "Yates thesis" has been accepted by various other historians. For
example, Hugh Trevor-Roper has said:

> The scientific revolution of the sixteenth and seventeenth centuries, it is
> now generally agreed, owed more to the new Platonism of the Renais-
> sance, and to the Hermetic mysticism which grew out of it, than to any
> mere "rationalism" in the modern sense of the word. Ficino, with his
> "natural magic," Paracelsus for all his bombast, Giordano Bruno in spite
> of his "Egyptian" fantasies, did more to advance the concept and investi-
> gation of a regular "Nature" than many a rational, sensible, Aristotelean
> scholar who laughed at their absurdities or shrank from their shocking
> conclusions. (EWC, 132)

In particular, Yates argues, this third tradition helped shape the formation of
the Royal Society, which to a great extent determined the nature and direction of
what we now call "science." For one thing, this tradition had a deep influence upon
Francis Bacon, whose writings formed much of the basis for the first phase of the
Royal Society. Also, John Wilkins, one of the Society's pillars, published in 1648
a book titled *Mathematicall Magick,* which drew extensively upon the writings of
John Dee and Robert Fludd, the two Englishmen who most influentially wrote
from the perspective of this tradition at the beginning of the seventeenth century
(Yates RE, 118–51, 184). Elias Ashmole, who in 1652 published *Chemical Britannic,*
a collection of alchemical writings, was another of the founding members. This
side of the origins of the Royal Society have not been generally known, Yates
adds, because by 1667, when Thomas Sprat published his official *History of the
Royal Society*, it was important for political reasons to put it in opposition to all

"magical" philosophies. The account of Bacon and others is "cleaned up" by being disassociated from the Dee-Fludd tradition (Yates RE, 186–88).

This magical tradition has been seen as influencing not only these earlier scientists, furthermore, but also "the great Newton" himself. This side of Newton has become increasingly known since the philosopher-economist John Maynard Keynes, having acquired Newton's manuscripts, concluded that he should be classified as "a magician" (EB, 313). J. E. McGuire and P. M. Rattansi argued in 1966 (NPP) that Newton thought of causation within a Neoplatonic frame-work. Likewise, Frank Manuel argued that Newton saw more than a metaphori-cal affinity among all forms of action at a distance, such as magnetism, gravity, telepathy, and healing at a distance. In relation to Newton's discovery of the law of attraction or gravity, Manuel said: "Newton lived in an animistic world in which feelings of love and attraction could be assimilated to other forces" (PIN, 85). Newton's interest in alchemy, Manuel pointed out, is shown by the fact that his library contained 175 books on it, that he left 650,000 words on it, that he combed over Ashmole's *Theatrum* time and time again, and that he performed alchemical experiments in his laboratory for many weeks a year, often working through the night (PIN, 162–63, 170–72). On the basis of these studies, Hugh Kearney in 1971 classified Newton as "the great amphibian" to emphasize his allegiance to both the magical and the mechanistic traditions (SC, Ch. 6). More recently, full-length works, such as Betty Jo Dobbs's *The Foundation of Newton's Alchemy* (1975) and Richard Westfall's monumental biography, *Never at Rest* (1984), have extensively discussed the extent to which Newton was indebted to the magical tradition, even though he argued against it.

The thesis that the magical tradition exerted a significant influence on modern science in a positive way, contributing to its methods and discoveries, is controversial (Westman and McGuire HSR; Vickers OSM). It will surely be some time before a consensus is achieved. What is not in doubt, however, is that this tradition was strongly opposed by the mechanistic tradition, which won the battle of the worldviews, becoming the framework in which most science would be carried out in the following centuries. I turn now to the understanding of this mechanistic tradition that has been developed in recent decades by viewing it in the light of its opposition to the Neoplatonic-magical-spiritualist tradition, an opposition that was largely rooted in theological ideas.

The Legal-Mechanical Tradition

In *Religious Origins of Modern Science,* Eugene Klaaren provides an insightful description of the theological side of this opposition in terms of a clash between the third tradition, which he calls the *spiritualist* tradition, and the tradition known as *voluntarism* (sometimes called *nominalism*). The latter's view of na-ture Klaaren felicitously calls the "legal-mechanical" view (RO, 147–48), because it combines the mechanism originating from the Greek atomists, especially Democritus, with legal ideas of nature based upon the voluntarist theologians, such as Scotus, Occam, and the Protestant Reformers, especially Calvin. Calling the view "legal" emphasizes the idea that the order of Nature is due entirely to divinely imposed "laws" (RO, 38, 165).

The voluntarist tradition emphasized God's absolute freedom. The world exists, with the order it has, entirely because of the (arbitrary) decision of God. Neither the world itself, nor any aspect of its order, has a shred of necessity about it. It is God's free creation, totally external to God. God is therefore not the soul of the world, as some spiritualists held, but entirely transcendent over it. The worst intellectual error, according to this view, is to confuse God and the world in any way.

Therefore, Robert Boyle, used by Klaaren as the main bridge from the voluntarist theological tradition to the legal-mechanical philosophy of nature, rejects the spiritualists' view that God is present in all things, saying that God *cannot* be united with matter (RO, 99–100, 158–59). Boyle rejected, furthermore, the view, common to spiritualists and Aristotelians, that creatures have "internal principles of motion," saying that these "vulgar" views make nature "almost divine." Motion is not inherent in matter, he insists: Matter could exist without motion (RO, 149, 162, 166). Creatures do not need the power to move themselves, because an external, omnipotent agent can do everything directly. We must distinguish clearly between matter, on the one hand, and the laws of motion, on the other, realizing that the latter exist only as imposed by God (RO, 132, 163). All of this imposed motion, furthermore, is local motion (locomotion), as opposed to internally motivated (appetitive) becoming. Of course, God causes some things to act *as if* they had appetition, but they do not in reality (RO, 178, 166–67, 171–72). In short, the world's order contains no inherent rationality to be discovered; it is completely imposed by the arbitrary *fiat* of God (RO, 147, 165, 177, 189). Only this view, Boyle held, respects the absolute difference between the Creator and the created and reflects the perfect transcendence, freedom, and power of the former.

Closely related to the spiritualists' error of confusing God and the world, Boyle believed, was their mixing of theology and natural philosophy (RO, 102, 150–51, 181). Overagainst them, he agrees with Francis Bacon that God's two books—Scripture and Nature—are to be kept separate. Theology is to be based on Scripture alone, being undiluted by any admixture of natural philosophy (RO, 103, 132). Theologians, likewise, should not, on the basis of scriptural revelation, interfere with natural philosophy, which is to be based on experiment. Boyle, therefore, sided on this issue with the two-level approach of the Aristotelian theologians, whose refusal to develop a "Christian philosophy" had been so severely criticized by the "pansophistic" spiritualists.

On substantive issues, the clash between the two traditions was nowhere stronger than on the idea of attraction at a distance. Blaise Pascal wrote that he could not understand how the natural magicians could claim that "the sympathy and antipathy of natural bodies are efficient causes, responsible for many effects as if inanimate bodies were capable of sympathy and antipathy" (Easlea WH, 120). With regard to Gilles Roberval's proposal that particles of the same kind have an attraction for each other, Descartes declared that "nothing is more absurd than the assumption . . . that a certain property is inherent in each of the parts of the world's matter and that, by the force of this property, the parts are carried towards one another and attract each other" (Easlea WH, 121). According to Descartes' alternative doctrine, "there exist no occult forces in stones or plants, no amazing and marvellous sympathies and antipathies, in fact there exists nothing in the whole of nature which cannot be explained in terms of purely corporeal causes, totally devoid of mind and thought" (Easlea WH, 111). The implication was the central denial of the mechanists, that there is no causal influence at a distance. As Richard Westfall says, "the fundamental tenet of Descartes' mechanical philosophy of nature [was] that one body can act on another only by direct contact" (NR, 381).

Of course, explaining the phenomena in question, especially gravity, while eschewing the "attraction" posited by the magical tradition was one of the mechanists' greatest challenges, for which they offered various hypotheses. Descartes had his theory of vortices. Christiaan Huygens suggested that, in Easlea's summary statement, "fluid matter surrounding the earth rotates 'in part' in circular motions *in all directions* so that the *net* movement of a gross body is downwards towards the centre of the earth." As Easlea comments, "The audacity of the mechanical philosophers was nothing short of impressive" (WH, 122–23).

The mechanical philosophers had trouble not only with the phenomenon of apparent attraction at a distance, but also with a wide range of other phenomena, from organic form and growth to the mind-body problem. As historians compare the mechanical school with the third tradition, it is not self-evident that the doctrines of the former were superior. Of course, most of the doctrines of the magical-spiritualist tradition, judged in the light of today's knowledge, seem ridiculous. But most of the doctrines of the seventeenth-century mechanists, if judged by the same standard, seem at least equally ridiculous. The question arises, then, as to why the legal-mechanical tradition won such a decisive victory, so that its basic stance has, from the eighteenth century to the present, generally been identified with *the* scientific worldview. Historians of the late seventeenth century have been busy providing answers to this question, to which we now turn.

REASONS FOR THE VICTORY OF THE LEGAL-MECHANICAL VIEW

Brian Easlea, who has written one of the best books on the topic, states clearly the view that is now a commonplace among historians of the period:

[T]he victory of this extraordinary [mechanical] philosophy over its equally extraordinary rival cannot be understood in terms of the relative explanatory successes of each basic cosmology but rather in terms of the fortunes of the social forces identified with each cosmology. (WH, 89–90)

It must be immediately added, however, that this *sociological* explanation is not opposed to the thesis of the present chapter, which is that *theological* issues lay behind the victory of the mechanistic view of nature. There is no opposition because the "cosmology" of each tradition was theological through and through. The point is that the theological cosmology of the legal-mechanical school was perceived to be supportive of the convictions and social interests of the ruling elite—the church, the state, and the wealthy—while the theological cosmology of the third tradition was perceived to be threatening to these convictions and interests. I will discuss this contrast in terms of five issues: political stability, the existence and transcendence of God, life after death, miracles, and witchcraft.[2]

Political Stability

If we ask why the magical-spiritual cosmology was not adopted by the leading opinion makers of the seventeenth century, the underlying answer is rather obvious: These opinion makers, almost by definition occupying positions of power, wealth, or (at least) comfort, favored social stability—the status quo—whereas the magical-spiritual cosmology was perceived to be supportive of a social transformation, if not rebellion. This perception was clearly accurate in the case of some of the representatives of the magical-spiritualist tradition, such as Paracelsus, who supported the peasants' uprising of 1525. Connecting this point with the earlier point about accusations of demon invocation, Easlea says:

> In the case of Paracelsian natural magic, suspicion of diabolical inspiration was reinforced by Paracelsus' social heterodoxy, by his obvious sympathy for the poor and dispossessed, and by his advocacy of a general spiritual and social reformation. (WH, 109)

Summarizing the work of historians of the era, Morris Berman writes that although the "ties between occult and revolutionary thought can be seen in a whole spectrum of leading radicals," we do not know the extent to which lower classes and radical groups held occult and "enthusiastic" beliefs. Nevertheless,

---

2. For reasons of space, I have deleted a sixth issue, the domination of nature, which has been much discussed in recent decades (e.g., Merchant DN).

there is little problem in demonstrating that such an association was made in the public (especially middle-class) mind of the time. . . . [T]he popular impression that communism, libertinism, heresy, and Hermeticism were part of some vast conspiracy is amply documented in the numerous statements made on the subject by clergymen. (RW, 123)

This "popular impression," which had a basis in at least some of the leading thinkers of the third tradition, was sufficient, we can see in retrospect, to have virtually guaranteed that this tradition would be relegated to the status of not only a protest movement but a despised heresy. The perceived danger of this philosophy from the perspective of the comfortable—that it was a threat to the whole social order, with its social, political, economic, and ecclesiastical hierarchies—underlay, as we will see, most of the theological reasons for preferring the legal-mechanical view.

The Existence and Transcendence of God

One of the most fundamental of these reasons was that the legal-mechanical view seemed to protect belief in a transcendent deity in relation to the dangers to this belief posed by the third tradition. The main danger came from the third tradition's animistic idea of matter as self-moving. On the basis of this idea, some members of this tradition had denied the need for a transcendent creator of the world. Being comprised of self-moving parts, they argued, the world could be a self-organizing whole. Sometimes the resulting worldview was atheistic, thereby confirming the widespread association of animistic ideas of matter with the atheism of Epicurus (Klaaren RO, 163); sometimes the resulting worldview was pantheistic (as in the case of Giordano Bruno and John Toland); and sometimes it was closer to what would today be called panentheism. In any case, the voluntarists' idea of a wholly transcendent deity who had created the world out of nothing, and therefore exercised absolute dominion over it, was denied.

This denial was, of course, threatening to the established order, as it was at least sometimes meant to be. The resulting worldview had no place for a God who, having created a heaven with arbitrary admission standards, had then with equal arbitrariness given the "keys to the kingdom" to a particular institution, delegating to it the power to guarantee admission to its faithful, obedient members while consigning others to hell. The immanent deity of the pantheists and panentheists, furthermore, had not arbitrarily decided to be accessible only through the hierarchy of the institutional church, but could be directly experienced by the masses. The rejection of the transcendent, voluntaristic creator would thereby undermine the basic assumptions upon which the authority of the church rested. We can understand why this worldview was threatening to those who favored the status quo, given the almost universal assumption that the authority of the government depended upon the support of the church with its power over people's extramundane status. To put the issue most starkly, what would prevent rebellion if the church were no longer seen to possess such power?

In this situation, the idea of matter as inert could seem a godsend. Because the mechanistic view of matter later became incorporated into a fully materialistic, atheistic worldview, many have assumed that this idea was originally adopted in the seventeenth century for anti-religious or at least extra-religious reasons. Nothing could be further from the truth. The idea that matter is essentially inert—that motion is not inherent in it—was used as one of the primary arguments for the existence of a transcendent creator. This apologetic point was stated clearly by Boyle in a writing published in 1686:

> [S]ince motion does not essentially belong to matter, as divisibility and impenetrableness are believed to do; the motions of all bodies, at least at the beginning of things, . . . were impressed upon them, either by an external immaterial agent, God; or by other portions of matter (which are also extrinsical impellers) acting upon them. (W, IV: 394)

Boyle's point in the last clause, of course, is that the motion of those "other portions of matter" would finally have to be explained by reference to God, the only "intrinsical impeller." Newton also uses the notion of inert material particles to prove the existence of God. Having pointed out that inertia is merely a passive principle, he says: "By this Principle alone there never could have been any Motion in the World. Some other Principle was necessary for putting Bodies into Motion" (Koyré FCW, 216).

The necessity of thinking of matter as moved by "certain active Principles," Newton says, leads to the conclusion that there is "a powerful ever-living Agent" who "in the Beginning form'd Matter in solid, massy, hard, impenetrable, moveable Particles" (Koyré FCW, 219, 217). This was surely one of the principles of his philosophy that he had in mind when he said in a letter to Richard Bentley, in an oft-quoted passage:

> When I wrote my treatise about our system, I had an eye upon such principles as might work with considering men for belief in Deity, and nothing can rejoice me more than to find it useful for that purpose. (Thayer NPN, 46)

Newton, however, argued for God's existence not only from the denial of self-motion to matter, but also from the denial of another power attributed to matter by the magical-spiritual tradition: the power to exercise and receive influence at a distance. This gave Newton an argument for God based upon the principle with which his name was most associated, gravity. It also gave him an opportunity to defend himself from the suspicion with which he was viewed by the Cartesian mechanists because of his use of the language of "attraction." Describing the context in which this suspicion arose, Westfall says that, besides banishing life, color, and other qualities from nature,

> the mechanical philosophy also banished . . . attractions of any kind. No scorn was too great to heap upon such notions. From one end of the century to another, the idea of attractions, the action of one body upon another with which it is not in contact, was anathema to the dominant school of natural philosophy. Galileo could not sufficiently express his amazement that Kepler had been willing to entertain the puerile notion, as he called it, that the moon causes the tides by action upon the waters of the sea. In the [16]90s, Huygens and Leibniz found similar ideas just as absurd for the same reasons. . . . An attraction was an occult virtue, and "occult virtue" was the mechanical philosophy's ultimate term of opprobrium. (IAN, 147)

The central role this point played in evaluating Newton is illustrated by the fact that Huygens, writing to a fellow Cartesian about Newton, said: "I don't care that he's not a Cartesian as long as he doesn't serve us up conjectures such as attractions" (Westfall NR, 464). By explicitly rejecting attraction at a distance as an inherent power of matter, Newton could clear himself of the suspicion that he was a "magician" while also providing a distinctive argument for the existence of God.

In his *Principia* of 1687, he states that gravity—unlike extension, hardness, impenetrability, moveability, and inertia—is not an essential or inherent property of matter, and he says that the words "attraction" and "impulse" are used only to refer to the "quantities and mathematical proportions" of the forces by which bodies approach each other, being neutral with regard to all theories attempting to explain this force (Koyré FCW, 173–77). Newton later, nevertheless, had to make his meaning clearer. In a letter of 1692 to Bentley, who was preparing public lectures, he wrote: "You sometimes speak of gravity as essential and inherent to matter. Pray do not ascribe that notion to me" (Koyré FCW, 178). In his next letter to Bentley, Newton is even more explicit, saying:

> It is inconceivable that inanimate brute matter should without mediation of something else which is not material, operate upon and affect other matter without mutual contact, as it must be if gravitation, in the sense of Epicurus, be essential and inherent in it. And this is one reason why I desired you would not ascribe innate gravity to me. That gravity should be innate, inherent, and essential to matter, so that one body may act upon another at a distance through a *vacuum,* without the mediation of anything else, . . . is to me so great an absurdity that I believe no man who has in philosophical matters a competent faculty of thinking can ever fall into it.

Having thereby disassociated himself from the Epicurean view of matter, with its atheistic associations, Newton then alluded to the theistic implications of his own view, adding:

> Gravity must be caused by an agent acting constantly according to certain laws, but whether this agent be material or immaterial I have left to the consideration of my readers. (Koyré FCW, 178–79)

Of course, as Westfall points out (NR, 506), Newton, by arguing that matter is inanimate so that it cannot exercise agency, had left his readers no choice but to conclude that the agent must be immaterial. Bentley got the point, arguing in his Boyle Lecture of 1692 that "mutual gravitation or spontaneous attraction can neither be inherent and essential to matter, nor ever supervene to it, unless impressed and infused into it by a divine power." Gravitation, Bentley concluded, provides "a new and invincible argument for the being of God, being a direct and positive proof that an immaterial living mind doth inform and actuate the dead matter and support the frame of the world" (Koyré FCW, 183, 184).

When Newton finally stated his views publicly, which he did in the Queries to the 1706 edition of the *Opticks,* he used gravitational attraction as one of the arguments for the existence of nonmaterial agency. In this context, Newton made clear the intent behind his famous statement that natural philosophy should not "feign hypotheses," which has often been interpreted in a positivistic sense, as if he left gravitational attraction without any explanation whatsoever, merely describing its force mathematically. That this was not his meaning is made clear by his criticism of those philosophers who refuse to attribute gravity "to some other Cause than dense Matter," instead "feigning Hypotheses for explaining all things mechanically." Against them, Newton says that the

> main business of natural Philosophy is to argue from Phaenomena without feigning Hypotheses, and to deduce Causes from Effects, till we come to the very first Cause, which certainly is not mechanical.

Then, after listing about a dozen questions that natural philosophy should chiefly be trying to resolve, he concludes:

> And these things being rightly dispatch'd, does it not appear from Phaenomena that there is a Being incorporeal, living, intelligent, omnipresent. . . . (Koyré FCW, 208–09)

We see, therefore, that Newton is not opposed to every kind of "hypothesis." What he rejected were the hypotheses of the Greek atomists and the Cartesians, which try to explain everything in mechanistic terms. Such hypotheses, he says, "are not only false; they lead straightaway towards atheism" (Koyré FCW, 234).

Newton's philosophy, as he told Bentley, was designed to lead straightaway to the opposite hypothesis—although he evidently thought of theism as more a *deduction* from the phenomena than a mere *hypothesis.*

Newton also makes clear that the deity required by his system transcends it, rather than being a soul of the world. There were, indeed, aspects of his position that suggested such a view. He had referred to space as eternal and as the "sensorium" of God. And he had said that God, "being in all places, is more able by his Will to move the Bodies within his boundless uniform Sensorium, and thereby to form and reform the Parts of the Universe, than we are by our Will to move the parts of our own Bodies" (Koyré FCW, 219). However, lest this sound too close to the ideas of the spiritualists, who seemed to the voluntarists to confuse God and the world, Newton hastened to add:

> And yet we are not to consider the World as the Body of God, or the several Parts thereof, as the Parts of God. He is an uniform Being, void of Organs, Members, or Parts, and they are his Creatures subordinate to him, and subservient to his Will. (Koyré FCW, 219)

Even more forcibly, in the famous *General Scholium* added to the second edition of the *Principia*, he said of God:

> This Being governs all things, not as the soul of the world, but as Lord over all; and on account of his dominion he is wont to be called *Lord God pantocrator* or *Universal Ruler;* for *God* is relative word, and has a respect to servants; and *Deity* is the dominion of God not over his own body, as those imagine who fancy God to be the soul of the world, but over servants. (Koyré FCW, 225)

Newton thereby asserted that his system, built upon the idea of matter devoid of the power to move itself or exert influence at a distance, pointed to the existence not only of a divine being, but also of one with the power to ordain, reveal, and enforce rules.

Life After Death

The authority of the church over the extramundane life of the soul depended not only on belief in the existence of a God who could have given it this authority, of course, but also on the belief that the soul *will have* an extramundane existence. This authority would be equally challenged, therefore, if philosophies denying life after bodily death were to spread. The third tradition's view of matter was again perceived as threatening, as the idea of matter as self-moving could be used to undermine the main argument for the immortality of the soul.

Given the belief in an omnipotent deity, to be sure, life after death should not have depended upon the soul's immortality: God could simply resurrect us, body and soul. The dominant view, however, was that this resurrection of the body would not happen until the end-time: The soul would survive until that time by virtue of its God-given immortality. In any case, whether or not the defenders of the faith logically should have been concerned about views that undermined the argument for the soul's immortality, they were, and the mechanistic view of matter again seemed to support their cause.

The main argument for the immortality of the soul was based upon Plato's idea that the soul alone is a self-moving thing. The fact that the body decomposes at death, therefore, provides no reason to suspect that the same fate awaits the soul, because the soul, by virtue of its capacity for self-motion, is different in kind from the matter of which the body is composed. This argument was undermined, of course, if matter itself was viewed as self-moving, as it was by many of those belonging to the third tradition. Not all or probably even most of them, to be sure, actually denied life after death. But some of them pointed back to Pietro Pomponazzi (1462–1525), who argued in a treatise *On The Immortality of the Soul* that this doctrine, which is valuable for maintaining the social order, is not supported by reason but can be affirmed only on the basis of faith (Cassirer et al. RPM, 277). Some of the thinkers of the third tradition, furthermore, explicitly affirmed *mortalism*, according to which when the body dies, so does the soul. Given the fact that the body, which is made of self-moving stuff, decays at death, there is every reason to believe, they argued, that the soul would too (Mosse PRE; Hill WTU). For some thinkers, such as Henry Stubbe, this argument was part and parcel of a comprehensive attack on traditional Christianity (Jacob BA, 218–19).

Against this mortalistic heresy, the mechanistic doctrine of matter, by portraying the human mind or soul as different in kind from the matter of which the body is composed, provided renewed support for the traditional argument for the soul's immortality based upon its unique power for self-motion. This point, along with the issue of an omnipotent deity, was evidently central to the conversion of Walter Charleton, a Royalist physician, from the magical to the mechanistic tradition. Charleton had published three books reflecting magical, alchemical ideas. But later, saying that atheists plotted to "undermine the received belief in an omnipotent *eternal being,* to murder *the immortality of the Soul* (the basis of all religion) and to deride the *Compensation of good and evil actions after death,*" Charleton renounced his earlier perspective in favor of Cartesianism (Easlea WH, 135).

Descartes himself was not indifferent to the support given to the belief in "compensation of good and evil actions after death" by his radical dualism between mind and matter. Having stated that "present institutions are practically always more tolerable than would be a change in them," Descartes asks why people rebel against divinely appointed rulers. His answer is that after the error of those who deny God . . . , there is none that leads weak minds further from the straight path of virtue than that of imagining that the souls of beasts are of the same nature as ours, and hence that after this present life we have nothing to fear or to hope for, any more than flies or ants. (PWD, I: 46)

Although many mechanists were uncomfortable with Descartes' particular version of dualism, the more general argument—that the idea of matter as insentient and inert provides a basis for arguing that we have something in us that, being different in kind, is not subject to decay—was very widespread. The fact that the mechanical philosophy was thereby seen as supporting belief in life after death, whereas the magical tradition was regarded as undermining it, was one of the reasons for the victory of the former over the latter. But there were still more reasons.

Miracles

Life after death and divine omnipotence were certainly necessary presuppositions of the claim that Christianity is the one true religion. But they were not sufficient. They, after all, could be used to support other religions, such as Judaism and Islam. The evidence that Christianity, alone among the religions on the face of the earth, has been ordained by God as the vehicle of truth and salvation was supported by the Christian miracles. For Protestants, this meant the miracles of the New Testament; for Roman Catholics, it meant those plus the miracles in the lives of the saints down through the history of the church. In either case, the Christian miracles, understood as events produced by God's supernatural intervention in the world, were God's signal of the singular truth and saving power of the Christian religion. Without this signal, the church's claim to possess the keys to the kingdom would be a groundless assertion. If the miracles did not happen, or if they could be given a naturalistic interpretation, the church's authority, and thereby the social order, would still be in peril, in spite of the legal-mechanical tradition's proofs for life after death and a transcendent creator.

Once again, the third tradition's view of matter provided a threat. At least some strands within this tradition, as we saw, attributed to matter, as one of its inherent, natural capacities, the power to exert and receive influence at a distance. Many defenders of the faith considered this idea dangerous, because it could be used to argue that, although the miracles of Jesus and the apostles were extraordinary occurrences, there were fully natural events, not different in kind from events that have occurred in other traditions (Easlea WH, 94–95). This argument was made, for example, by Henry Stubbe in a letter to Boyle published in 1666. Boyle responded by arguing that Stubbe's attempt to give a naturalistic explanation for the biblical miracles depended upon his false view of an "animated and intelligent universe," which, unlike the mechanical philosophy, fails to recognize the distinction between God and nature (Jacob BA, 218–19).

Given the importance of this issue, we can understand why Boyle insists so strongly and so often, overagainst the "vulgar philosophy," that matter interacts only by contact (W, III: 453, 457; IV: 416). Against those "enemies of Christianity" who, "granting the truth of the historical part of the New Testament, (which relates to miracles) have gone about to give an account of it by . . . natural complexions," he recommends the virtues of the mechanical philosophy. Because no natural explanation of miracles is possible consistent with its principles, Boyle says, people who accept it will "frankly acknowledge, and heartily believe, divers effects to be truly miraculous, that may be plausibly ascribed to other causes in the vulgar philosophy."

Boyle's response to Stubbe in England was a replay of an earlier argument by Fr. Marin Mersenne, the predecessor, along with Pierre Gassendi, of Descartes in introducing the mechanical philosophy into France. In 1623, Mersenne published a criticism of the Hermetic-Cabalistic-Paracelsian philosophy, dealing especially with Giordano Bruno ("one of the wickedest men whom the earth has ever supported . . . who seems to have invented a new manner of philosophizing only in order to make underhand attacks on the Christian religion") and Robert Fludd ("Bruno's vile successor and principal enemy of Christian religion" [Crombie M, 317]). When Fludd replied, Mersenne, realizing that he needed an alternative system to defeat Fludd's Cabalistic philosophy, appealed for help to Gassendi, who introduced Mersenne to the Democritean, mechanistic philosophy, which had recently been revived in Italy by Galileo (Easlea WH, 108). Mersenne thereby became a major figure in the ascendancy of the mechanistic philosophy of nature, as indicated by the title of Robert Lenoble's book, *Mersenne ou la naissance du méchanisme* ("Mersenne or the Birth of Mechanism").

According to Lenoble, Mersenne came to see the magical tradition as "public enemy number one" by virtue of the fact that it, by affirming universal magic, denied the supernatural character of the miracles upon which the Catholic Church was built (MNM, 9, 120, 157). By regarding all wonders as natural, the magical philosophy in effect denied laws of nature, and without laws of nature to be broken, there could be no miracles in the true sense (MNM, 133). Desiring a stable scientific order with real laws, Mersenne regarded Aristotelianism, which in its pure form denied influence at a distance, as superior to Hermetic naturalism. But he found Galilean mechanism even better, because it made absolutely clear that certain miraculous effects are naturally impossible (MNM, 133, 157–58, 375). Emphasizing the fact that the mechanistic philosophy, rather than being introduced to undermine the belief in miracles, as older studies suggested, was used to support it, Lenoble says:

> Nature reduced to an interplay of mechanical forces was the salvation of science: it was, for religion, a new guarantee of its transcendence and its

dignity. The belief in miracles, far from opposing this evolution, favored it instead; miracles could be produced only in a nature submitted to laws. (MNM, 381; my translation)

I have thus far pointed out the way in which Protestants, such as Boyle, and Catholics, such as Mersenne, employed the mechanistic philosophy in the same way, to support the truly supernatural character of the Christian miracles. Many Protestant thinkers, however, had an additional reason for rejecting the magical view of nature. The fact that post-biblical miracles were used to support Catholicism led Protestants to say that the age of miracles was over. As Keith Thomas describes the Protestant attitude: "Miracles were the swaddling-bands of the early Church, necessary for the initial conversion of unbelievers, but redundant once the faith had securely established itself" (RDM, 124–25). In their polemics, the Protestants accused the Catholic Church of being riddled with magical superstitions and practices (RDM, 51–57, 65, 68, 75–76). Many Protestant thinkers were intent on taking all magic out of Christianity. In such circles, the Neoplatonic-magical-spiritual tradition was a victim of guilt by association: To reject it was to strike a blow against the despised Catholicism. The acceptance of a totally ordered world, devoid of self-moving powers and occult forces, was regarded as a natural consequence of the Reformation view that monotheism, in the true sense, means belief in a single, all-directing Providence and thereby a disenchanted world (RDM, 470–71, 657). Protestant thinkers were even more predisposed by their Christian faith than were Catholics, therefore, to favor the mechanistic over the magical-spiritual view of nature.

Witchcraft

We have thus far looked at evidence suggesting that the mechanistic view of nature, far from being accepted primarily because of empirical evidence or its greater explanatory fruitfulness, was accepted by the opinion makers of the late seventeenth century because it supported theism of a particular sort, life after death (as a precondition for extramundane rewards and punishments), and the supernatural character of the Christian miracles, all of which undergirded the church's authority and thereby the stability of the social order, especially in the face of the threat of uprisings of the poor. It was also used, as we have just seen, to bolster the Protestants' claim that their form of Christianity was superior to the Catholic form. We have yet to consider, however, one of the most important of the theological-sociological reasons for the victory of the legal-mechanical over the magical worldview: the relationship of these worldviews to witchcraft. Brian Easlea considers this factor so important, in fact, that his book on the era is titled *Witch Hunting, Magic, and the New Philosophy*. In a statement summarizing the previous point and providing a transition to the present one, Easlea says:

> Natural magic . . . posed a threat to Christianity. If nature is occult and extraordinary phenomena have a natural . . . explanation, then the miracles of Christ may either have been natural phenomena or the work of an exceptional magician, not necessarily the Son of God. The response of orthodox Christians . . . was to declare natural magic inefficacious and all so-called magical feats either illusions or the work of demons. Natural magic therefore existed always under the cloud of Satanism. *And Satan's powers were daily proved by the necessity of witch-hunts.* (WH, 109)

As this statement suggests, the magical worldview, rightly or wrongly, became associated with witchcraft. The other point of the present subsection, argued by Easlea and other historians of the era, is that the mechanical philosophy came to be regarded as the antidote to witchcraft—both the practice and the accusations thereof.

Although we may know that "witches" were at one time both feared and persecuted and that this "witch craze" reached its height in the sixteenth and seventeenth centuries, it is surely hard for us to realize the extent to which this phenomenon characterized the period. Intellectuals tend to think of the period primarily in terms of the achievements of Copernicus, Kepler, Galileo, Bacon, Harvey, Descartes, Boyle, and Newton, and thereby as the period in which "modern science and rationality" replaced "medieval superstition." Relatively few people, however, participated in these activities, compared with those influenced by the witch craze (which, incidentally, included Kepler, whose mother was accused of witchcraft). In Easlea's words, the phenomena associated with the witch craze "made living in the sixteenth and seventeenth centuries, if a time of creative achievement for a few, a misery and rack for the many" (WH, 2). Another generally under-appreciated feature of the period is the degree to which intellectuals themselves were preoccupied with the witch craze. In their *Witchcraft in Europe 1100–1700,* Alan Kors and Edward Peters say: "Many contemporary observers looked upon manifest witchcraft as quantitatively and qualitatively the single greatest threat to Christian European civilization" (WE, 5).

For many other intellectuals, however, the problem was not "manifest witchcraft" but the *belief in* and *fear of* witchcraft, which led to the hysterical persecution and execution of large numbers of innocent people. The "trials" were usually travesties of justice. In many countries, once a person was accused, it was virtually impossible for her—the vast majority were women—to prove her innocence. In France, Thomas reports, the acquittal rate was sometimes as low as five percent (RDM, 520, 542). This persecution, furthermore, was primarily the product not of the most ignorant, but of the most educated, members of society. "Indeed," says Hugh Trevor-Roper in *The European Witch Craze of the Sixteenth and Seventeenth Centuries,* the more learned a man was in the traditional scholarship of the time, the more likely he was to support the witch-doctors. The most ferocious of witch-burning princes, we often find, are also the most cultured patrons of contemporary learning. (EWC, 154)

Encyclopedias of witchcraft were produced by many of the leading minds of the age (such as the renowned Jean Bodin of del Rio, who knew nine languages and had by the age of nineteen produced an edition of Seneca, citing 1100 authorities). Trevor-Roper says:

To read these encyclopaedias of witchcraft is a horrible experience. Each seems to outdo the last in cruelty and absurdity. Together they insist that every grotesque detail of demonology is true, that scepticism must be stifled, that sceptics and lawyers who defend witches are themselves witches, that all witches, "good" or "bad," must be burnt, that no excuse, no extenuation is allowable, that mere denunciation by one witch is sufficient evidence to burn another. (EWC, 152)

The more we learn about the extent and the horrors of the witch craze, the more we can understand why sensitive, humane thinkers of the time would want to put an end to it, which would mean overcoming its root cause or causes.

A phenomenon as complex and long-lasting as the witch craze was surely fed by several causal streams. One factor, emphasized especially by feminist commentators on the period, is related to the fact, mentioned above, that almost all of the victims were women. Whatever else the witch craze was, it was a misogynist holocaust, which was part of a more general movement to masculinize thought and society. The persecution of women, as many commentators have said, was surely not unrelated to the (successful) attempt to develop a more fully masculine science of nature (Merchant DN, Chs. 5–7). Another dynamic behind the witch craze is suggested by the fact that most of the victims were poor. That is, the witch-huntings can be understood as part of the attempt by the establishment to keep the poor and dispossessed in their place. Pointing out that the persecutions were correlated with famine, price rises, and related crises, Easlea suggests that (male) peasant rebellion and (female) peasant witchcraft could be seen as having the same objective: to turn the (social) world upside down (WH, 37). Commenting that "the witch-hunters were simultaneously suppressing peasant rebellions," Easlea suggests that the witch mania was an attempt by the privileged to keep the world, from their perspective, right side up. For support, Easlea (WH, 37–38) quotes the following observation by Marvin Harris:

The witchcraft mania . . . dispersed and fragmented all the latent energies of protest. It demobilized the poor and the dispossessed [and] filled them with mutual suspicions. . . . . In so doing, it drew the poor further and further away from confronting the ecclesiastical and secular establishment with demands for the redistribution of wealth and the levelling of rank. . . . It was the magic bullet of society's privileged and powerful classes. (CP, 239–40)

These psychological and sociological factors were surely central. But intellectual factors also played a role. A thorough discussion of the various ways in which philosophical and theological beliefs contributed to the belief in witchcraft and to the witchhunts would, of course, need to be lengthy and complex. However, because our question here is how the revulsion against the witch craze contributed to the victory of the mechanistic view of nature over the magical view, we can limit our inquiry to the question of how the magical and mechanistic views were generally *perceived* to be related to the belief in witchcraft.

Most writers agree that "witchcraft" as understood in this period had two aspects. One was the old idea of a witch as one who could produce certain magical effects by sorcery. The magic was "benevolent" or "malevolent" depending upon whether these effects were beneficial or harmful to the recipient. Although the church officially held that all magic was of the devil (because only the devil had the power to simulate supernatural effects), throughout most of the Middle Ages it was indulgent of benevolent magic. And it treated malevolent magic merely as a form of *maleficium,* deliberately causing harm to others: Those convicted of practicing it were, Jeffrey Russell reports, treated like any other criminals who cause harm to others (WMA, 13). In the late Middle Ages, however, theologians added a second element, the idea that the witch owed her special powers to a *deliberate pact* with the devil (as distinct from a *tacit* pact, which was commonly thought by medieval theologians to be implied by dabbling in magic). This deliberate pact made witchcraft into a Christian heresy— in fact, the worst of all heresies: explicit devil-worship. From then on, the *maleficium* involved was, for the theologians, a secondary concern (Thomas RDM, 438–39; Trevor-Roper EWC, 116). In 1468, the pope declared witchcraft an "exceptional crime," thereby in effect, Trevor-Roper says, removing all limits to the application of torture (EWC, 118).

The actual relationship of this "low magic" of witchcraft to the "high magic" of the Renaissance Neoplatonists is a complex issue. On the one hand, the belief in witchcraft understood as a pact with the devil was *undermined* by the Neoplatonic-magical worldview: As we have seen, all sorts of effects that had to be considered "supernatural" from the viewpoint of the less flexible Aristotelian view (which denied the possibility of natural influence at a distance) could, from the perspective of the Neoplatonic-magical view, be considered "natural." Therefore, alleged phenomena that by the Aristotelian clergy had to be attributed to Satan could be explained by the Neoplatonists without recourse to that hypothesis. As Trevor-Roper says:

Nature, to the neo-Platonists, might be filled with "deomons" and charged with "magical" forces, operating by sympathies and antipathies. It might not exclude the existence of "witches"—creatures who, by arcane methods, contributed to short-circuit or deflect its operations. But at least it had no need of such vulgar mechanisms as particular satanic compacts, with their ridiculous concomitants of carnal intercourse, "imps," broomsticks and the witches' Sabbat. It is no accident that "natural magicians" like Agrippa and Cardano and "alchemists" like Paracelsus and his disciples were among the enemies of the witch-craze, while those who attacked Platonist philosophy, Hermetic ideas and Paracelsian medicine were also, often, the most stalwart defenders of the same delusion. (EWC, 132)

As Trevor-Roper indicates, besides the logical fact that the Neoplatonic-magical philosophy could do without the hypothesis of the pact with the devil, which gave the impetus to the witch persecutions, it is a historical fact that many of the chief critics of these persecutions were adherents of this philosophy, including John Webster, Johann Weyer, and Reginald Scot, who were the writers most often attacked by those who thought it important to uphold the belief in witchcraft in the strong sense (Trevor-Roper EWC, 132, 143–49; Thomas RDM, 578–79; Kocher SR, 67–70, 131–37).

On the other hand, the idea of witchcraft was widely associated in the public mind with the Neoplatonic-magical worldview. For one thing, although Plotinus himself had condemned the invocation of demons, this practice became associated with Neoplatonic cosmology in the early modern period (Easlea WH, 56). Although the "invocation of demons" within this cosmology was very different from the "pact with the devil" within the Christian-Aristotelian framework, the two notions did have in common the possibility of contact with disembodied spirits (Thomas RDM, 437; Kocher SR, 69–70). In the public mind, this similarity often overrode the crucial differences. The widely *perceived* implication of the Neoplatonic-magical worldview, in other words, was often different from its logical implication. As Trevor-Roper puts it:

The gulf between the neo-platonic demons, which filled and animated all Nature, and the diabolic hierarchy of the inquisitors might be very deep and logically impassable, but to the common eye—and even to some uncommon eyes—it was also very narrow and could be jumped. When Ficino and Pico della Mirandola, Reuchlin and Cardano, Copernicus and Paracelsus, Giordano Bruno and Campanella all believed, or seemed to believe, that men, by arcane knowledge, might make angels work for them and so control the movements of heaven, it was not unreasonable for ordinary men to suppose that witches, by a baser acquisition of power, might make devils work for them and so interfere with events on earth. (EWC, 134)

Contributing to this association was the fact that not all Neoplatonists rejected the belief in witchcraft. Probably the two most prominent exceptions were Henry More, the Cambridge Platonist, and Joseph Glanvill, a member of the Royal Society, who both agreed with the widespread argument that the denial of witches was the first step toward materialism and therefore atheism. In opposition to the advocates of the Neoplatonic-magical worldview who were offering non-demonic explanations of the powers of witches, they argued that the phenomena of witchcraft prove the truth of immaterial substances. Witchcraft, wrote More, proves "that there are bad Spirits, which will necessarily open a Door to the belief that there are good ones, and lastly that there is a God" (Prior JG). More and Glanvill believed that extraordinary phenomena should be a subject of scientific study (which is today called "psychical research" or "parapsychology"), and Glanvill even called upon the Royal Society to use the scientific method to study spirits (Trevor-Roper EWC, 132–33; Thomas RDM, 577–78). Although this circumspect approach was a far cry from that of those who favored witch beliefs in order to support witch persecutions, the prominence of More and Glanvill probably helped reinforce the widely held association between Neoplatonism and belief in witches.

Therefore, although both logically and historically the Neoplatonic-magical worldview primarily worked to *counter* the kind of witch beliefs that led to the witch craze, it was widely perceived as *contributing* to those beliefs. The result was that many humane intellectuals, looking for a way to put an end to the witch beliefs and thereby the persecutions, thought that the magical worldview needed to be overcome. For some of these intellectuals, the mechanistic worldview seemed best suited for achieving this victory. This aim was one of Mersenne's motives, Lenoble argues, for favoring the mechanistic worldview (MNM, 18, 89–96). In any case, there is general agreement that the mechanistic worldview of Mersenne and Descartes was in fact instrumental to ending the witch persecutions in the latter part of the seventeenth century (Kors and Peters WE, 14, 311–12; Trevor-Roper EWC, 180–81; Thomas RDM, 571–79).

SUMMARY AND CONCLUSION

Summarizing the effect of the triumph of the legal-mechanical philosophy, Keith Thomas says:

> It involved the rejection both of scholastic Aristotelianism and of the Neoplatonic theory which had temporarily threatened to take its place. With the collapse of the microcosm theory went the destruction of the whole intellectual basis of astrology, chiromancy, alchemy, physiognomy, astral magic and their associates. The notion that the universe was subject to immutable natural laws killed the concept of miracles, weakened the

belief in the physical efficacy of prayer, and diminished faith in the possibility of direct inspiration. The Cartesian concept of matter relegated spirits, whether good or bad, to the purely mental world; conjuration ceased to be a meaningful ambition. . . . The triumph of the mechanical philosophy meant the end of the animistic conception of the universe which had constituted the basic rationale for magical thinking. (RDM, 644)

As the statement by Thomas suggests, the mechanical philosophy undermined the intellectual bases not only for "magic" (both "natural" and "spiritual"), but also for various religious beliefs and practices, such as divine inspiration, prayer for healing, and even the existence of "spirits," with this denial of spirits eventually leading, as Glanvill and More had predicted, to the denial of God and life after death. Indeed, as we saw in Chapter 2, the contemporary form of "scientific naturalism," which is based on the mechanical philosophy, is completely antithetical to any significantly religious view of reality. Perhaps the greatest irony of modern intellectual history is the fact, recently revealed by historians of the seventeenth century, that this mechanical philosophy was then adopted primarily as a weapon to defend a supernaturalistic form of Christian faith. What is clear now, in any case, is that if there ever is to be a harmony between science and religion, the attachment of the scientific community to the mechanistic worldview will have to be transcended.

As subsequent chapters will point how, however, there are other reasons, beyond the desire for a harmony of religious and scientific beliefs, for this transcendence of the mechanistic worldview on the part of the scientific community. For example, the view of nature as comprised of inert, insentient bits of matter, originally adopted partly to support belief in the immortality of the human soul by showing it to be different in kind from the matter of which the body is composed, has led to an insoluble mind-body problem, with the result that "science" seems to stand opposed to our inescapable presuppositions about the reality, freedom, and efficacy of our conscious experience (Chapter 6). Likewise, this mechanistic view of nature, which was adopted partly because it made natural influence at a distance inconceivable, has led to a wholly *a priori,* nonempirical treatment of data suggesting that such influence does occur (Chapter 7). What currently passes as the "modern scientific worldview," furthermore, is unable to account for the reality of time (Griffin PAP; PUST) or do justice to the evident facts of the evolution of our world (Chapter 8).

However, the idea that the mechanistic philosophy of nature is somehow uniquely the "scientific" view has had such a stranglehold upon the modern mind that previous demonstrations of its complete inadequacy have, at least until very recently, led only a relatively few scientists and philosophers of science to adopt an alternative philosophy. The dominant attitude has evidently been that, whatever problems the mechanistic view creates, we must live with them, trusting that somehow they will eventually be resolved within the mechanistic framework—even if this "resolution" means rejecting beliefs, such as the belief in our own freedom, that we cannot help presupposing in practice. The assumption that the mechanistic worldview simply *is* the scientific worldview still evidently makes it difficult for many scientists and philosophers to look rationally at alternative philosophies, to see if they might be more adequate.

It is in relation to this mind-set that the new historical understanding of the rise of the "modern scientific worldview" can be especially helpful. Whereas we had previously assumed that the mechanistic view of nature was adopted primarily for rational-empirical reasons, we have now learned that it was adopted primarily for sociological and theological reasons. As this realization sinks in, intellectuals will probably find it easier to take seriously the idea that an alternative philosophy of nature might provide a better worldview for the scientific community.

Besides providing a reason for considering alternatives to mechanism, this new historical perspective on the seventeenth century even provides some support for the Whiteheadian philosophy in particular, insofar as it can be considered a postmodern version of the worldview that, in its premodern version, was defeated by the legal-mechanical tradition. That is, the Whiteheadian philosophy is also a "pansophist" philosophy, being both religious and scientific. And, like at least some of the philosophies in the Neoplatonic-magical-spiritualist tradition, it (1) regards the human being as a microcosm, therefore not different in kind from the rest of the universe; (2) regards each natural unit as self-moving; (3) attributes to each natural unit the capacity to exert influence at a distance; (4) affirms the reality of a nonsensory mode of perception; (5) regards the world as present in a divine reality, which is the soul of the universe; and (6) regards this divine reality as naturally present and active in the world. These ideas have been widely regarded as suspect, even heretical, in both religious and scientific circles. We have learned, however, that they were rejected not because they were antiscientific. They, in fact, belonged to a worldview that promoted scientific attitudes and many scientific discoveries. Rather, their disreputable status came about because they were part and parcel of a naturalistic religious outlook. The rejection of these ideas in the name of a wholly supernaturalistic version of Christianity led to a complete divorce of science and religion, to the detriment of both.

We cannot undo the past. But by accepting a postmodern version of the alternative rejected in the seventeenth century, we may enable future generations to enjoy the kind of harmony between their scientific and their religious presuppositions that has been increasingly impossible since the rise of the "modern scientific worldview." In Part II, we will examine some crucial respects in which Whitehead's postmodern philosophy could bring about a new harmony between scientific and religious beliefs.

Part II

Harmonizing Science
and
Religion Within
Whiteheadian Naturalism

6

SCIENCE, NATURALISM, AND THE
MIND-BODY PROBLEM

The central problem for modern philosophy has been the mind-body problem. Having touched on it in previous chapters, we will in the present chapter examine this problem in detail, seeing how recent discussions within the philosophical community support the thesis that an integration of our scientific intuitions with our religious and ethical intuitions requires a naturalistic worldview based upon panexperientialism. Crucial to this discussion will be the idea that any acceptable solution must do justice to our "hard-core commonsense notions," as discussed in Chapter 4.

THE MODERN MIND-BODY PROBLEM

The problem is how we can explain the relation of our conscious experience to our bodies, especially our brains, so as to do justice simultaneously to two sets of beliefs: our *science-based beliefs* about the world, including ourselves, and our *commonsense beliefs* about ourselves. Some of the crucial beliefs in this second category are (1) that we have conscious experience; (2) that this conscious experience, while influenced by our bodies, is not wholly determined thereby but involves an element of self-determining freedom; and (3) that this partially free experience exerts efficacy upon our bodily behavior, giving us a degree of responsibility for our bodily actions. It is been widely thought that science implies the falsity of one, two, or even all three of these commonsense beliefs. Many scientists and science-based philosophers, accordingly, have sought to eliminate these beliefs from their worldviews. As John Searle says, "the general form of the mind-body problem has been the problem of accommodating our commonsense and prescientific beliefs about the mind to our general scientific conception of reality" (MBW, 215). The problem, as Searle points out, is that this "accommodation" often results in the outright *elimination* of our commonsense beliefs. Indeed, one of the major movements in philosophy in the latter part of the twentieth century has been "eliminative materialism." Recognizing that these three commonsense beliefs do not fit within the reductionistic, materialistic worldview that has become associated with the science, some materialists recommend their elimination from the repertoire of respectable beliefs.

Problems Created by Eliminative Materialism

One problem that this recommendation creates is that these commonsense beliefs are presupposed by a religious-ethical viewpoint. We presuppose that other human beings, like ourselves, have conscious experiences and thereby intrinsic value, value for themselves. This presupposition stood behind, for example, Kant's dictum that we are to treat other human beings as ends in themselves, not merely as means to our ends.[1] The religious and ethical life also presupposes that we have a significant degree of freedom. It presupposes, further, that our freedom to make choices exerts causal efficacy upon our bodies, so that people are, at least generally, responsible for their bodily behavior. We have a serious conflict between science and religion, therefore, if science stands in conflict with these beliefs, as many believe. For example, Francis Crick, in *The Astonishing Hypothesis: The Scientific Search for the Soul,* says:

> The Astonishing Hypothesis is that "YOU," your joys and your sorrows, your memories and your ambitions, your sense of personal identity and free will, are in fact no more than the behavior of a vast assembly of nerve cells and their associated molecules. As Lewis Carroll's Alice might have phrased it: "You're nothing but a pack of neurons." . . . The scientific belief is that our minds—the behavior of our brains—can be explained by the interactions of nerve cells (and other cells) and the molecules associated with them. (AH, 3, 7)

---

1. The less anthropocentric, more ecological ethic that has been emerging in our time is based in part on the assumption that experience and thereby intrinsic value are not limited to human beings. I have elsewhere (WDEW) proffered a reconciliation of the Whiteheadian worldview, with its hierarchy of intrinsic value, and deep ecology, with its egalitarianism of inherent value.

This position, which stands, says Crick, in contrast with "the religious concept of a soul," puts science "in a head-on contradiction to the religious belief of billions of human beings alive today" (AH, 7, 261).

The problem created by the recommendation that we eliminate our threefold belief in the reality, self-determination, and causal efficacy of conscious experience cannot, however, be reduced to simply a problem between science and religion, which would allow the typical modernist response: So much the worse for religion! This facile solution is not possible, because this threefold belief, rather than belonging uniquely to a religious outlook on life, is part of our hard-core commonsense beliefs, which are presupposed in practice by all human beings in all their activities—including their scientific activities. This threefold belief cannot be eliminated from our repertoire of beliefs because, even if we deny it verbally, we will inevitably continue to presuppose it in practice. For example, the recommendation by a philosopher that we eliminate this threefold belief presupposes (1) that we consciously understand the recommendation, (2) that we can freely choose to accept the recommendation, and (3) that our bodily actions, such as our "speech acts," will henceforth be guided by this free choice. The denial of this threefold belief is, therefore, irrational, because it involves a performative self-contradiction.[2]

Eliminative materialists are recommending that we replace commonsense beliefs, or what they like to call "folk beliefs," with what they consider scientific beliefs. "Folk astronomy," for example, held the Earth to be the center of the universe and thought of the sun and all the stars as revolving around the Earth. Included in this folk astronomy was the idea of literal "sunrises" and "sunsets." This commonsense or folk belief has now been replaced by a more scientific understanding. By analogy, they suggest, "folk psychology" thought in terms of conscious experiences, beliefs, volitions, and so on, but it is now time to replace these commonsense beliefs with truly scientific conceptions, according to which the whole universe, including human activity, is to be explained in terms of a materialistic framework, in which there is no room for such beliefs.

The problem with this suggestion is that it subsumes two entirely different kinds of beliefs under the rubric of "commonsense beliefs." Searle points to this equivocation in saying that eliminative materialists "claim that giving up the belief that we have beliefs is analogous to giving up the belief in a flat earth or sunsets" (RM, 48). The eliminative approach to reconciling our science-based beliefs with our (hard-core) commonsense beliefs, accordingly, does not work, even apart from the question of the relation between science and religion. It does not work because it involves a self-contradiction between two types of beliefs: those contained in the theory and those inevitably presupposed in practice, including the practice of advocating eliminative materialism. Eliminative materialism is no more adequate for science than it is for religion.

---

2. See note 6 of Ch. 4.

The Modern View of Matter as the Source of the Problem

In spite of the fact that eliminative materialism is self-contradictory, its emergence as one of the major responses to the mind-body problem is a significant development. The recent period has been very important for the discussion of the mind-body problem for two reasons. First, partly due to the idea that the relation of a computer to its "soft-ware" provides a model for thinking of the relation between a brain and its mind,[3] there has been a veritable explosion of interest in the mind-body problem. Second, arguably the most important result of this ferment has been a growing realization that, given the modern assumption about the nature of matter, no solution that does justice to our commonsense assumptions is possible.

The modern assumption about matter is that it, at least in its most elementary forms, is wholly devoid of experience (sentience) and spontaneity. Given this assumption, there are two options: dualism and materialism. According to dualism, the mind is both (1) numerically distinct from, and (2) ontologically different in kind from, the body. The dualist's mind, being distinct from the brain, provides a locus for conscious freedom. But dualists have never been able to explain, at least if they reject supernaturalist answers, how mind and body, being different in kind, could interact.

Materialism, according to which nothing nonmaterial or nonphysical exists, says that the mind is in some sense identical with the brain. While readily admitting that this "identism" could not be reconciled with our commonsense beliefs about freedom, most materialists in earlier decades were confident that they could at least explain how conscious experience can be identical with the insentient neurons comprising the brain. More recently, however, an increasing number of materialists have been gravitating toward either eliminative materialism, which, by denying the reality of experience, denies that there is anything to explain, or agnostic materialism, which says that, although we cannot deny the reality of experience, we will never be able to explain how it arises. Both are admissions of defeat, saying that we cannot reconcile our scientific with our commonsense beliefs.

The Theological Basis for the Modern View of Matter

The assumption behind this whole discussion is that the modern view of matter, according to which its ultimate units are devoid of experience and spontaneity, is a "scientific" belief. As we saw in Chapter 5, however, this mechanistic view of matter is not a "scientific" belief in the sense of being based upon empirical

---

3. Although the term "functionalism," which has become the standard term for this view, was coined by Hilary Putnam, he has more recently (WL) rejected this view as grossly inadequate.

facts. It had its origin, instead, in theory, especially theological theory, as this idea was used to support the existence of a supernatural deity, the view that Christian miracles were genuinely miraculous because requiring supernatural intervention, and the immortality of the soul.

It is this last use of the mechanistic idea of matter that is especially relevant to our present concerns. The argument, it will be recalled, was that, given the view of matter as inert, wholly devoid of any experience and spontaneity, it is obvious there must be something in us that is different in kind from matter, which we call our soul. Given the fact that this soul is different in kind from the matter composing the body, the fact that the body decays at death is no reason to think that the soul will decay. This argument is used to this day. For example, John Hick, in a chapter defending dualism as a basis for belief in life after death, says:

> Intuitively, it seems odd that of two realities whose careers have been carried on in continuous interaction, one should be mortal and the other immortal. But it also seems, intuitively, odd to deny that of two independent realities of basically different kinds, one might be capable of surviving the other. (DEL, 126)

This dualistic solution to the threat of mortalism, of course, created another problem, which was how mind and body, if they be wholly different in kind, could interact. Matter was said to occupy space in an impenetrable way, mind was not. Matter was said to operate entirely by efficient causation, by pushing and being pushed, whereas mind was said to operate by final causation, being moved by intangible things such as ideas and ideals. How minds could affect bodies composed of matter was as mysterious as how such bodies could affect minds. The unintelligibility of mind-body interaction is usually considered the basic problem of Descartes' dualism, to which he admitted having no solution. However, in a recent reinterpretation of Descartes' dualism, Gordon Baker and Katherine Morris have convincingly argued that the fact that the relation between mind and body was rationally unintelligible was not something he grudgingly admitted. Rather, it was part and parcel of his position, which was that mind and body can influence each other only because this relationship has been ordained by God (DD, 167–70).

This appeal to divine omnipotence to render the mind-body relation somewhat intelligible was more explicitly made by subsequent thinkers. Malebranche, bringing out explicitly the "occasionalism" or "occasionalist interactionism" that Baker and Morris show to be Descartes' real position, said that mind and matter could not really interact. They appear to interact because God causes them to do so: Upon the occasion of my hand being on the hot stove, God causes me to feel pain; then, upon the occasion of my decision to move my hand, God causes my hand to move. Closely related is the doctrine of "parallelism," according to which God, in creating the world, had preordained that mind and body would run along parallel with each other, only *appearing* to interact. Some theists, such as Thomas Reid, held the more straightforward view that God, being omnipotent, could simply make unlike things interact.

In the seventeenth century, the fact that dualism required an appeal to God to explain the (at least *apparent*) interaction of mind and body could be considered a point it its favor, because it provided evidence for the existence of a supernatural deity. Speaking of Descartes' version of this appeal to God, Baker and Morris say:

> It is ironic that one common criticism of Descartes' dualism points to the *a priori* impossibility of his *explaining* mind-body interaction. This is precisely his doctrine, not a *problem* for it. (DD, 153–54)

The Predictability of the Present Impasse

Although historically they are surely correct, from the standpoint of more recent sensibilities it seems unfortunate to have to appeal to supernatural power to explain something that seems so natural. William James illustrated the change in sensibilities by commenting derisively, in relation to Descartes' view: "For thinkers of that age, 'God' was the great solvent of all absurdities" (SPP, 195). This new sensibility, with its unwillingness to accept supernaturalist solutions to philosophical problems, began developing in France in the eighteenth century, with the result that the mind-body *relation* became the mind-body *problem.* In his article on the soul-body relation in the *Encyclopaedia,* for example, Denis Diderot said of soul-body interaction that it is "a fact which we cannot put in question but whose details are completely hidden from us" (Lorimer S, 105). This agnostic admission of inexplicability would become increasingly prevalent.

The fact that the mechanistic doctrine of matter led to this impasse, however, has not led many thinkers, especially in the English-speaking world, to question it. The fact that this view of matter was originally adopted primarily for theological reasons was forgotten. It was simply accepted as "the scientific view" and thereby as beyond question. Other views were impatiently dismissed with such epithets as "mystical," "anti-scientific," and "implausible."

However, given our present historical understanding, we can see that the dead-end reached by both dualists and materialists was predictable. The fact that materialism's attempt to overcome dualism has failed should be no surprise, because the mechanistic view of matter was formulated partly to show the need for a mind or soul different in kind from the bodily constituents. Likewise, the fact that the interaction of conscious experience and brain processes cannot be rendered intelligible within a naturalistic framework should be no surprise, because the mechanistic view of matter was formulated partly to show the need for a supernatural agent.

The above points will now be illustrated by some recent discussions of the mind-body problem. In the next section, I look at dualism as exemplified by four prominent dualists, Karl Popper, John Eccles, H. D. Lewis, and Geoffrey Madell. In the third section, on materialism, I examine the positions of John Searle and Colin McGinn.

DUALISM'S DIFFICULTIES

The strength of dualism, in comparison with materialism, is that, by speaking of the mind as numerically distinct from the brain, it provides a basis for explaining two major features of our experience: its unity and its freedom. With regard to the unity of consciousness, some materialists candidly admit this to be a problem for their identification of the mind with the brain. Thomas Nagel, for example, says that "the unity of consciousness, even if it is not complete, poses a problem for the theory that mental states are states of something as complex as a brain" (VW, 50). Eliminative materialists try to avoid this problem by saying that the unity of experience is an illusion. Daniel Dennett, for example, says that the head contains billions of "miniagents and microagents (with no single Boss)" and "that's all that going on" (CE, 458, 459). If that were the case, however, even the *appearance* of unity would be a mystery. John Searle is more candid: Besides including "unity" as one of the "structures of consciousness" (which he illustrates by pointing out that one can have experiences of a rose, a couch, and a toothache "all as experiences that are part of one and the same conscious event"), he admits: "We have little understanding of how the brain achieves this unity" (RM, 130). Dualists avoid this problem by saying that the mind is a full-fledged actuality, numerically distinct from the brain. John Eccles, for example, says that "the unity of conscious experience is provided by the self-conscious mind, not by the neural machinery" (HS, 22).

This same doctrine likewise provides a locus for the freedom that we all presuppose in practice. Materialists, by contrast, find it difficult to affirm freedom: If there were simply billions of microagents, but no overall "Boss," we could not make self-determining responses to the influences upon us. Dualism, however, says that, besides the billions of microagents constituting the brain, there is another agent, distinct from the brain, which we call the mind or soul. Because it is said to be a single individual, the freedom that we seem to exercise can be affirmed.

This freedom, furthermore, is usually not limited to the power to determine our own mental states. Dualists generally also attribute to the mind the power to influence its body. Those dualists who do not are known as epiphenomenalists. They maintain that, although the body can influence the mind, the mind is simply a nonefficacious by-product of the brain, with no power to influence the brain in return. Although epiphenomenalism was at one time quite prevalent, most philosophers nowadays reject it, saying that the efficacy of conscious experience for bodily behavior is too obvious to deny. For example, William Seager observes that this efficacy "presents the aspect of a datum rather than a disputable hypothesis" (MC, 188). And John Searle, who includes "the reality and efficacy of consciousness" among obvious facts about our minds, endorses "the commonsense objection to eliminative materialism" that it is "crazy to say that . . . my beliefs and desires don't play any role in my behavior" (RM, 54, 48). Most dualists agree, as shown by the fact that "dualism" is usually equated with "interactionism," according to which mind and body act on each other.

Popper and Eccles

At this point, however, dualists encounter difficulties: Although they may *affirm* interactionism, they have to admit that they cannot *explain* how mind and body can influence each other. An example is provided by Karl Popper, one of the twentieth century's most influential philosophers of science. At one time, Popper assumed that an explanation would be forthcoming, saying: "What we want is to understand how such nonphysical things as *purposes, deliberations, plans, decisions, theories, tensions,* and *values* can play a part in bringing about physical changes in the physical world" (OCC, 15). Later, however, in *The Self and the Brain: An Argument for Interactionism,* he, in effect, admitted failure by trying to minimize the importance of the once-urgent problem: "Complete understanding, like complete knowledge," he said, "is unlikely to be achieved" (SB, 105). Insofar as Popper did try to explain how the mind's efficacy upon its body is imaginable, he said: "I think that the self in a sense plays on the brain, as a pianist plays a piano" (SB, 494–95). Popper, however, had affirmed the self to be different in kind from the matter comprising the body, even accepting the pejorative description of dualism as belief in "a ghost in the machine" (SB, 16–17, 494–95). He surely realized that a physical-physical (finger-piano key) relation provides no help in understanding the possibility of a mental-physical relation.

More recently, the coauthor of *The Self and the Brain,* John Eccles, has claimed to have solved the problem in a book entitled *How the Self Controls Its Brain.* He is able to make this claim, however, only because he identifies *the* problem to be solved as that of avoiding a violation of the law of the conservation of energy (HS, 23, 72, 140, 168). Although Eccles recognizes that "[s]elf-brain dualism demands primarily two authentic orders of existents with completely independent ontologies" (HS, 167), he ignores the conceptual problem of how two such orders of existents could interact.

H. D. Lewis

For another example of agnostic dualism, we can examine the position of H. D. Lewis, as articulated in *The Elusive Self* and *The Elusive Mind*. There is no doubt that Lewis, besides making a numerical distinction between the self (or mind) and the brain, thinks of this distinction as involving an ontological dualism. He speaks of the importance of showing "how radical is the difference we must draw between mental states or processes, on the one hand, and material or physical states, on the other, including one's own bodily states," adding that the "finality of this distinction seems to me to be the essence of what is usually understood by the term dualism" (ES, 1). Explicitly aligning himself with Descartes, for whom spatial extension constituted the essence of the physical as distinct from the mental, Lewis speaks of a "non-spatial purpose" as able to "bring about a physical change" (ES, 34). He says that "the obvious divide from which dualism takes its course" is that between things with and without sentience (ES, 4).

That absolute divide was not so problematic in Descartes' time, when it was assumed that sentient and insentient things had both been created by God *ex nihilo*. Given the evolutionary perspective of our time, however, this idea of an absolute line between sentient and insentient beings raises new problems. One of these is the question of exactly *where to draw this absolute line*. The most extreme view would be to draw the line at human beings, saying that all other animals are mere machines, with no feelings—the view that is usually attributed to Descartes. Most dualists draw the line much further down. Wherever it be drawn, however, it is arbitrary. Some say that sentience exists when there is a central nervous system; having such a system, however, is not an all-or-none affair, but a matter of degree. Some suppose that sentience emerges with the rise of life, but where exactly is that? Are we to say that eukaryotic cells are sentient but that prokaryotic cells are not? Or, if we include the latter, what is the reason for excluding the virus, which has some of the properties traditionally used to characterize "living" things? And, if we extend sentience to the virus, then why exclude macromolecules such as DNA and RNA, which, the early view of them as little machines notwithstanding, have been found to have remarkable organismic properties?

Both the evolutionary perspective and the empirical observation of the world, in short, suggest continuity, whereas dualism presupposes that at some point an absolutely new type of actuality emerged: one with experience. Dualism says: Up to that point a purely mechanistic, externalist account is adequate, but at that point a radically new principle of behavior suddenly comes into existence. Lewis affirms this view, while recognizing its difficulty.

> In the long history of our planet, . . . a point must have been reached where, out of dispositions of non-sentient physical matter, there emerged—how or why need not concern us now—an entirely new ingredient of

sentient existence. . . . The question just when . . . is one we must leave mainly to the scientist. . . . It may in some cases be exceptionally difficult to draw the line. . . . The philosopher as such cannot settle this. He can only affirm . . . that at some point response and behavior ceases to be reasonably explicable without recourse to some element of at least sentient existence. . . . [I]t is new and incapable of being accounted for plausibly in the same terms as the physical explanation which was exhaustive up to that point. (ES, 4)

This same sudden appearance of sentience that occurred at some point in the evolutionary process, Lewis says, must also be posited in relation to every human being: "One has no reason to suppose that an unfertilised ovum or the sperm which reaches it has any kind of sentience. . . . But at some point, presumably before actual birth, it must happen" (ES, 4).

Besides the problem of exactly where to draw the absolute line between the insentient and the sentient, there is the insoluble problem of how such an emergence could occur, which Lewis admits, saying: "The mystery of how the change comes about is another matter" (EM, 26).

In addition to these problems involving emergence, furthermore, there is the original problem of how, once the mind or sentience has emerged, it can interact with its wholly insentient bodily members. Like Popper, Lewis seeks to belittle this problem, which he does by appealing to Hume's understanding of causation. Hume had said that, given an empiricist approach to defining terms, we have no understanding of causation as real influence, according to which one thing or event actually brings about another. Our notion of causation, he said, is exhausted by the idea of "constant conjunction," meaning that one kind of event, which we call the "effect," constantly comes right after another kind of event, which we call the "cause." Given this understanding of what causation means, Hume added, we have no basis upon which to dictate *a priori* what kinds of events can be linked in cause-effect relations. Lewis seeks to exploit this point to minimize the damage done to dualism by the fact that it speaks of causal interaction between unlike kinds of things. He first states the problem.

How does the mind send its message to the body, and how does the body instruct the mind? These are, of course, questions to which no answer is ever given, the alleged "transactions" "remain mysterious," as Ryle puts it. . . . The conclusion we are expected then to draw is that the influence of distinct mental processes on physical ones, and vice versa, is a wholly fictitious one. (EM, 26–27)

"There is, however," Lewis then adds, "little in these arguments or their implications to cause us serious anxiety." With an appeal to Hume, he says: "We find that things behave in a certain way" (EM, 27). Pointing out that the difficulty of understanding dualistic interaction is based on the acceptance of the assumption that "like is caused by like," Lewis says:

> Descartes himself could get out of the difficulties presented by his special views about causation by insisting that God could cause anything he liked to be related. There is not, in any case, any need for us to follow Descartes at all points in order to accept the substance of his interactionist theory. (ES, 33)

Lewis can avoid this resort to supernaturalism, he believes, by his resort to the Humean analysis of causation, which allows us to avoid the assumption that cause and effect must share "a common nature."

> We do not in the last resort explain causal relations, except in the sense of unfolding in greater detail the way things do in fact behave. To seek for an explanation of causal relations . . . , to try to pass to some level beyond that of the way in which we find in fact that things do behave, is to follow the wildest will o' the wisp; and no philosopher should be so led astray today. . . . All we can say is that the state of our minds influences our bodies in certain ways, and that the state of my body affects my mind. Why this should happen we do not know. . . . We must be contented to accept what we find. (ES, 33)

On this basis, Lewis claims that causal relations between mind and body create no more problems than do causal relations in general. "The alleged influence of mind on body, then, is indeed remarkable. But in a way so are all other causal relations" (EM, 28–29). Therefore, he suggests, "there is nothing in the last resort more perplexing or astonishing about my mental processes affecting the movements of my body than about a flame consuming the paper to which it is applied" (ES, 38–39). In response to Bernard Williams' criticism of Descartes' position, Lewis says:

> We may indeed admit that there is "something deeply mysterious about the interaction which Descartes's theory required between two items of totally disparate natures. . . ." But it is no more mysterious than many other things which we find in fact to be the case, and it is somewhat unfair for this reason to speak of "the obscurity of the idea that immaterial mind could move *any* physical thing." "Obscurity" . . . suggests that there is something which should be made plain. But there is a limit to explanation and a point where we just have to accept things as we find them to be. (ES, 33)

At most, Lewis says, the mystery of how mind can influence body is, in comparison with the mystery involved in other causal relations, greater only in degree (ES, 33). The mystery involved in mind-brain interaction, he concludes, is no reason to reject dualism.

There are, however, three problems in Lewis' argument. The first is that, in comparing the mystery involved in dualism's mind-body interaction with other mysteries, he fails to distinguish between natural and artificial mysteries. The world is full of *natural* mysteries, such as how our universe began, how evolution proceeded (expressed in the question as to which came first, the chicken or the egg), and how spiders know how to spin their webs. These may be permanent mysteries, to which human beings will never know the answers. They are different in kind, however, from *artificial* mysteries, which are created purely by human conceptions or definitions. The traditional problem of evil, for example, is created purely by the human idea that the Creator of our universe is not only perfectly good but also all-powerful, in the literal sense of essentially possessing all the power. The modern mind-body problem, likewise, is created by the conception of the physical world, including our physical bodies, as comprised of "matter," the ultimate units of which are devoid of experience and spontaneity. The fact that many natural mysteries may permanently exceed our capacities for understanding is no excuse for resting content with insoluble mysteries of the artificial type.

The second problem with Lewis' position is that he arguably does not heed his own advice. He rightly asks, rhetorically: "would it not be better for philosophers, rather than trying to explain away or discredit extra-ordinary facts of experience, to stop and wonder at them and their possible further implications?" (ES, 38–39). This is precisely one of the reasons for affirming panexperientialism: We realize that mind and body do interact. We realize, further, that if they were different in kind, this interaction would be impossible—at least apart from, in Whitehead's words, "an appeal to a *deus ex machina* . . . capable of rising superior to the difficulties of metaphysics" (SMW, 156). The "further implications" of these realizations would seem to be that the mechanistic view of matter must be untrue and that, insofar as we rightly reject the idealist view that the physical world is not really real, panexperientialism must be true.

Lewis, by contrast, takes the "further implications" to be the truth of the Humean view that wholly unlike things, with no common nature, must be capable of causal interaction. This conclusion, however, brings us to a third problem: *Although Lewis affirms the Humean conclusion,* that there is no good reason to stipulate *a priori* that only things with a common nature can causally interact, *he rejects the Humean understanding of causation upon which this conclusion was based.* Hume, as mentioned earlier, defined causation as nothing but regularity of succession. He thereby denied that we have any empirical basis for thinking of efficient causation in terms of a *real influence* of the cause upon the effect. Lewis, however, does not accept this Humean understanding. He speaks instead of "the *influence* of distinct mental processes on physical ones" and of minds as having the "*power* of affecting physical things" (EM, 26, 124; emphasis added). Lewis is right to reject the Humean view of causation: We do experience causation as the real influence of one thing upon another— specifically, the influence of our bodies upon our experience and of our decisions upon our bodies. In rejecting Hume's analysis of causation, however, *Lewis forfeits the right to retain its implication that cause and effect need have nothing in common.*

Lewis' treatment of dualism, in sum, is important in two respects. He illustrates the point that dualists, if they eschew the appeal to supernatural causation, must regard several aspects of the mind-body relation as wholly mysterious, especially when and how mind emerged out of purely physical processes and how, once it did emerge, it could interact with the body. Also, like Popper, Lewis provides no good reason to say that these insoluble mysteries should not count against the truth of the dualistic metaphysic.

Geoffrey Madell

More recently, another dualist, Geoffrey Madell, has been more willing to admit this. Assuming—as do virtually all dualists and materialists—that dualism and materialism constitute the only two real options, he begins his *Mind and Materialism* by saying: "Sympathy with the underlying motivation behind materialism must rest in part on an appreciation of the difficulties which any dualist position confronts" (MM Preface [n.p.]). Then, in speaking more concretely of these difficulties, he says that "it is admitted on all sides that the nature of the causal connection between the mental and the physical, as the Cartesian conceives of it, is utterly mysterious" (MM, 2). Explicitly rejecting the way of softening this problem attempted by Lewis, Madell says that even the Humean view of causation cannot dissolve "the oddity of the claim that it is in the brain *and in the brain only* that perfectly ordinary physical processes have the power to produce something utterly unlike themselves, namely, immaterial states" (MM, 140).

Madell, like Lewis, includes the first emergence of mind as part of this mystery, saying that "the appearance of consciousness in the course of evolution must appear for the dualist to be an utterly inexplicable emergence of something entirely new, an emergence which must appear quite bizarre." Like Lewis, furthermore, Madell says the same for the emergence of sentience in each human: "A parallel emergence occurs, the dualist claims, in the course of the development of the embryo, but it is an event of equal inexplicability" (MM, 140–41).

Madell refers to his book rightly as "a limited and qualified defence of dualism" (MM, 9). Seeing the difficulties he has acknowledged, however, one might well wonder why he defends it at all. The answer, to which most of his book is devoted to supporting, is that whatever the problems of dualism, those of materialism are far worse. This current orthodoxy, he says, involves a doctrine that is "totally mysterious where it is not simply incredible." On that basis, he concludes that "interactionist dualism looks to be by far the only plausible framework in which the facts of our experience can be fitted" (MM, 135). His argument for dualism, in other words, is primarily negative: Materialism is false, therefore dualism must be true. After his statement that he is offering only "a limited and qualified defence of dualism," he says: "The very factors which, to my mind, make materialism impossible to accept, point strongly towards some sort of dualist position" (MM, 9). The assumption, again, is that only these two positions are live options.

Madell does briefly mention panpsychism,[4] but he dismisses it in one sentence, saying that it does not have "any explanation to offer as to why or how mental properties cohere with physical" (MM, 3), a claim that at best applies only to certain versions of panpsychism. Madell shows no signs of having examined the writings of any actual advocates of panpsychism, referring instead, like most other contemporary philosophers, only to the version that Thomas Nagel describes, which is a Spinozistic version, according to which all things, including all aggregational things such as rocks, have a mental aspect as well as a physical aspect. One would think that, given the manifold difficulties that Madell has pointed out in both dualism and materialism, this third alternative should be examined more seriously. There is, in fact, a version of it that (unlike the Spinozistic version) agrees with Madell that the falsity of materialism points to the truth of "some sort of dualist position." That is, if by "dualism" one is referring simply to the numerical distinction between mind and brain, then the Whiteheadian version of panpsychism, or panexperientialism, is a "sort of dualist position,"[5] involving interaction between mind and brain. It thereby avoids the problems caused by materialism's equation of mind and brain.

We have seen, in sum, that there is a growing acknowledgment among dualists themselves that dualism contains insoluble problems, at least within a naturalistic framework, and that the only reason for continuing to hold it is the supposition that materialism, which has even greater problems, is the only serious alternative. As we will see in the next section, contemporary materialists are increasingly coming to the same position, except for their judgment that the difficulties in dualism are even greater than those in materialism.

---

4. Although I sometimes employ the term "panpsychism," because others use it, the term "panexperientialism" is better for the Whiteheadian position, because the term "psyche" has two connotations that do not apply: It suggests that all individuals have *conscious* experience (which was evidently Whitehead's reason for eschewing the term), and it suggests that the ultimate units are enduring things, rather than momentary events.

5. I will argue below, however, that this position should not be called "dualistic," because this term inevitably suggests the Cartesian position.

Materialism's Mysteries

In examining recent discussions of the mind-body problem by materialists, we will examine the adequacy of materialism in terms of the same three commonsense beliefs: the existence of conscious experience, its efficacy for bodily behavior, and its freedom, beginning with the issue of freedom. One of the strengths of dualism, as we saw, is that it provides a basis for taking at face value our presupposition that we exercise a significant degree of freedom. One of the problems created by materialism is that, by identifying the mind with the brain or some aspect thereof, it removes that basis. We will examine this issue in the writings of philosopher John Searle, who candidly points out that his position cannot do justice to the freedom he presupposes in practice.

Searle on the Scientific Impossibility of Freedom

In a book entitled *Minds, Brains, and Science,* Searle describes the mind-body (or mind-brain) problem as "the question of how we reconcile a certain traditional mentalistic conception that we have of ourselves with an apparently inconsistent conception of the universe as a purely physical system" (MBS, 8). The problem, more specifically, is this: "We think of ourselves as *conscious, free, mindful, rational* agents in a world that science tells us consists entirely of mindless, meaningless, physical particles" (MBS, 13). Particularly important for our purposes is Searle's claim that the problem is caused by what "science tells us." He does not, in other words, think of the late modern worldview, with its materialistic reductionism, merely as a philosophical interpretation that has been associated with science for contingent historical reasons. In spite of all the recent reflection upon the fact that theories are always "underdetermined" by the empirical evidence, he thinks that this late modern worldview is itself dictated by the scientific evidence. Because of this conviction, he believes science to be in conflict with common sense.

As we saw earlier, Searle is critical of the way in which some of his fellow materialists have accommodated common sense to science. Insofar as *eliminative* materialism rejects the existence of conscious beliefs and the efficacy of those beliefs for bodily behavior, he rejects it as obviously absurd. Searle believes that these two (hard-core) commonsense beliefs can be reconciled with the scientific worldview, as he understands it. Many fellow materialists do not think that Searle's efforts in this regard are successful, as we will see below. For now, however, we will focus on freedom, the third of our hard-core commonsense beliefs, which even Searle does not find reconcilable with materialism, which he calls "the scientific worldview" or simply "science." Searle states this conflict forthrightly:

> Our conception of ourselves as free agents is fundamental to our overall
> self-conception. Now, ideally, I would like to be able to keep both my
> commonsense conceptions and my scientific beliefs. . . . [W]hen it comes
> to the question of freedom and determinism, I am . . . unable to reconcile
> the two. (MBS, 86)

In calling our conception of ourselves as free agents a "commonsense conception," Searle uses the term in exactly the sense I employ it in speaking of
"hard-core commonsense beliefs." He says, for example, that no matter how
many arguments against it may be marshaled by philosophers, including himself, it is "impossible for us to abandon the belief in the freedom of the will"
(MBS, 94). Using the ideas of "sunsets" and a flat Earth as examples, he points
out that it *is* possible to give up *some* commonsense convictions (namely, those
belonging to what I have called our *soft-core* common sense), "because the
hypothesis that replaces it both accounts for the experiences that led to that
conviction in the first place as well as explaining a whole lot of other facts that
the commonsense view is unable to account for." But, Searle points out, "we
can't similarly give up the conviction of freedom because that conviction is
built into every normal, conscious, intentional action" (MBS, 97). Spelling out
this point, he says:

> Reflect very carefully on the character of the experiences you have as you
> engage in normal, everyday ordinary human actions. You will sense the
> possibility of alternative courses of action built into these experiences.
> Raise your arm or walk across the room or take a drink of water, and you
> will see that at any point in the experience you have a sense of alternative
> courses of action open to you. (MBS, 95)

The commonsense belief in freedom, accordingly, is different in kind from the
kind of common sense involved in the belief that the Earth is flat:

> We don't navigate the earth on the assumption of a flat earth, even though
> the earth looks flat, but we do act on the assumption of freedom. In fact
> we can't act otherwise than on the assumption of freedom, no matter how
> much we learn about how the world works as a determined physical sys-
> tem. (MBS, 97)

To understand Searle's position, it is important to see that he is speaking
of freedom in the real, or libertarian, sense of the word, not in the Pickwickian
sense accepted by many philosophers, according to which freedom is said to
be compatible with physical determinism. The distinction can be clarified in
terms of Searle's statement, quoted above, that "you have a sense of alternative
courses of action open to you." According to the compatibilist rendering of

freedom, you may have a *sense* that you have alternative courses, but you do not, really: Although you may *think* that you made a genuine "decision," cutting off alternative possibilities, *in fact* the antecedent conditions dictated exactly the course of events that ensued. For Searle, by contrast, the freedom that we all presuppose in practice implies an *affirmative* answer to the question, "Could we have done otherwise, all other conditions remaining the same?" (MBS, 89). This point is important, Searle stresses, because "the belief that we could have done things differently from the way we did in fact do them . . . connects with beliefs about moral responsibility and our own nature as persons" (MBS, 92).

However, although Searle's position on freedom as a (hard-core) commonsense belief is the same as mine, he does *not* take commonsense beliefs in this sense as the ultimate criteria for a philosophical theory. He gives that role, instead, to the contemporary scientific conception of the world, which he simply calls "science." Because of this, he concludes that our belief in freedom must be an illusion, in spite of its being ineradicable (MBS, 5, 94, 98). To see why he is led to this paradoxical conclusion, we must see what it is, in his view, that "science tells us" about the world.

We got a glimpse of Searle's view about this in statements quoted at the outset of this section, in which he refers to "the universe as a purely physical system" and of "a world that science tells us consists entirely of mindless, meaningless, physical particles" (MBS, 8, 13). These statements reflect the standard modern conception, shared with dualists, according to which the ultimate units of the world are "purely physical" in the sense of being devoid of experience or sentience ("mindless"). The dualist's position is rejected, however, by the insistence that the world consists "entirely" of such particles. Indeed, Searle endorses what he calls "naive physicalism," defined as "the view that all that exists in the world are physical particles with their properties and relations" (RM, 27). Pointedly rejecting the dualist's distinction between the brain and the mind, Searle says of the human head that "the brain is the only thing in there" (RM, 248). Indeed, Searle regards this as an item of "knowledge," one of those things that, thanks to science, "we know for sure" (RM, 248, 247). Far from being a property of a distinct mind, "consciousness is just an ordinary biological, that is, physical, feature of the brain" (RM, 13).

Given this conception of the world, it follows that, if we are to consider freedom to be real, we must be able to attribute it to the brain. That, however, is impossible: "Science," he says, "allows no place for freedom of the will" (MBS, 92). Searle himself raises our next question: "Why exactly is there no room for the freedom of the will on the contemporary scientific view?" (MBS, 93)

The first part of Searle's argument is the claim that the world consists entirely of physical particles. From this it follows that everything that happens must be caused by these particles, which implies that there is nothing that really acts freely:

> Since nature consists of particles and their relations with each other, and
> since everything can be accounted for in terms of those particles and their
> relations, there is simply no room for freedom of the will. (MBS, 86)

Searle's argument, more exactly, is that according to science, all explanation is
in terms of bottom-up causation: "Our basic explanatory mechanisms in physics
work from the bottom up" (MBS, 93). And the scientific worldview, Searle
maintains, entails that this mode of explanation is to be used not only for the
objects of physics in the narrow sense, but for all phenomena. This point rules
out freedom:

> As long as we accept the bottom-up conception of physical explanation,
> and it is a conception on which the past three hundred years of science are
> based, the psychological facts about ourselves, like any other higher level
> facts, are entirely causally explicable in terms of . . . elements at the fun-
> damental micro-physical level. Our conception of physical reality simply
> does not allow for radical freedom [by which Searle means freedom in the
> libertarian, noncompatibilist sense]. (MBS, 98)

Searle, incidentally, is factually in error in saying that this has been the conception "on which the past three hundred years of science are based." As we have
seen, this totally reductionistic view came to be associated with science only in
the latter half of the nineteenth century (except in France, where this transition
occurred a century earlier). Prior to that, science was associated with the dualistic view, which emphatically did not accept a reductionist view of human
beings, according to which their behavior was to be explained wholly in bottom-
up terms. Searle's point would be largely correct, however, if it were modified
to speak of "the past one hundred and fifty years of science."

In any case, one might argue that Searle seems to be presupposing an
outmoded view of physics, according to which it reveals absolute determinism
at the micro-level, whereas quantum physics now speaks of indeterminacy, which
arguably betokens an element of spontaneity at the lowest level of nature. It is
possible, one could suggest, that this indeterminacy at the quantum level becomes magnified in the human brain, with its extreme complexity, accounting
for that freedom that we all presuppose in practice. Searle's retort to this suggestion constitutes a second point in his argument. Quantum indeterminacy is
irrelevant, he says, because "the statistical indeterminacy at the level of particles
does not show any indeterminacy at the level of the objects that matter to us—
human bodies, for example" (MBS, 87). Searle here is referring to what is often
called "the law of large numbers," according to which indeterminacy at the level
of the elementary units gets canceled out in objects comprised of large numbers
of these units. For example, although there may be some indeterminacy in the
electrons, protons, and neutrons of which a billiard ball is composed, these

indeterminacies cancel each other out, so that the billiard ball itself operates in a completely deterministic, predictable manner. A human body should operate in equally deterministic manner.

At this point, however, Searle's critic might claim that Searle is presupposing a false analogy, as if a human being were organized in the same way as a billiard ball. The critic could begin by pointing out that Searle begs the question by speaking of "a human body." There is no doubt but what a human body, in the sense of a corpse, is analogous to a billiard ball. The question at issue, however, should be whether a living, conscious human being is analogous, and here the answer is clearly No: The human being, by virtue of its conscious experience, is able to be a *self-determining* organism, thereby different in principle from a billiard ball or a corpse. And, Searle's critic could continue, it is the human being's mind or conscious experience that makes indeterminacy at the quantum level relevant to the question of freedom. We are not to suppose that the indeterminacies of the trillions of particles in the brain somehow magically combine on their own to produce the high-level freedom that we know ourselves to have. Rather, the relevance of quantum indeterminacy is that, because the elementary particles comprising the brain are not rigidly determined, the mind can influence them. In this way, we can understand how we are responsible, as we presuppose, not only for our decisions, but also for our bodily behavior.

Part of Searle's response to this objection has already been anticipated by citing his denial that there is anything in the head except the brain: By rejecting the existence of a mind distinct from the brain, he is implying that a conscious human being *is* structurally no different from those aggregational objects in which the macro-properties are fully determined by the behavior of particles at the micro-level. He, in fact, says this explicitly: Directly after the above-quoted comment that "our basic explanatory mechanisms in physics work from the bottom up," he continues:

> That is to say, we explain the behaviour of surface features of a phenomenon, such as the transparency of glass or the liquidity of water, in terms of the behaviour of microparticles such as molecules. And the relation of the mind to the brain is an example of such a relation. (MBS, 93)

Conscious experience, in other words, is an "emergent property" of the brain in the same way that liquidity, solidity, and transparency are emergent properties of water, ice, and glass, respectively. In each case, the emergent properties "have to be explained in terms of the causal interactions among the elements" (RM, 111).

Therefore, everything about the mind, including each of its seemingly free decisions, is fully determined by the behavior of the molecules in the brain. Searle, in discussing the mind as emergent from the brain, specifically denies that the mind, once it emerges, has any degree of freedom to determine its own states, such as its decisions, with which it could then influence its body:

> A feature F is emergent2 iff [meaning "if and only if"] F is emergent1 and F has causal powers that cannot be explained by the causal interactions of *a, b, c*. . . . If consciousness were emergent2, then consciousness could cause things that could not be explained by the causal behavior of the neurons. The naive idea here is that consciousness gets squirted out by the behavior of the neurons in the brain, but once it has been squirted out, it then has a life of its own. . . . [O]n my view, consciousness is emergent1, but not emergent2. (RM, 63)

It might seem that this denial that consciousness contains an element of spontaneity, which then guides the body, is in contradiction with Searle's statement, which we cited earlier, that it is "crazy to say" that our beliefs play no role in our behavior (RM, 48) and also with his positive affirmation that "one's desires, as mental phenomena, conscious or unconscious, are real causal phenomena" (MR, 65). His position, however, is that, although there really is top-down causation from the mind to the brain, everything in the mind, in terms of which it exercises causation upon the body, had been previously determined by bottom-up causation from the body. His position, therefore, is not explicitly self-contradictory (although see below). His affirmation of top-down causation does not, however, mean an affirmation of freedom:

> [T]he top-down causation works only because the mental events are grounded in the neurophysiology to start with. So, corresponding to the description of the causal relations that go from the top to the bottom, there is another description of the same series of events where the causal relations bounce entirely along the bottom, that is, they are entirely a matter of neurons and neuron firings at synapses, etc. As long as we accept this conception of how nature works, then it doesn't seem that there is any scope for the freedom of the will because on this conception the mind can only affect nature in so far as it is a part of nature. But if so, then like the rest of nature, its features are determined at the basic microlevels of physics. (MBS, 93)

In saying that the mind is "a part of nature," Searle means that it is a physical state of the brain. This allows him to make two points at once: that the mind is totally determined by the particles in the brain, and that the mind is part of the chain of physical causes and effects (MBW, 227). If one were, by contrast, to try to allow some freedom to the mind by saying that it is not simply a physical state of the brain but an emergent mental entity that is distinct from the brain, Searle would bring out the standard argument against dualistic causation:

> [I]f our thoughts and feelings are truly mental, how can they affect any-thing physical? How could something mental make a physical difference? Are we supposed to think that our thoughts and feelings can somehow produce chemical effects on our brains and the rest of our nervous system? How could such a thing occur? Are we supposed to think that thoughts can wrap themselves around the axons or shake the dendrites or sneak inside the cell wall and attack the cell nucleus? (MBS, 117)

Searle also rejects the suggestion that quantum indeterminacy would somehow allow a mind distinct from the brain to influence it:

> [I]t doesn't follow from the fact that particles are only statistically deter-mined that the human mind can force the statistically-determined particles to swerve from their paths. Indeterminism is no evidence that there is or could be some mental energy of human freedom that can move molecules. (MBS, 87)

Searle, accordingly, has closed every possible opening for conceiving our felt freedom as genuine. He is left with an irreconcilable contradiction between science (as he conceives it), on the one hand, and our ineradicable commonsense conviction of freedom, which is presupposed in our morality, on the other hand. Faced with this choice, he assumes that "the contemporary scientific view" is more to be trusted, so that our feeling of freedom must be an illusion.[6] He realizes, however, that this solution is unsatisfactory, because our conviction of freedom is so strong that "neither this discussion nor any other will ever con-vince us that our behavior is unfree" (MBS, 98). This unsatisfactory result leads him to say that he is "confident that in our entire philosophical tradition we are making some fundamental mistake, or a set of fundamental mistakes in the whole discussion of the free will problem" (MBP, 145). I believe, of course, that Searle is absolutely right at this point. Before exploring the nature of these mistakes, however, we need to look further at the mysteries to which materi-alism leads.

---

6. Searle's position on this issue is widely shared. Thomas Nagel's analysis is the same: Although we necessarily presuppose freedom in practice, no coherent account of freedom seems possible (VN, 110–17, 123). Colin McGinn says that "it is much more reasonable to be an eliminativist about free will than about consciousness" (PC, 17n). There is, in fact, a virtual consensus among advocates of materialism that it cannot be reconciled with our belief in freedom. The only materialists who think otherwise are those, like William Lycan, who define freedom so as to make it compatible with causal determinism (C, 113–18), which amounts, as Nagel and Searle point out, to defining it away.

Kim and Searle on the Efficacy of Mentality for Bodily Behavior

We have been asking whether materialism can do justice to three of our hard-core commonsense beliefs: the reality of conscious experience, the efficacy of this experience for bodily behavior, and the freedom of this experience and thereby of the resulting bodily behavior. Searle's analysis has shown that materialism cannot do justice to our commonsense belief in freedom. With regard to the second belief, involving the efficacy of conscious intentions for bodily behavior, Searle and virtually everyone else agrees that a position, to be adequate, must do justice to this belief (see Charles Birch's discussion of the centrality of purposes [OP]). However, after years of effort devoted to this issue, Jaegwon Kim has concluded that materialism, in regarding the microlevel studied by physics as causally sufficient for all phenomena, cannot really avoid epiphenomenalism, even though this means a *reductio ad absurdum* of the position (SM, 102–07, 348–60, 367). Because the form of materialism that Kim was seeking to defend, which insisted that all vertical causation must be upward (SM, x–xv, 76–77, 353–54), is the same as that articulated by Searle, Kim's analysis suggests that Searle's position does not really allow efficacy to be ascribed to mental causation, so that it is at least implicitly self-contradictory on this point.

The Emergence of Consciousness: Searle's Position

Although materialism cannot do justice to freedom and the efficacy of consciousness, most materialists have believed that it can at least do justice to the most obvious of our three commonsense beliefs, the fact that conscious experience itself exists. As Searle says, "if your theory results in the view that consciousness does not exist, you have already simply produced a *reductio ad absurdum* of the theory" (RM, 8). Most materialists have, furthermore, believed that eliminativism could be avoided on this point. Searle is one of these.

Searle believes that, given the view that the world consists entirely of insentient physical particles, he can explain how consciousness has emerged. His answer, as we saw earlier in passing, is that conscious experience emerges out of certain states of the brain in the same way that liquidity emerges out certain states of H_2O molecules, solidity out of different states, and so on. However, other philosophers, including many fellow materialists, do not believe that Searle's argument works. One problem is that his analogies are not valid. To see this, it will help to have one of Searle's concise statements of these analogies before us:

> Consciousness is a higher-level or emergent property of the brain in the utterly harmless sense of "higher-level" or "emergent" in which solidity is a higher-level emergent property of H_2O molecules when they are in a

lattice structure (ice), and liquidity is similarly a higher-level emergent property of H_2O molecules when they are, roughly speaking, rolling around on each other (water). Consciousness is a mental, and therefore physical, property of the brain in the sense in which liquidity is a property of systems of molecules. (RM, 14)

Searle sometimes uses the fashionable term "supervenience" in place of "emergence" to express this analogy, saying that consciousness is supervenient on certain neurophysiological states just a liquidity is supervenient on certain molecular states.

Fellow materialist William Seager[7] points out that these two relations do not have the similarity that Searle claims. Having distinguished between *constitutive* supervenience and merely *correlative* supervenience, Seager explains the difference:

Roughly speaking, in cases of constitutive supervenience the dual evidence provided by a knowledge of a system's basic components and their link to its behavior is decisive for ascription of the supervenient property. . . . [I]t makes credible the idea that the joint activity of the various components, through their own causality, could reasonably be claimed to produce the system's overall behavior. (MC, 179)

For example, given our scientific knowledge of H_2O molecules, we can understand why they would produce liquidity at one temperature and solidity at another temperature. We can understand, in other words, how liquidity and solidity can be constituted out of these molecules. The supervenience is *constitutive*. Nothing about the scientific knowledge of neurons, however, gives the slightest clue as to why they, when combined together in a brain, should produce consciousness. This is merely *correlative* supervenience.

In response, Searle agrees that, whereas the supervenience involved in all the analogies he employs is constitutive, all we have in the mind-brain relation is correlative (or what he calls *causal*) supervenience, in which we say that the brain causes conscious experience to arise (RM, 125). Although Searle apparently does not realize it, this concession is in effect an admission that his analogies, through which he hoped to show that the emergence of conscious experience is simply one more example of a familiar phenomenon, provide not even the

---

7. At the time Seager gave his critique, he still held a conventional version of materialism, while being frank about its problems (MC). However, more recently, after becoming aware of the Whiteheadian version of panexperientialism, he has sketched what he calls a panpsychist version of materialism (CIP), which is particularly helpful in bringing out the connection between panexperientialism and the notion of "information," a connection only implicit in my discussion.

slightest intuitive understanding of how conscious experience could have emerged out of a brain comprised of insentient particles.

The Emergence of Consciousness: Nagel's Position

Searle's position illustrates Thomas Nagel's complaint that "much obscurity has been shed on the [mind-body] problem by faulty analogies" (MQ, 202). In a more complete statement of his point, Nagel has said:

> Every reductionist has his favorite analogy from modern science. It is most unlikely that any of these unrelated examples of successful reduction will shed light on the relation of mind to brain. But philosophers share the general human weakness for explanations of what is incomprehensible in terms suited for what is familiar and well understood, though entirely different. (MQ, 166)

It has especially been Nagel who has driven home to fellow materialists the difficulty of making sense of the emergence of conscious experience. The difficulty arises not simply because of the fact, mentioned above, that the *scientific* knowledge of neurons provides no basis for understanding how a brain comprised of them could produce conscious experience. The most serious difficulty arises because of the *philosophical* assumption, widely assumed to be vouchsafed by science, that these neurons are devoid of all sentience or experience whatsoever. Using the French term *pour soi* for that which, having experience, is something "for itself," and the term *en soi* for that which, having no experience, is merely something "in itself," Nagel has famously said:

> One cannot derive a *pour soi* from an *en soi*. . . . This gap is logically unbridgeable. If a bodiless god wanted to create a conscious being, he could not expect to do it by combining together in organic form a lot of particles with none by physical properties. (MQ, 189)

Although I do not know why Nagel used the term "bodiless god," we can take it as directly relevant to our chapter's theme, which is whether we can make sense of the mind-body relation within a naturalistic context. Assuming that by "bodiless god" Nagel meant a creator without "hands" in the sense of the capacity to *manipulate* the world, we can take the expression to refer to a creator without supernatural powers. A conscious being could be created out of insentient particles only by a *deus ex machina* who, in Whitehead's words quoted earlier, is "capable of rising superior to the difficulties of metaphysics" (SMW, 156).

McGinn on Mind-Matter Mystery

Whether or not this point was in Nagel's mind, it has become a central point in the thought of Colin McGinn, whose reflections upon "the problem of consciousness" were given their direction by Nagel (PC, viii). McGinn has, however, taken Nagel's point about the "logically unbridgeable gap" even more seriously, arguing that the problem is insoluble in principle. Indeed, drawing upon Noam Chomsky's distinction between "problems," which human minds are in principle able to solve, and "mysteries," which in principle elude our understanding, McGinn says: "The mind-body problem is a 'mystery' and not merely a 'problem'" (PC, 29).

As I have pointed out, it is not merely scientific ignorance of the brain's workings, which may be overcome in time, but a philosophical doctrine of the brain, according to which it is comprised of insentient neurons, that makes the mind-body relation unintelligible in principle. The fact that McGinn's pessimism is based upon this philosophical conception is made clear in a formulation of the issue on the first page of his book: Having asked, "How is it possible for conscious states to depend upon brain states?", McGinn then shows that he assumes this question to be identical with the quite different query: "How could the aggregation of millions of individually insentient neurons generate subjective awareness?" (PC, 1). That McGinn simply assumes matter to be insentient is shown in a passage in which he is discussing the problem of the rise of consciousness in an individual human being:

> [T]he human sperm and ovum are not capable of consciousness, and it takes a few months before the human foetus is. So when consciousness finally dawns in a developing organism it does not stem from an immediately prior consciousness: it stems from oblivion, from insensate (though living) matter. (PC, 46)

This assumption that the elementary units of matter are insentient (or insensate), in the sense of being devoid of experience of any sort, leads McGinn to conclude that the rise of conscious experience is unintelligible in principle. Besides saying that "our understanding of how consciousness develops from the organization of matter is non-existent" (PC, 19), he says that a characterization of the properties of matter that lie behind the rise of consciousness is "something we do not have, even as a glint in the theoretician's eye" (PC, 85). And, indicating that this is not a lack that might be overcome in time, he says:

> The difficulty here is one of principle: we have no understanding of how consciousness could emerge from an aggregation of non-conscious

elements. . . . [I]t remains a mystery . . . how mere matter could form itself into the organ of consciousness. (PC, 213)

"I think the time has come," adds McGinn, "to admit candidly that we cannot resolve the mystery" (PC, viii).[8]

One of the virtues of McGinn's discussion is that, besides simply pointing out that we cannot explain the emergence of experience from nonexperiencing entities, he further specifies the nature of the difficulty. McGinn does this in terms of Descartes' characterization of the difference between consciousness and matter, according to which matter is, and consciousness is not, essentially characterized by spatial extension. However, rather than regarding this difference as ontological in nature, descriptive of what consciousness and matter are in themselves, McGinn poses the problem in terms of our *conceptions* of consciousness and matter. The problem is that our conceptions are necessarily based upon our perceptions (PC, 13, 29, 60) and that our perception of matter gives us a purely spatial conception of it—a conception that provides no clue as to how consciousness, which we necessarily conceive in nonspatial terms, could arise from it.

[T]he senses are geared to representing a spatial world; they essentially present things in space with spatially defined properties. But it is precisely *such* properties that seem inherently incapable of resolving the mind-body problem. . . .

No property we can ascribe to the brain on the basis of how it strikes us perceptually . . . seems capable of rendering perspicuous how it is that damp grey tissue can be the crucible from which subjective consciousness emerges fully formed. (PC, 11, 27)

On the basis of this notion, that we necessarily regard physical things as purely "spatial entities," McGinn rejects the usefulness of the kinds of analogies to which Searle appeals. While agreeing that just as liquids have a hidden structure that accounts for their macro-properties, such as liquidity, there must be a hidden structure that accounts for consciousness, McGinn does not believe that the former type of emergence provides a helpful model for thinking of the latter:

Principally, this is because the relationship between molecules and liquids is one of spatial composition, whereas the emergence of consciousness from neural aggregates appears to be nothing of the kind. (PC, 79n)

---

8. McGinn is by no means alone in this judgment. For example, in *Brains and People,* William S. Robinson, while accepting a materialist approach, has said that there is no "imaginable story" leading from talk of neurons in the brain to "our seeing why *such* a collection of neurons has to be a pain," and that this absence of understanding "is not merely a temporary limitation" (BP, 29).

The impossibility of understanding how consciousness could arise out of entities necessarily understood in purely spatial terms is what leads McGinn to the conclusion that the mind-body relation constitutes not simply a problem but a terminal mystery, a human solution to which is not possible in principle—at least within a naturalistic worldview.

McGinn is aware of the approach "that says that nothing merely natural could do the job, and suggests instead that we invoke supernatural entities" (PC, 2). This solution, however, is not one that McGinn himself can adopt: "It is," he says, "a condition of adequacy upon any account of the mind-body relation that it avoid assuming theism" (PC, 17n). He does say of the supernaturalist solutions that they "at least recognize that something pretty remarkable is needed if the mind-body relation is to be made sense of" (PC, 2). He fully admits, furthermore, that he cannot, by producing a naturalistic account of the emergence of consciousness, show the supernaturalistic account to be unnecessary. For example, having said that the theory of evolution, by showing how our world could have come about naturalistically, undermines the claim of creationists that a supernatural creator is necessary to explain the existence of our world, McGinn adds:

> In the case of consciousness the appearance of miracle might also tempt us in the "creationist" direction, with God required to perform the alchemy necessary to transform matter into experience. . . . We cannot, I think, refute this argument in the way we can the original creationist argument, namely by actually producing a non-miraculous explanatory theory. (PC, 17n)

McGinn returns repeatedly to this point, that the only possible constructive solution to the mind-body problem would be a supernaturalistic solution. For example, in a passage in which McGinn speculates that consciousness must have arisen "when some of the fancier models of mollusc took up residence in the oceans, or when fish began to roam the depths," he says:

> we do not know how consciousness might have arisen by natural processes from antecedently existing material things. Somehow or other sentience sprang from pulpy matter, giving matter an inner aspect, but we have no idea how this leap was propelled. . . . One is tempted, however reluctantly, to turn to divine assistance: for only a kind of miracle could produce *this* from *that*. It would take a supernatural magician to extract consciousness from matter, even living matter. Consciousness appears to introduce a sharp break in the natural order—a point at which scientific naturalism runs out of steam. (PC, 45)

This thought—based on the equation of "scientific naturalism" with materialism—leads McGinn to wonder if any theists have used the mind-body problem.

I do not know if anyone has ever tried to exploit consciousness to prove the existence of God, along the lines of the traditional Argument from Design, but in this post-Darwinian era it is an argument with more force than the usual one, through lack of an alternative theory. It is indeed difficult to see how consciousness could have arisen spontaneously from insentient matter; it seems to need an injection from outside the physical realm. (PC, 45n)

Swinburne's Supernaturalism

The answer is Yes, the mind-body problem *has* been exploited in this way, most notably by Richard Swinburne, who provides an "argument from conscious-ness," which he introduces with these words:

I do not know of any classical philosopher who has developed the argu-ment from consciousness with any rigour. But one sometimes hears theo-logians and ordinary men saying that conscious men could not have evolved from unconscious matter by natural processes. I believe that those who have said this have been hinting at a powerful argument to which philoso-phers have not given nearly enough attention. (EG, 161)

This "powerful argument" is the one to which McGinn refers. Believing, like McGinn, that most events in the world, including "brain-events," are purely physical in the sense of being devoid of experience, Swinburne argues that there is "no natural connection between brain-events and correlated mental events," with the result that this correlation is inexplicable apart from appeal to the will of an omnipotent deity (EG, 172–73). Swinburne summarizes the argument thus:

[S]cience cannot explain the evolution of a mental life. That is to say, . . . there is nothing in the nature of certain physical events . . . to give rise to connections [to mental events]. . . . God, an omnipotent, omniscient, perfectly free and perfectly good source of all, would need to be postu-lated as an explanation of many diverse phenomena in order to make his existence probable. But the ability of God's actions to explain the other-wise mysterious mind-body connection is just one more reason for postu-lating his existence. . . . God, being omnipotent, would have the power to produce a soul thus interacting, to produce intentionally those connections which, we have seen, have no natural connections. (ES, 198–99)

McGinn, of course, cannot accept this solution. "Naturalism about con-sciousness," he holds, "is not merely an option. It is a condition of understand-ing" (PC, 47). Although I would not argue for it in those terms, I agree with

McGinn about the need for a naturalistic account of the mind-body relation. This is one of the main reasons for rejecting the assumption, shared by dualists (like Swinburne) and materialists (like McGinn), that the ultimate units of nature are devoid of experience and spontaneity.

McGinn's Fideism

McGinn believes that he can avoid this option—that he can provide a solution of sorts to the mind-body problem while retaining the modern notion of matter. While insisting that no constructive solution is possible (without supernaturalism), he believes that he has provided *a naturalistic but nonconstructive solution*. It is not a constructive solution, because it does not "specify what it is about the brain that is responsible for consciousness." It is naturalistic, however, because it insists that "whatever it is it is not inherently miraculous" (PC, 2). McGinn's nonconstructive solution, in other words, consists of the assertion that, although it is hidden from us in principle, some fully natural property of the brain must account for consciousness:

> In reality, there is no metaphysical mind-body problem; there is no *onto-logical* anomaly, only an epistemic hiatus. The psychophysical nexus is no more intrinsically mysterious than any other causal nexus in the body, though it will always strike *us* as mysterious. This is what we can call a "nonconstructive" solution to the problem of how consciousness is possible. (PC, 31)

We can well ask, however, whether this is a solution of *any* sort. The goal of traditional materialists was to show how consciousness could be reduced to material states of the brain. Realizing that this program failed, some materialists decided simply to deny the existence of consciousness. For example, one eliminativist, Paul Churchland, has said: "If we do give up hope of reduction, then elimination emerges as the only coherent alternative" (OS, 507–08). McGinn, however, believes that there is another alternative, which involves a combination of agnosticism and gnosticism. Although he agrees with eliminativists that no reduction of consciousness to matter can be given, this does not prevent his "knowing that a reduction *exists*" (PC, 31n). But how is such knowledge possible? McGinn makes this claim to knowledge again in distinguishing existential from effective naturalism. An *effective naturalism*, in which we are "able actually to specify naturalistic necessary and sufficient conditions for the phenomenon in question," now exists, McGinn says, with regard to reproduction, the weather, and the movements of the planets. "No hint of the divine need now be recognized in these areas; the miraculous has given way to the mechanistic." Defining *existential naturalism* as "the thesis, metaphysical in character, that

nothing that happens in nature is inherently anomalous, God-driven, an abroga-
tion of basic laws—whether or not *we* can come to comprehend the processes
at work," McGinn says that "we can be in position to know that existential
naturalism is true of consciousness without being in a position to convert this
into effective naturalism" (PC, 87, 88).

This metaphysical thesis, however, sounds much more like faith than knowledge. In places, McGinn admits this, saying that existential naturalism, when it
cannot be turned into effective naturalism, is "an article of metaphysical faith"
(PC, 87). In fact, calling his position "agnostic realism," McGinn compares
himself to an "agnostic theist." Although McGinn continues to speak in terms of
knowledge—saying that we "can know that something exists without knowing
its nature. We can assert that a gap is filled without being able to say how it is
filled" (PC, 119)—he certainly does not believe that the agnostic theist, who
claims to know that God exists without being able to answer various objections
such as the problem of evil, *knows* theism to be true. Likewise, Geoffrey Madell,
while admitting that he does not know *how* dualism is true, nevertheless seems
to think that he knows *that* it is true, but McGinn would not agree with this
claim to knowledge. Being in the same position, McGinn cannot plausibly claim
to know materialism to be true.

The Neglected Alternative: Panexperientialism

McGinn believes that he knows this because he believes that he knows all the
alternatives to materialism to be false. I share McGinn's belief that we can be
confident of the falsity of idealism, according to which what we call the "physical world" does not actually exist. Given the assumption that some form of
realism must be true, we are left with three alternatives: Some version of dualism, some version of materialism, and some form of panpsychism or panexperientialism. I share McGinn's confidence, as I have already indicated, that dualism
is too problematic to be considered true. We agree, accordingly, that the choice
must be between the two forms of pluralistic monism: materialism and
panexperientialism. We also agree, however, that materialism is extremely problematic. This would seem to imply that we should look carefully at the various
kinds of panexperientialism, to see if one of them is less problematic than all the
versions of dualism and materialism. McGinn, however, does not do this.

McGinn's treatment of this third type of realism is superficial in the extreme. This is surprising, given his recognition that panexperientialism, unlike
dualism, would meet his criterion that any acceptable solution be naturalistic.
Regarding the elementary units of the world as having experience, he concedes,
"is not supernatural in the way postulating immaterial substances or divine interventions is." He dismisses it, however, as "extravagant"—on the same page, incidentally, on which he says that "something pretty remarkable is needed if the

mind-body relation is to be made sense of" (PC, 2). This casual dismissal is all the more surprising in the light of his statement that, if we credit neurons with proto-conscious states, "it seems easy enough to see how neurons could generate consciousness" (PC, 28n). In the light of his realization that it is impossible to understand how neurons could do this if we continue to regard them as wholly devoid of experience, one would think McGinn would recommend that we examine all the extant versions of panexperientialism, to see if one of them can help us finally solve this centuries-old problem. Instead, however, McGinn simply says that he will "here be assuming that panpsychism, like all other extant constructive solutions, is inadequate as an answer to the mind-body problem" (PC, 2n).

Although McGinn fails to mention it in this context, the "here" in this statement reflects the fact that he had examined panpsychism in a previous book. This examination, however, was limited to a couple pages (CM, 31–33). In this brief treatment, besides not exploring various versions of panpsychism, he did not even explore an extant version as presented by an advocate. Instead, he looked only at the Spinozistic version portrayed by Thomas Nagel, which differs from the version advocated in the present volume in at least two crucial respects. In my Whiteheadian-Hartshornean version, a distinction is made between experience as such, which even the lowest-level individuals are said to have, and *conscious* experience, which emerges only in very high-level individuals. Also, this position is best called not simply "panexperientialism" but "panexperientialism with organizational duality," because a distinction is made between two ways in which low-level individuals can be organized: into "compound individuals," in which a higher-level of experience (which might be conscious) emerges, and merely "aggregational societies," such as rocks and telephones, in which no higher-level experience emerges. A rock as such, therefore, has no experience whatsoever; the highest experiences in the rock are those of the billions of molecules. In McGinn's imagined panpsychism, by contrast, neither of these distinctions obtains, so that "elementary particles enjoy an inner life" and "rocks actually have thoughts" (CM, 32). It is on the basis of *this* understanding of what it entails that McGinn makes not only his aforementioned claim that panpsychism is "extravagant" but also the stronger claims that it is "outrageous" and "absurd."

Not having really explored panexperientialism, however, McGinn is hardly in a position to claim that he knows, by having eliminated all the alternatives, materialism to be true. We have ample reason, furthermore, to conclude that materialism must be false, because it cannot do justice to our hard-core commonsense assumptions about the reality, the efficacy, and the freedom of consciousness. This threefold failure, it must be stressed, means that materialism is inadequate not only for morality and religion but also for science, because scientists, precisely in their scientific activities, presuppose all of these beliefs. No worldview can be adequate for "science" that has no room for the activities of scientists!

In a more extensive discussion (UW), I have given several reasons to consider panexperientialism not to be obviously absurd and to see it, in fact, as providing a far more adequate and self-consistent worldview than either dualism or materialism. One of the most important reasons for this conclusion is the thesis of this chapter: that panexperientialism alone, among realistic philosophies, can avoid an insoluble mind-body problem within a naturalistic framework. Having argued the negative part of this thesis—that dualism and materialism cannot solve the mind-body problem—I turn now to the positive presentation of a panexperientialist solution based on Whitehead's philosophy.

TEMPORALITY AND PANEXPERIENTIALISM

A central thesis of the present volume is that much of the apparent conflict between science and religion has been due to the acceptance of a worldview in which the fundamental nature of temporality has been overlooked, and that a necessary dimension to revealing the essential harmony between science and religion is the adoption of a worldview in which temporality is given its due. I will here show how this thesis applies to the mind-body problem.

One of the merits of McGinn's discussion is the fact that it shows the difficulties of dualism and materialism to be rooted in the impossibility of understanding how the physical world, if it be conceived to be wholly spatial, could give birth to, and otherwise interact with, conscious experience. To see why McGinn speaks of impossibility, we need to bring out some premises. First, in saying that we necessarily conceive of matter as *wholly* spatial, McGinn implies that we do *not* think of it as also essentially temporal. This means that we necessarily conceive of matter as wholly different from conscious experience, which we conceive to be nonspatial but temporal. A second, closely related presupposition behind the mind-body problem, for dualists and materialists alike, is that we cannot attribute experience to that which is conceived to be purely spatial, because experience is necessarily conceived as essentially temporal.

With regard to matter, McGinn's twofold claim is that not only must our conception of it be based upon our *perception* of it, but also that this perception is *sensory* perception, which presents matter as essentially and purely spatial. McGinn's claim that conceptions, to be meaningful, must be based upon perceptions is certainly correct, as is his claim that sensory perception (especially vision) presents things as purely spatial. What is dubious, however, is his claim that our conception of what we call "matter" or the "physical world" must be based exclusively upon sensory percepts of it.

Panexperientialism is based upon the contrary supposition, that we can and should think about the units comprising the physical world by analogy with our own experience, which we know from within. The supposition, in other words, is that the *apparent* difference in kind between our experience, or our "mind," and the entities comprising our bodies is an illusion, resulting from the fact that we know them in two different ways: We know our minds from within, by identity, whereas in sensory perception of our bodies we know them from without. Once we realize this, there is no reason to assume them really to be different in kind. This solution to the mind-body problem, interestingly enough, was suggested by Immanuel Kant, who, in a discussion of "the communion of soul and body," said:

> The difficulty peculiar to the problem consists . . . in the assumed heterogeneity of the object of inner sense (the soul) and the objects of the outer senses, the formal condition of their intuition being, in the case of the former, time only, and in the case of the latter, also space. But if we consider that the two kinds of objects thus differ from each other, not inwardly but only in so far as one appears outwardly to another, and that what, as thing in itself, underlies the appearances of matter, perhaps after all may not be so heterogeneous in character, this difficulty vanishes. (CPR, 381 [B, 428])

Although McGinn quotes this passage, he rejects Kant's suggested solution. Panexperientialism, by contrast, accepts and develops this suggestion, saying that our perception of our own conscious experience can provide data for conceiving the entities forming the body, such as the neurons comprising the brain. We can, therefore, think of these entities as having not only spatial extension but also temporal duration—*which makes it possible to conceive of their having a low-level degree of experience.*

A crucial event in the history of this solution to the mind-body problem, as Pete Gunter has stressed (HB, 137–41), occurred between the first and second books published by Henri Bergson. In his first book, *Time and Free Will* (1889), Bergson articulated an absolute dualism between physical nature, which was described as spatial but nontemporal, and the mind, which was described in terms of temporal duration. Bergson soon realized, however, that this made the interaction of mind and matter, which he presupposed, unintelligible. In his next book, *Matter and Memory* (1896), he overcame this dualism by attributing a primitive memory and thereby temporal duration to that which we, from without, call matter. The ultimate units of the universe, in other words, are not purely spatial bits of matter but spatial-temporal events, with temporal as well as spatial extension. With the dualism overcome, mind-body interaction could be understood as a purely natural occurrence.

This Bergsonian solution was developed more fully by Whitehead. The fundamental units of the actual world, the fully actual entities, were said by Whitehead to be "actual occasions," by which he meant that they were temporally as well as spatially extensive (PR, 77). Each ultimate unit of nature, thus

conceived, is an event, "with time-duration as well as with its full spatial dimensions" (RM, 89). It takes time, in other words, to be actual. Events at the subatomic level may take less than a billionth of a second to occur, but this makes
them qualitatively different from matter as traditionally conceived, according to
which it can exist in an "instant," meaning a slice of space-time with no duration. According to that traditional view, as characterized by Whitehead, "if material
has existed during any period, it has equally been in existence during any portion
of that period. In other words, dividing the time does not divide the material"
(SMW, 49). This means that "the lapse of time is an accident, rather than of the
essence, of the material. . . . The material is equally itself at an instant of time"
(SMW, 50). In matter thus conceived, there is no internal becoming. The only
motion is external motion, or locomotion: the motion of a bit of matter from one
place to another. One side of the mind-body problem can be phrased in terms
of the question: How could the locomotion of bits of matter in the brain give rise
to the internal motion, or becoming, in our experience? Whitehead's view avoids
this question, saying that there is no "nature at an instant," that even the most
primitive units of nature have temporal duration, during which internal becoming, analogous to that in our own experience, occurs.

In fact, once temporal duration with its internal becoming has been attributed to all actual entities, it is natural to attribute experience to them, because
we have no way to conceive of this internal duration except by analogy with the
duration we know in our own experience. Whitehead, therefore, also referred to
actual entities as "occasions of experience," which is the basis for referring to
his philosophy as panexperientialism. Once this move is made, the emergence of
conscious experience out of the neurons in the brain is no longer a complete
mystery (as McGinn admits). The same is true for the more general interaction
of mind and body. Regarding temporality as fundamental is, accordingly, central
to overcoming the modern mind-body problem.

A possible objection, at this point, might be phrased thus: "We can perhaps grant that this attribution of temporality and thereby experience to all actual
entities can solve the mind-body problem, but is there any justification for this
attribution?" One justification is provided by the very fact that this move *can*
solve the mind-body problem. We know, in other words, that our conscious
experiences and our bodies interact. The panexperientialist view of the body is
justified by the very fact that it, and evidently it alone, can explain this interaction within a naturalistic, realistic framework.

A second justification is provided by the fact that our conscious experience
appears to be as much a part of "nature" as anything else. We should, therefore,
use our privileged vantage point on our own experience to understand what
"natural entities" are in themselves. That is, our own experiences are the only
events in the world that we are able to view from within. If we take seriously
the idea that our experiences are fully natural, rather than being supernatural
additions to nature, we should generalize what we know about our own experi-

ences to all other events (with less sophisticated experience being attributed, of course, to less complex types of individuals).

A third basis for attributing experience to what are usually called "natural entities" is provided by our direct experience of our own bodies, which provides our most direct observation of nature. At this point, Whitehead explicitly contravenes the conventional viewpoint, which McGinn exemplifies. In responding to the question, "How do we observe nature?", Whitehead says: "The conventional answer to this question is that we perceive nature through our senses" (MT, 158). If we accept this answer, we end up with McGinn's view that we must think of natural entities as purely spatial, because sensory perception does indeed, Whitehead agrees, "spatialize" nature (SMW, 50; PR, 209). However, our intellectual conception of nature need not be based entirely upon this type of perception, as McGinn supposes, because we have another way to perceive nature, which is, in fact, a more direct form of perception. In opposition to the conventional approach, in which all direct observation has been identified with sensory perception, Whitehead says:

> All sense perception is merely one outcome of the dependence of our experience upon bodily functionings. Thus if we wish to understand the relation of our personal experience to the activities of nature, the proper procedure is to examine the dependence of our personal experiences upon our personal bodies. (MT, 159)

If we pay attention to this dimension of our experience, we are led to a quite different view of the entities comprising nature: "[A]mong our fundamental experiences," Whitehead points out, is the "direct feeling of the derivation of emotion from our body" (MT, 159–60). This is our primal relation to our body and thereby to what we call the physical world:

> The primitive form of physical experience is emotional—blind emotion— received as felt elsewhere in another occasion and conformally appropriated in subjective passion. In the language appropriate to the higher stages of experience, the primitive element is *sympathy,* that is, feeling the feeling *in* another and feeling conformally *with* another. (PR, 162)

If this is the truth about the relation of our experience to our bodies, the implication is that the body is not comprised of "vacuous actualities," meaning things that are actual and yet wholly devoid of experience (SMW, 58; PR, 29, 167). Rather, if we derive emotions and other feelings from our bodily members, these members must themselves have feelings of their own, even if feelings of a lowlier sort. As Charles Hartshorne has pointed out, it is Whitehead's panexperientialist view of matter, not the view of the dualists and materialists, that is rooted in immediate experience:

The "ocean of feelings" that Whitehead ascribes to physical reality is not only thought; so far as our bodies are made of this reality, it is intuited. What is not intuited but only thought is nature as consisting of absolutely insentient stuff or process. No such nature is given to us. (SC, 13)

The three justifications I have already given for panexperientialism are empirical in one sense of the term: They are based upon appeals to our immediate experience. A fourth justification is empirical in the sense of appealing to science: As pointed out in Chapter 4, natural scientists have recently been finding evidence for experience lower and lower down the phylogenetic scale. If bacteria and other prokaryotic cells give evidence of making decisions on the basis of memories, then it is hardly far-fetched to suppose that the much more complex eukaryotic cells in our brains have experience. Although many people may be willing to accept the idea that neurons have experience, thereby overcoming the major source of the mind-brain problem, they might still boggle at the idea of panexperientialism, according to which even electrons and other subatomic particles have a primitive degree of experience. The mystery, they would say, is now how the jump was made from insentient particles to sentient cells.

Two aspects of twentieth-century physics, however, provide reason to attribute experience all the way down. I refer, first, to the indeterminacy of quantum physics. As mentioned in Chapter 4, this indeterminacy, interpreted realistically, is suggestive of an element of spontaneity, or self-determination, at the quantum level, and spontaneity is in turn suggestive of experience. The second aspect returns us to the theme of the fundamental nature of temporality: Physics now speaks of space-time, or time-space, which means that space and time are inseparable. It is almost universally accepted, furthermore, that time and space are not absolute containers that would exist apart from spatial and temporal actualities. Rather, time exists because temporal processes occur, and space exists because these events or processes are spatially extended. Now, if time and space are inseparable, so that we must always speak of space-time, or time-space,[9] the implication is that the ultimate units of nature are spatiotemporal occurrences, which returns us to Whitehead's point that the ultimate actual entities of the world are actual occasions, with internal duration as well as spatial extensiveness. The final step in this fifth argument is that it is impossible to conceive of internal duration except as experience.

A sixth justification, also involving time, is the argument that we run into insuperable paradoxes if we assume that time arose at some time in the evolutionary process. The only nonparadoxical view is pantemporalism, according to which time stretches all the way back. We can make sense of pantemporalism,

---

9. See Milič Čapek's "Time-Space rather than Space-Time" (NAT).

however, only by assuming panexperientialism. I have laid out this argument elsewhere (PAP; PUST).

There are, accordingly, many reasons to adopt panexperientialism, several of which are scientific. Indeed, whereas earlier I argued only that science, contrary to widespread contemporary opinion, does not necessarily support materialism, I now add the point that, insofar as science supports any position, it is panexperientialism. As I indicated above, however, one of the most important of the reasons to favor panexperientialism is the fact that it, and apparently it alone, can solve the mind-body problem. In the final section, I will develop this point more fully by showing how panexperientialism supports the position needed to take account of our hard-core commonsense beliefs: nondualistic interactionism.

Nondualistic Interactionism

A confusion that runs through most discussions of the mind-body problem is the equation of "interactionism" with "dualism." Cartesian dualism involves two distinct theses, a numerical thesis and an ontological thesis. The *numerical* thesis simply says that mind and brain are not numerically identical but are *two things*: The mind is one actuality, the brain is another actuality (or, really, an aggregational society of billions of actualities). The *ontological* thesis adds the additional point that the mind and the brain are *two ontologically different kinds of things:* The mind is one kind of actuality (a mental one), the brain is another kind of actuality (a physical one). A position should be called "dualistic" only if the ontological thesis as well as the numerical thesis is affirmed. If a position affirms merely the numerical thesis, saying that the words "mind" and "brain" do not refer to the same entity, it should *not* be called "dualistic," because the word "dualism" inevitably suggests the Cartesian idea that mind and brain are ontologically different types of things, which is the idea that creates the problem of dualistic interactionism.

Of course, although the numerical and the ontological theses are distinguishable, they were not really separable for Descartes, due to his view of matter, and thereby of the brain, as devoid of both experience and spontaneity. *Given the Cartesian view of matter, the numerical thesis implies the ontological thesis.* Because almost all scientists, philosophers, and theologians have retained that early modern view of matter, they perpetuate the assumption that any position that distinguishes (numerically) between mind and brain is *ipso facto* "dualistic." For example, Daniel Dennett's argument that materialism must be true is based on his assumption that the only alternative is dualism, according to which "conscious thoughts and experiences cannot be brain happenings, but must be . . . something in addition, made of different stuff" (CE, 29). The transition from the numerical thesis ("something in addition") to the ontological thesis ("made of different stuff") occurs without comment. Dualists no less than materialists tend to assume that the first thesis implies the second. For example, John Hick begins a chapter on "Mind and Body" in this way:

> We have the two concepts of body and mind, and various rival views of the relation between them. According to the . . . mind/brain identity theory the two concepts refer to the same entity. This is the monistic option; all the others are dualist, regarding body and mind as *distinct entities*, and indeed *entities of basically different kinds.* (DEL, 112; emphases added)

As the words I italicized in this quotation show, Hick here, while making a quick transition from the numerical to the ontological thesis, at least distinguishes between them. A few pages later, however, he says: "In rejecting the mind/brain identity, then, we accept mind/brain dualism. We accept, that is to say, that mind is a reality of a different kind from matter" (DEL, 120). Here Hick, like virtually all other writers on the topic, simply equates the two theses.

The importance of this equation of the numerical thesis with the ontological thesis, or at least the assumption that the first implies the second, cannot be overestimated. It leads to the conclusion that interaction between mind and brain is unintelligible, because it builds a Catch-22 into the discussion, implying that interaction between mind and brain is inconceivable whether one distinguishes between them or not. On the one hand, the very notion of "interaction" implies two distinct things that can causally influence each other. So, if we, with identists, do not (numerically) distinguish the mind from the brain, we cannot affirm interactionism. On the other hand, the assumption that mind and brain are (ontologically) different in kind implies that although they are (numerically) distinct, interaction between them is inconceivable. One of the most vicious effects of the modern view of matter is that, by apparently forcing a choice between numerical identism and ontological dualism, it has made the task of making sense of interactionism, and thereby our hard-core commonsense assumptions about the mind-body relation, impossible.

Panexperientialism, by allowing for interactionism without dualism, finally enables us to make sense of these assumptions. Because, according to this view, the mind is a unified actuality, distinct from the billions of neurons comprising the brain, the unity that characterizes our experience is intelligible. Because the mind is distinct from the brain, but not ontologically different in kind from the cells making up the brain, our presupposition about their interaction—that the body influences our conscious experiences and these experiences in turn influence our bodies—is not unintelligible. Because the mind at any moment is a full-fledged actuality—as actual as the brain cells and their constituents—we can take at face value our assumption that our decisions are not fully determined by causal forces coming up from the body, but that they are truly "decisions," in which our minds exercise genuine *self*-determination. Finally, our hard-core commonsense conviction that our bodily behavior is significantly guided by these decisions made by the mind is intelligible because of two features of this position. The first—that the mind, being of the same ontological type as the cells comprising the brain, can intelligibly exercise causal influence ("downward causation") upon them—has already been stated. The second feature is the idea that the mind, besides being as fully actual as the individual cells comprising the brain, has far more power than they do. This idea explains why the mind can exercise a dominating influence over the body, providing that overall coordination that so radically distinguishes the behavior of a living, conscious person from that of a corpse.

Panexperientialist Emergence

Whereas many of the ideas in the previous paragraph simply summarize ideas that have been explained earlier, some additional explanation will be helpful with regard to some of them. One distinctive feature of panexperientialism of the Whiteheadian-Hartshornean type is its view of emergence. Given the mechanistic view of the ultimate units of nature as vacuous bits of matter, it was impossible to do justice to evolutionary emergence. There was no way to think of the emergence of higher types of actualities out of the organization of lower types. A living cell, for example, could not be thought to have any ontological unity based in a higher type of actual entity in which the distinctively living properties of the cell could inhere. In the panexperientialist view, by contrast, an actual entity is an occasion of experience, which arises from its prehensions of prior actual entities, especially those in its immediate environment with which it is contiguous.

In this framework, we *can* understand evolutionary emergence to involve, at least sometimes, the emergence of higher-order types of actual entities. For example, the atom need not be thought to consist only of its subatomic particles and the relations between them. It can be thought to involve, as well, distinctively atomic occasions of experience, more complex than the electronic, protonic, and neutronic occasions of experience. In this way, the holistic behavior of the atom—as manifested, for example, in the Pauli exclusion principle (as discussed by Ian Barbour [RAS, 104–05])—can be assigned to an inclusive actuality. A molecule, likewise, can be thought to involve, above and beyond atoms, distinctively molecular occasions of experience. Higher-level actual entities can also be supposed to exist in macromolecules, such as DNA and RNA, in organelles, and in cells. This philosophy does not dictate *a priori* that higher-level actual occasions emerge at each of these levels: That judgment is to be made on an empirical basis, in terms of whether (say) a water molecule or an organelle shows sufficient unity of response to its environment to merit positing a "regnant" or "dominant" member to account for this unified response. The point is that the panexperientialist philosophy allows us to posit the emergence of such a higher-level member if the evidence warrants it. A society having such a higher-level member, which gives the society as a whole a unity of experience and response, has been called a "compound individual" by Hartshorne (CI), because a higher-level individual has been compounded out of lower-level individuals. Although panexperientialism by definition rules out ontological dualism, it, unlike materialism, allows for an *organizational duality* between aggregational societies and compound individuals.

In the context of this distinction, the existence of "minds" or "souls" in human beings and other animals is not an evolutionarily unprecedented type of emergence, as generally supposed by materialists and dualists alike. Rather, the emergence of a mind out of that complex organization of cells that we call a brain is simply the highest-level example of a type of emergence that has been occurring throughout the evolutionary process. The human mind seems so unique to us because it is the only emergent actuality that we know from within. However, this emergence is (by hypothesis) only different in degree, not different in kind, from the emergence of living occasions in eukaryotic cells out of macromolecules and organelles. Accordingly, *contra* Searle, we do not need to say that there is nothing in the head but atoms, nor that these atoms are entirely reducible to their subatomic parts. There is the mind. There are the billions of living occasions in the neurons. And there are organellular, macromolecular, molecular, and atomic occasions of experience. We need not suppose, with reductionistic materialists, that our conscious experience somehow emerges magically as a property of insentient atoms, or, with dualists, that our minds have only insentient subatomic particles with which to interact.

Two Modes of Creativity

The difference between this monistic pluralism and the materialistic type of monistic pluralism can be characterized in terms of different views of the universal "stuff" embodied in all actual entities. Materialism assumes this stuff to be "energy" as described by contemporary physics. Whitehead, who had himself focused on mathematical physics before turning to philosophy, regarded the physicist's energy as an abstraction from the full-fledged "creativity" embodied at all levels of actuality, from electrons to human beings (AI, 186). This embodiment of creativity in each actual occasion involves two modes. In the first mode, it is embodied in the occasion's moment of *subjectivity,* during which the occasion enjoys its own experience. This mode has two poles: the "physical pole," during which the occasion receives the causal influence from the past, and the "mental pole," during which it exercises its own final causation or self-determination. Following this subjective mode of existence, the occasion exists in its *objective* mode, which means that it is an object for subsequent subjects. In this mode, its capacity to exercise self-determination is over, but it can now exercise efficient causation upon others. From this perspective, the physicist's "energy" involves a twofold abstraction: It deals entirely with an event's objective mode of existence, during which it exercises efficient causation, thereby abstracting entirely from its subjective mode, during which it is an "occasion of experience" for itself. Even with regard to the objective mode, furthermore, the physicist's energy involves only the quantitative aspect of the occasion's causal efficacy, ignoring its qualitative dimensions, such as the transference of emotional tone (SMW, 153).

The Relation Between Efficient and Final Causation

Alongside the concept of a compound individual, which has a dominant member, the understanding of the mind or soul as a *personally ordered society* of dominant occasions of experience, oscillating between self-determination and causal influence on others, is the other central element in the Whiteheadian explication of the freedom we all presuppose in practice. This idea explains one of the most difficult of all philosophical problems: that of freedom and determinism, or, more precisely, how final and efficient causation are related. The traditional view of the mind, according to which it is simply an enduring mental substance, made it seem as if it must either be totally determined from without, by the efficient causation from the body, or else totally self-determined from within, by its own final causation. Some philosophers who accepted the second view developed the previously mentioned doctrine of parallelism, according to which mind, while not really interacting with the body, ran along in parallel with it. In the thought of some advocates of parallelism, such as Leibniz, the synchronization of our perceptions and decisions with the events going on inside and outside our bodies was given a supernatural explanation, in terms of a harmony preordained by God. For other advocates of parallelism, this synchronization was left an unexplained mystery. Either way, however, parallelism was too incredible to attract much of a following. Most philosophers and scientists, therefore, gravitated toward determinism.

However, if the mind is constituted by a series of momentary experiences, which first exist subjectively and then objectively, there is a constant oscillation between final and efficient causation. Each occasion of experience begins by receiving causal influence from prior occasions; it then exercises its own final causation or self-determination in terms of a goal; and then it becomes one of the many efficient causes upon subsequent occasions. In this way we can do justice to our three presuppositions (1) that we are heavily influenced by the causal power of the past; (2) that we do, nevertheless, exercise a degree of freedom in each moment; and (3) that this free decision influences our bodily actions.

Panexperientialism's view of the body then explains how the third of these three presuppositions can be true, namely, that our mental decisions influence our bodily behavior. Materialists and dualists, as we have seen, have had great difficulty in explaining this influence. That is partly because they think of the bodily cells as vacuous actualities, which could not conceivably be affected by thoughts, feelings, and decisions. But it is also partly because they think of these cells, along with their constituents, as operating in terms of inflexible "laws of nature." From the viewpoint of panexperientialism, however, these so-called laws are abstractions, being simply the widespread *habits* of nature. They are, more especially, the habits of (say) molecules *in inorganic environments,* such as rocks or test tubes. Each occasion of experience is internally constituted by its appropriation of influences from its environment, especially its contiguous environment. A molecule in a living cell in the brain of a conscious human being, accordingly, will be subject to influences very different from those that influenced it when it was in the soil, before it was (say) absorbed by a carrot that was then eaten by the human being. From this perspective, there is, contrary to the opinion of John Searle and most other materialists, nothing "unscientific" about supposing that the free decisions of the mind cause the molecules in the body to act otherwise than they would if they were in a different environment.

In sum: Because it regarded mind and body not only as numerically distinct but also as different in kind, dualism's distinction between mind and body could not be sustained. As a result, science became associated with materialism, which, besides having most of dualism's problems, also found it impossible to make sense of the freedom, the efficacy, and even the reality of conscious experience—ideas that are inevitably presupposed in practice. These consequences of materialism's identification of mind and brain, especially the denial of freedom, have created an apparent conflict between the presuppositions of science and the presuppositions of religion and morality. But this appearance is doubly superficial. On the one hand, it is not "science" that denies freedom, but only the materialistic form of naturalism with which science has recently been identified. On the other hand, this materialistic worldview, with its deterministic implications, is as contrary to the presuppositions of scientific practice as it is to moral and religious practice. Panexperientialism, by providing the basis for the type of solution to the mind-body relation that is required by science as well as by religion and morality, shows that the conflict between science and religion with regard to this relation is merely apparent.

Besides overcoming the apparent conflict between the freedom presupposed by religion and the determinism allegedly presupposed by science, this solution to the mind-body problem does much more, making this chapter foundational for the remainder of the book. By supporting the numerical distinction between mind and brain and the reality of nonsensory prehensions, this chapter provides a position in terms of which we can take seriously the evidence for "extrasensory perception" as usually understood, which in turn provides a basis for taking testimonies to distinctively religious experience more seriously. This position also provides a basis from which the evidence for life after death, almost universally ignored by modern liberal theologians, can be taken seriously. The panexperientialism of this position, finally, provides a basis for again thinking of divine influence in individuals at all levels, from humans to quarks, thereby for thinking of divine influence throughout the evolutionary process.

7

PARAPSYCHOLOGY, SCIENCE, AND RELIGION

One of the main ways in which the present book differs from most books on science and religion is by including a chapter on the branch of science originally called "psychical research" but now usually called "parapsychology." The fact that parapsychology is usually ignored in the science-and-religion literature requires explanation, especially given the fact that parapsychology has generally been regarded by its proponents as the major bridge between science and religion. That it would become this was clearly the view of those who founded the Society for Psychical Research (SPR) in 1882 as a new branch of scientific study. This society was founded, the charter says,

> to investigate that large body of debatable phenomena designated by such terms as mesmeric, psychical and spiritualistic . . . in the same spirit of exact and unimpassioned enquiry which has enabled Science to solve so many problems. (Gauld FPR, 138)

One of those founders, Frederick Myers, declared in his presidential address of 1900 his hope that the SPR would provide a scientific preamble to all religions by being able to "demonstrate that a spiritual world exists, a world of independent and abiding realities, not a mere 'epiphenomenon' or transitory effect of the material world" (PA, 117).

Those who were not as hopeful as Myers of providing scientific grounding for a positive creed at least hoped to undermine the materialistic worldview, which, partly through the influence of Darwinism, was increasingly thought to be scientifically validated (Gauld FPR, 141–42). J. B. Rhine, the best-known American parapsychologist from the 1940s through the 1960s, said that those nineteenth-century founders "sought to confront the onslaught of the natural sciences on religion by marshalling in its defense the forces of a new kind of science" (CP, 7). Indeed, Rhine himself conceived of parapsychology as religion's science, meaning that, whereas most of the other branches of science were either neutral to religion or had served to undermine it, parapsychology provided positive support. "The relationship of parapsychology to the field of religion," Rhine suggested, should be "much the same as that of physiology to medicine, or that of physics to engineering" (RM, 209). J. Shoneberg Setzer developed this point in an article entitled "Parapsychology: Religion's Basic Science" (1970). Charles Tart, a well-known psychologist, has more recently testified to the help that parapsychology can provide:

> I happened upon a partial resolution of my personal (and my culture's) conflict between science and religion. Parapsychology validated the existence of basic phenomena that could partially account for, and fit in with, some of the spiritual views of the universe. (PSI, xii–xiii)

This proffered help has for the most part, however, been refused by Christian philosophers of religion and theologians, including those who have focused on the issue of science and theology. As to why, we must distinguish between conservative-to-fundamentalist thinkers, on the one hand, and liberal thinkers, on the other. A major reason for the rejection of parapsychology on the part of the former was explained in Chapter 5: Parapsychology, by treating events involving influence at a distance as natural (albeit extraordinary) occurrences, undermines the belief that the "miracles" of Christian history were actually supernatural events, requiring special divine agency, and thereby the belief that these miracles provide divine testimony to the unique truth of Christian faith. A positive employment of parapsychology to provide empirical support for various Christian doctrines, accordingly, would undermine the method of authority upon which conservative-to-fundamentalist theologies are based.

Our primary concern here, however, is with liberal religious thinkers, who have rejected the method of authority in favor of basing claims for truth on reason and experience. One might suppose that such thinkers would welcome whatever empirical support parapsychology can supply. William James, who was the leading figure of the American branch of the SPR, hoped that it would. In the same paragraph in which he said, "As the authority of past tradition tends more and more to crumble, men naturally turn a wistful ear to the authority of reason or to the evidence of present fact," James made his famous declaration about religion and empiricism:

> Let empiricism once become associated with religion, as hitherto, through some strange misunderstanding, it has been associated with irreligion, and I believe that a new era of religion as well as of philosophy will be ready to begin. (ERE, 270)

The kind of empiricism James had in mind was a "thicker and more radical empiricism," which included an examination of "the phenomena of psychic research so-called" (ERE, 270, 271).

Virtually all liberal philosophers of religion and theologians, however, have ignored the parapsychological study of these phenomena, refusing the possible aid that it might give. The primary reason for this refusal is their acceptance of the suspicion in intellectual circles in general of paranormal claims and the study thereof, a suspicion that is generally conjoined with a suspicion that parapsychology is not really a science, only a "pseudoscience." Coexisting in a relationship of mutual reinforcement with these suspicions is usually an almost total ignorance of the nature of parapsychology and the phenomena that it has studied. This ignorance exists partly because parapsychology is held in such disrepute in intellectual circles that any investigation of it is discouraged. An even simpler explanation of this ignorance is the fact that, because of the virtual absence of serious discussions of parapsychology in most intellectual circles, philosophers of religion and theologians, like most other intellectuals, are not likely to be exposed to any such discussion in the books and journals they read or the conferences they attend.

The background to why paranormal claims are met in intellectual circles with such suspicion—which often takes the form of *a priori,* dogmatic rejection and even vilification of the proponents of these claims—was introduced in Chapter 5. As we saw, one of the motivations behind the adoption of the mechanistic view of nature was the desire to exclude the possibility of influence at a distance as a natural capacity. Given that restricted view of what is naturally possible, events unexplainable in terms of causality by contact could not occur except through supernatural intervention. In leading intellectual circles, as we have seen, this theistic supernaturalism was replaced with a noninterventionist deism in the eighteenth century and then with a completely atheistic view in the nineteenth century. From the perspective of this later worldview, the kinds of events in question could not happen, period. The result was that "science" sanctioned the *a priori* rejection of not only most of the "miraculous" events reported in the Bible (and, for Roman Catholic and Eastern Orthodox Christians, the history of the saints) but also of the various kinds of phenomena studied by parapsychology, because the common element running through these phenomena is that they all appear to involve some kind of influence at a distance.

This connection with influence at a distance is reflected in the names for many of them, such as *telepathy* (meaning "feeling at a distance"), *telekinesis* (the older word for psychokinesis), and *remote viewing* (the newer word for clairvoyance). This connection is also implicit in the term *extrasensory perception:* Sensory perception involves a chain of contiguous cause-effect relations from the object to the perceiver's mind, so that any perceptual process that did not use sensory channels would evidently involve perception at a distance. The transition from the scientific worldview of the seventeenth century to that of the nineteenth and twentieth centuries, accordingly, means that "science," with all its prestige and power, has seemed to stand opposed to all claims for the occurrence of events of this type, whether the claims be made in the name of religious supernaturalism or of a "deviant science" with a naturalism more permissive than that of the current scientific orthodoxy.

As we have seen, the transition from the early modern to the late modern worldview involved a transition not only from supernaturalism to atheism but also from dualism to materialism. This latter transition, which involved the rejection of the idea that the mind is an entity distinct from the brain with power of its own, closed off the possibility that the types of events in question might be explainable by appeal to the power of the mind, as parapsychologists usually suggest.

The fact that the worldview now dominant in intellectual circles rules out the phenomena studied by parapsychologists goes far toward explaining why parapsychology is generally rejected, or at least ignored, by liberal philosophers of religion and theologians. Although they may not accept a completely materialistic view of human beings, they tend to accept sensationism, according to which all the mind's perceptions come through the body's sensory organs. They tend even more strongly to accept what can be called "motorism," according to which all of the mind's influence upon the world beyond its body is through its body's motor-muscular system. The ideas of extrasensory perception and psychokinesis are, accordingly, ruled out almost as fully as by a completely materialistic standpoint. Also, the world of discourse of most liberal philosophers of religion and theologians generally shelters them from exposure to serious discussions of parapsychology and its data.

I am one for whom the usual protective mechanisms failed. While reading around on the mind-body problem, I stumbled onto an intelligent discussion of the philosophical implications of parapsychology by psychologist John Beloff (EM, Ch. 7). Thanks partly to the fact that I had already adopted a philosophical position that does not rule out paranormal phenomena *a priori,* and partly to the fact that I had time, being on sabbatical leave, I followed up this lead and began reading widely in the field. Through this extended exposure, I became impressed with the quantity of the work that had been done and with the quality of much of it. Since that initial exposure (during the 1980–81 academic year), I have become increasingly convinced not only that most of parapsychology's types of phenomena are genuine, but also that this fact is of utmost importance for philosophy of religion and theology. With regard to theology in particular, I believe that the incorporation of parapsychological data is crucial for the development of forms of postmodern liberal theology that can overcome the thinness, often verging on vacuity, of most forms of modern liberal theology. I agree with William James that a thicker, more radical empiricism, one that includes the empirical evidence of parapsychology, can provide the basis for a new era in religious thought. In particular, we will no longer be forced to choose between intellectual credibility and religious robustness. Thanks to the evidence provided by parapsychology, we can have a theology that, while fully naturalistic in worldview and liberal in method, is as robust religiously as any supernaturalistic, authoritarian theology. Whereas modern liberal theologies have achieved a reconciliation of science with *theology* at the expense of its religious content, parapsychology is crucial for a form of liberal theology that effects a reconciliation of science with *religion.*

For such an approach to be widely adopted, however, the obstacle that has hitherto excluded parapsychology from discussions of the relation between science and theology—the suspicion that parapsychology is not to be considered a genuine branch of science—must be overcome. I will approach this topic in the following section by discussing the nature of parapsychology and its distinctive subject matter. In the third section, I will examine the reasons that have been given for regarding parapsychology as a pseudoscience, showing that they do not stand up. In the fourth section, I will discuss some respects in which the panexperientialism articulated in the present volume shows parapsychological phenomena not to be as exceptional to ordinary processes as they are often thought to be. Finally, in the fifth section, I will look at some of the ways in which parapsychology, assuming the authenticity of its data, is theologically important.

PARAPSYCHOLOGY AND ITS DATA

Parapsychology involves the notion of *paranormal* phenomena, which involves the notion of causal influence at a distance. The term *psi* has come to be widely used as a synonym for paranormal influence. The notion of psi or the paranormal, more precisely, involves the notion of *causal influence at a distance to or from a mind.* Of course, given the mechanistic view of nature, even the idea of action at a distance that involves no mind, but only so-called inanimate nature, is controversial; but the study of such phenomena would not belong to parapsychology. For example, even if it is decided that gravitational attraction and the nonlocality associated with Bell's theorem involve some kind of causal influence at a distance, this conclusion would not place them among the phenomena studied by parapsychology. Its phenomena are limited, as the very words *psychical* research and para*psychology* imply, to interactions involving *psyches* or minds. Influence at a distance *to* minds, usually called "extrasensory perception" (abbreviated ESP), can also be called "receptive psi." Influence at a distance *from* minds, usually called "psychokinesis" (abbreviated PK), can also be called "expressive psi." Most of the phenomena studied by parapsychology can be placed under one or the other of these headings.

A more precise definition, however, is needed. Although I have thus far spoken of parapsychology as the scientific study of paranormal phenomena, it is more properly to be defined as the scientific study of *ostensibly* paranormal phenomena. With this definition, even those who do not believe in paranormal events, understood as events actually involving causal influence at a distance, can agree that parapsychology has phenomena to study, because there clearly are events that *prima facie* appear to involve influence at a distance to or from minds. With this definition, furthermore, one can be a parapsychologist without being a "believer," that is, without believing that any events are genuinely paranormal. There can be—and in fact are—skeptical parapsychologists. The first question for parapsychology, then, is whether any of the *ostensibly* paranormal events should be considered *truly* paranormal—whether they defy all attempts to explain them in terms of chains of contiguous cause-effect relations. Events are truly (not merely ostensibly) paranormal if they really involve influence at a distance.

Given this characterization of parapsychology, we will now look at some evidence showing that parapsychology has a distinctive subject matter—that is, that events occur that are at least ostensibly paranormal. We will at the same time examine reasons to suspect that some of these ostensibly paranormal events are truly paranormal. The evidence for paranormal occurrences consists of three broad types: (1) anecdotal evidence, (2) methodical investigations of reports of spontaneous events, and (3) experimental evidence.

Anecdotal Evidence

Anecdotal evidence consists of reports of apparently paranormal occurrences, whether from contemporaries or historical sources. In one sense, these anecdotal reports cannot be said to constitute evidence, insofar as they have not been methodically investigated, at least not by people employing standards and methods of documentation that would allow us to regard their reports as evidential. Parapsychologists, indeed, generally think of anecdotal reports less as real evidence for the paranormal than as the natural phenomena providing the occasion for their discipline. In another sense, however, these anecdotal reports do constitute evidence, because many of the same kinds of phenomena have been reported in virtually all societies from which we have records. This fact can be seen by reading Sir James Frazer's 1896 classic, *The Golden Bough*. Frazer himself regarded all these reports as superstitious. The similarity of the reports from different times and places, however, raises the question as to why superstition would so often take the same forms in unconnected places.

This point is at the center of a recent sociological, cross-cultural study of such phenomena, *Wondrous Events*, by James McClenon. Taking issue with what has been called the "cultural source theory," according to which religion and belief in paranormal events result entirely from cultural sources, McClenon stresses the degree to which wondrous events, by which he primarily means "psychic" or paranormal phenomena, are universal and thereby transcultural (WE, 1, 2, 6, 7, 36, 151, 240). Particularly problematic for the cultural source theory is the fact that these phenomena tend to fall into natural clusters, suggesting that they are "natural kinds," not artifacts of particular belief systems or of religious needs engendered by particular cultures:

> The cultural source theory cannot explain the existence of natural categories of anomalous experiences, such as apparitions, precognitions, extrasensory perceptions, night paralysis, out-of-body experiences, synchronistic events, and déjà vu, all of which have cross-culturally consistent features. (WE, 56)

The cultural source theory has been employed by some of the most vociferous critics of parapsychology. For example, philosopher Paul Kurtz, founder of the Committee for the Scientific Investigation of Claims of the Paranormal (CSICOP), suggests that beliefs in both religion and the paranormal, which he equates with "magical thinking," are due to succumbing to "the transcendental temptation," arising from discontent with the present world, to posit a transcendental realm of escape (TT, xi–xii, 22, 350). Another prominent member of CSICOP, who also equates parapsychology with magical thinking, psychologist James Alcock, uses B. F. Skinner's ideas to suggest that operant conditioning based on need reduction "underlies the development of belief in magic and religion" (PSM, 16).

On the basis of McClenon's cross-cultural examination of stories about wondrous events, by contrast, he says: "Most stories do not support the argument that they originated from religious needs or from the fear of death" (WE, 45). More generally, in response to the assumption "that wondrous perceptions are cultural products originating simply from religious need and scientific ignorance" (WE, 18), McClenon says that according to the actual evidence, the types of experiences in question, rather than being caused by religious beliefs, are themselves causes of such beliefs, accounting for the similarities of folk religious beliefs in disparate cultures (WE, 8, 12, 160, 166, 168, 248), a thesis that is reflected in the full title of McClenon's book, *Wondrous Events: Foundations of Religious Belief*. In any case, McClenon has provided well-documented support for the belief, which lay at the root of the founding of parapsychology as a scientific discipline, that the ubiquity of anecdotal reports of paranormal events provides strong *prima facie* evidence for their reality.

Evidence from Methodical Investigations of Spontaneous Cases

The second kind of evidence for paranormal phenomena consists of *methodical investigations* of reports of ostensibly paranormal spontaneous occurrences. These types of reported occurrences, most of which were mentioned in the quotation from McClenon about "natural categories" of paranormal events, include telepathy, clairvoyance, apparent precognition, apparitions, hauntings, poltergeist outbreaks, and out-of-body experiences. *Telepathy* is the direct (extrasensory) reception of information from another mind. *Clairvoyance* is the extrasensory reception of information about an objective state of affairs. *Apparent precognition* refers to an experience in which a person seems to have received information about an event prior to its occurrence.[1] *Apparitions* are more or less lifelike appearances of people who are not physically present. Although they may be of the living, most apparitions are of the dead, especially those who have *just* died. Apparitions are quite brief, usually lasting for less than a minute, and are one-time occurrences (although there are *collective* apparitions, which are witnessed by two or more persons at the same time and place, and *multiple* apparitions, which are witnessed by two or more people at about the same time at different places). *Hauntings*, which often involve apparitions, are recurrent phenomena, perhaps recurring for years or even decades, and are identified with a place (rather than with particular people). *Poltergeist outbreaks*, by contrast, are usually associated with a particular person, rather than a place. Like hauntings, they involve recurrent phenomena, but they usually run their course in a manner of weeks. Although the German term *poltergeist* means noisy ghost or spirit, most parapsychologists believe that the phenomena, which are often quite bizarre, are actually produced unconsciously by the person with whose presence they are associated. *Out-of-body experiences* (OBEs) are, as the term suggests, experiences in which persons seem to themselves to be consciously experiencing the world from a perspective at some distance from their body, which they may view from the outside. Whether the mind or soul is in fact out of the body is, of course, a matter of debate.

Although critics of parapsychology like to refer to all reports of spontaneous psi occurrences as merely "anecdotal" evidence, this term is misleading in

---

1. The word "apparent" is added to this category because I consider genuine precognition to be logically impossible, a point that will be discussed later.

relation to events that have been subjected to rigorous, methodical investigation.[2] The purposes of the investigation of reports of such phenomena are to see if the details of a given report can be corroborated by independent witnesses, to ascertain the credibility of the witnesses, and to judge whether the unusual occurrence could be explained by normal means, such as subliminal sensory information, trickery, misperception, faulty memory, or mere coincidence. Although critics have often charged psychical researchers with credulity in accepting the authenticity of some of these reports, the standards have generally been quite high. For example, one case investigated in the early days of the SPR involved a Mrs. Barter in England, whose husband was in India during the time of the Indian Mutiny. She dreamed that her husband, after being wounded in the leg, bound his leg with a puggaree (the cloth band wrapped around the crown of a hat or helmet). Although Mrs. Barter's dream was correct about most of the details of what had in fact happened, her case was rejected as evidence for genuine telepathy or clairvoyance for two reasons: Her husband had actually bound his leg with a silk necktie rather than a puggaree, and she had had another dream about him that did not correspond to any facts. Her dream, in other words, was dismissed as mere coincidence (Tyrrell A, 26).

Investigation of apparitions was one of the central activities of the SPR in its early years. The most interesting apparitions were those that seemed to be "veridical," meaning that some information was conveyed to the person that he or she had no normal way of knowing. A good percentage of such apparitions were "crisis apparitions," occurring at the time that the apparent (the person whose apparition was seen) was undergoing a crisis, perhaps death. The veridical information in an apparition occurring at the time of the death of the apparent might be simply the fact that he or she had in fact died, especially if the death occurred at a time when the witness had no particular reason to expect it. The veridicality in such cases is increased if the apparition conveys information as to the mode or cause of the apparent's death.

For example, in one well-known case, a young Englishman, himself later to become a military man, reported awakening in the middle of the night to see an apparition of his older brother, who was already in military service in another country. The apparition was kneeling and, when it turned its head, the younger brother saw a wound on the right temple with red streaming from it. Being disturbed by this apparition, he went to a friend's room and, telling the friend what he had seen, asked if he could sleep on the couch. The next morning he told others in the house what he had seen, including his father, who admonished him not to repeat such nonsense. About two weeks later, they learned that the

---

2. The claim that only experimental evidence for psi can be considered "scientific" would imply that the only genuine scientific studies of any sort must be experimental, thereby excluding most of geology, astronomy, paleontology, and many other sciences from the status of true science.

older brother had been killed on the night in question, that he had been struck by a bullet in the right temple, and that he had been found in a kneeling posture, as his body had been propped up by other dead bodies amid which it had fallen. Investigation of this case involved obtaining corroboration from others that the younger brother had indeed reported seeing the apparition that night, and with those details (Myers HP, 230–31). Whether such cases provide good evidence for survival of death is debatable, but they at least provide evidence for extrasensory perception. (For more on apparitions, see my discussion in PPS, Ch. 7.)

Cases seeming to involve only simple extrasensory perception have also been investigated. In 1970, psychiatrist Ian Stevenson published a book titled *Telepathic Impressions,* which contains 195 such cases. The criteria for inclusion were (1) that the percipient had an accurate feeling about a distant person while having had no normal way of knowing about that person's situation; (2) that the percipient then told someone else about it or took unusual but, in the light of the distant person's situation, appropriate action; and (3) that this response can be independently corroborated (TI, 10–11).

To illustrate: A case in 1949 involved Joicey Acker Hurth. Three months after she was married, she and her husband were staying with his parents in Cedarburg, Wisconsin, about a thousand miles from her home in Anderson, South Carolina. Although she had been completely happy, one night she woke up sometime after midnight feeling very sad and began to cry. When her husband awakened and asked why she was crying, she said that she felt that something was wrong. Not having slept the rest of the night, her appearance the next morning led her parents-in-law to ask if there had been a "lover's quarrel." Then, while fixing breakfast, Joicey suddenly exclaimed, "It's my father! Something is terribly wrong with my father!" Because her father had been very healthy and a letter from him only a few days earlier had mentioned no problems, she had, the others reminded her, no reason for her concern. A few minutes later, however, a telephone call from her aunt informed her that her father was in a coma, due to a drug he had taken for a backache without proper instructions. Immediately returning to South Carolina, Joicey got to the hospital just before her father died. This was, incidentally, not a once-in-a-lifetime event for Joicey Hurth: She reports having six such experiences, four of which were sufficiently verified to be included in Stevenson's book. (For more on telepathy, see my discussion in PPS, Ch. 2.)

In recent times, much attention has been given to reports of out-of-body experiences (OBEs). Such experiences have evidently occurred in most cultures: A recent survey by Dean Sheils (CCS) found beliefs in OBEs existing in fifty-one of fifty-four cultures studied. The experiences are also very similar from culture to culture, having about a dozen common elements. The most interesting of these features from the viewpoint of parapsychologists is the fact that some reports of OBEs contain evidence of "veridical observations" during OBE episodes. Several cases of this type, which often occur in hospitals in near-death situations, have been reported in *Recollections of Death* by cardiologist Michael Sabom (RD), who had been incredulous when he first read of such cases in Raymond Moody's first book (LAL). In these cases, the patient, upon reviving, will typically speak about details of the operation and the conversation among the nurses and doctors, sometimes adding details about unusual clothing or hairstyles of a doctor or nurse.

Although such reports are often impressive, critics can point out that hearing is the last of the senses to go and that, even though the patient was apparently unconscious, he or she could have reconstructed the scene imaginatively on the basis of auditory clues. More impressive, accordingly, are reports that include veridical information from other parts of the hospital or from even more distant locations. Some patients have, for example, correctly reported on conversations that relatives were having in a part of the hospital far away from the operating room; some have correctly reported the existence of a shoe on the ledge or the roof of the hospital; and some have correctly reported on activities of family members at home. Investigations of such reports involve verifying the accuracy of the reports by examining the doctors' and nurses' reports, interviewing the relevant people for corroboration, and checking to see if there was any way that this apparently paranormal information could have been learned through normal means. Verifying that the information was indeed obtained paranormally does not by itself settle the most interesting question raised by such reports, which is whether the mind was really out of the body. But it does at least add to the evidence for spontaneous extrasensory perception. (For more on out-of-body experiences, see my discussion in PPS, Ch. 8.)

Evidence for spontaneous psychokinesis can be investigated in poltergeist outbreaks. These outbreaks are reported in many cultures. For example, a recent book by Alan Gauld and A. D. Cornell reports on hundreds of poltergeist and haunting cases occurring since 1525 in such diverse locations as Canada, China, Europe, Iceland, India, Indonesia, Jamaica, Madagascar, Malaysia, Scandinavia, South Africa, Turkey, the USA, and the former USSR. This study showed that the poltergeist cases, in spite of occurring in different centuries and different cultures, had many elements in common. For example, the focal person, in whose presence the phenomena occurred, would usually be between twelve and fifteen years of age and would often have severe emotional problems. As the term "noisy ghost" suggests, unaccountable sounds—either explosive or percussive—are common. Although these occurrences produce by far the most bizarre form of psi phenomena, they are the most well-investigated of the spontaneous psi phenomena, because of their recurrent but concentrated nature. That is, in these cases the paranormal happenings tend to last only about two months, and during that period the phenomena are likely to occur daily.

For an illustration, we can look at the famous case in Rosenheim, Germany. In 1967, the law office of Sigmund Adam was suddenly afflicted with a rash of disturbances: Fluorescent lights on the ceiling went out repeatedly, sometimes following a loud bang; incandescent bulbs exploded and electrical fuses blew; all the telephones would ring simultaneously, although nobody would be on the line, and the telephone company registered a great number of calls that no one in the office had made. Adam called in electricians, telephone experts, and police, but no normal explanation could be found. Finally, Germany's leading parapsychologist, Hans Bender, and two physicists that he brought discovered that the phenomena occurred only when an eighteen-year-old secretary, who was a new employee, was present. After Bender expressed his belief that the phenomena were produced by (unconscious) psychokinesis, they intensified: Paintings turned over on their hooks; plates jumped off the wall; drawers opened by themselves; a 400-pound filing cabinet twice moved about a foot—all while Bender, the physicists, the police, and the engineers were watching, and some of the phenomena were even captured on videotape. After the secretary took another job, the disturbances in Adam's office ceased, although they occurred for a while, less intensely, at her new place of employment (Broughton PCS, 217–19).

Experimental Evidence

The third kind of evidence, which is the kind generally considered most important for discussions of the scientific status of parapsychology, comes from experiments, in which the possibility of normal explanations can be more fully ruled out. I will begin with experiments involving extrasensory perception.

Working on the assumption that people are probably more receptive to extrasensory influences while they are asleep, when there is less competition from sensory percepts, the "dream laboratory" at the Maimonides Medical Center in New York ran an interesting set of experiments. After the subject had gone to sleep and begun to dream (as indicated by brain waves and rapid eye movements), a person in another room would seek to "send" images from a (randomly selected) picture. The subject would then be awakened and asked to describe his or her dream. Outside judges, not knowing which of several pictures was the actual target, would then, by studying the transcription of the description, decide which of the pictures it most closely approximated. In one case, the target picture was *Zapatistas,* a painting by Carlos Orozco Romero depicting Zapata's Mexican-Indian followers, who traced their ancestry to the Mayans and Aztecs. The painting shows armed horsemen and marching men, with mountains and dark clouds in the background. The subject, a psychiatrist named William Erwin, said this in describing his dream:

> A storm. Rainstorm. It reminds me of traveling—a trip—traveling one time in Oklahoma, approaching a rainstorm, thundercloud. . . . I got a feeling of memory, now, of New Mexico, when I lived there. There are a lot of mountains around New Mexico, Indians, Pueblos.

The next morning, when asked if anything more came to mind, Erwin said:

> Here it gets into this epic type of thing . . . a DeMille super-type colossal production. I would carry along with it such ideas as the Pueblo going down to the Mayan-Aztec type of civilization. (Broughton PCS, 91)

The judges had no problem in agreeing that this was a clear "hit."

Turning from telepathy to clairvoyance: In the early days of the Duke University parapsychology laboratory, there were many card-naming tests, in which the subject, shielded from all sensory clues, tried to name the card on the top of the deck. The tests normally used "Zener" cards, which contain five types of cards in a deck of twenty-five. By chance alone, one should get an average of five correct per run. In some cases, however, subjects did significantly better. In tests for clairvoyance from buildings a hundred or more yards away, Hubert Pearce, a divinity student, once averaged, in twelve runs, 9.9 hits per run. Over the total of seventy-four runs, he averaged 7.1 hits, the odds against this occurring by chance being twenty-two billion to one (Broughton PCS, 69; Edge et al. FP, 162). In another kind of test, using ordinary playing cards, the subject tried to guess both the suit and the denomination. Although by chance one should get exact hits less than 2 percent of the time, Bill Delmore, a Yale law student, got them 6 percent of the time, the odds against chance being calculated at a million trillion trillion to one. Especially impressive were "confidence calls," in which he was most confident of being correct: Making twenty such calls, he was correct fourteen times (Edge et al. FP, 171).

Since those early days of parapsychological experimentation, experiments of the card-guessing type have been superseded by many new kinds of experiments. One type of test that has become popular, called "remote viewing," can be considered either telepathy or clairvoyance or a combination of the two. One recent test involved Marilyn Schlitz, an anthropologist, and Elmar Gruber, a German parapsychologist. Gruber, while in Rome, was to go to randomly selected places, while Schlitz, back in Michigan, was to try, at the pre-arranged time, to "see" where Gruber was. On one of the ten trials, he was sent to a hill close to the airport, from which he could see the terminal and the planes taking off. The hill had many holes, dug by people looking for coins. Schlitz wrote:

> Flight path? Red lights. . . . A hole in the ground, a candle-shaped thing. . . . Something shooting upward. . . . The impressions that I had were of outdoors and Elmar was at some type of—I don't know if institution is the right word—but . . . a public facility. He was standing away from the main structure, although he could see it. . . . I want to say an airport but that just seems too specific. There was activity and people but no one real close to Elmar. (Broughton PCS, 121)

According to the five judges who evaluated the ten trials, Schlitz made a direct hit on six of them (including this one).

Another recently popular type of experiment, called the *ganzfeld,* can also be considered a combination of telepathy and clairvoyance. The subject's sensory input is minimized by various means, such as placing halved table-tennis balls over the eyes and playing white noise through headphones while the subject is sitting in a reclining chair. After the subject is thoroughly relaxed, the "sender" in another room seeks to impress images from a picture (randomly selected by a computer program) on the subject's mind. The subject then describes the images he or she is having, after which judges are asked to say to which of four pictures (one of which was the target picture) the description best corresponds. In one case, the other three pictures were of a Chinese nobleman, flowers, and a parking lot with rows of snowbound automobiles, while the target picture had a pickup truck traveling down a curving country road. The subject said:

> There was a road . . . a hard-packed pebble road. . . . And, there was a very fleeting image of being inside of a car and I could see just the rearview mirror. . . . A feeling of going very rapidly. . . . Also, at one point, the feeling of being out in the country and wide open spaces. . . . I had the feeling of driving out in the country.

Upon seeing the target picture after the session, the subject said, "That's it—that's what I've been seeing" (Broughton PCS, 101).

Not all sessions, of course, produced such clear hits. But after a long series of experiments, it was determined that 45 percent of them showed significant results, compared with the 5 percent that would be expected from chance alone (which is, of course, an extremely significant result). A further discovery was that a certain type of person—one who has practiced some form of mental discipline, has had at least one "psychic" experience, and is a Feeling and Perceiving type according to the Myers-Briggs Type Indicator—will have an even higher success rate, in one case 64 percent (Broughton PCS, 113). In any case, the *ganzfeld* experiments, which have been examined very closely by critics, led to the inclusion of a section on "Psi Phenomena" in the tenth edition of the standard *Introduction to Psychology* (long known as "Hilgard and Atkinson"). Taking a purely empirical approach, the authors (Atkinson et al.) criticize those who reject psi phenomena on the grounds of impossibility, saying that "such a priori judgments are out of place in science; the real question is whether the empirical evidence is acceptable by scientific standards" (IP, 235).

To turn now from tests for extrasensory perception or receptive psi to those for psychokinesis or expressive psi: Just as the founders of the parapsychology laboratory at Duke wanted to make studies of ESP rigorously scientific by using card-guessing experiments, they sought to do the same thing for PK by measuring the ability of people to affect falling dice. Although statistical analyses sometimes showed signs of PK, they were very weak. Recently, however, a new process that has become widely accepted in the social sciences, called "meta-analysis," has been employed on these old tests taken together (which involved over 2.5 million throws of the dice), showing that the results, although usually weak in each test, were so consistently positive that the odds against chance are 10^{70} to 1 (Broughton PCS, 295). More recently, subjects have been asked to try to introduce a bias in random number generators. A report published in *Foundations in Physics* in 1989 dealt with 597 experimental studies (and 235 control studies) carried out by 68 investigators in the 1970s and 1980s. The odds against chance that the positive results could have been due to chance alone were calculated at about 1 in 10^{35}. The subjects involved in these experiments, like those involving dice, were usually ordinary people, with no history of paranormal experiences (Broughton PCS, 290). What these tests taken cumulatively suggest, accordingly, is that such people, while incapable of producing dramatic effects, can rather consistently produce slight effects in the behavior of physical things, the reality of which can be revealed by statistical analysis.

Some individuals, however, seem capable of producing results that can be directly observed—sometimes called "conspicuous PK." One form this capacity occasionally takes is the ability of some people to impress images on undeveloped photographic film. One such person, Ted Serios, was studied extensively by Denver psychiatrist Jule Eisenbud, whose writings about Serios include some of the paranormal photographs produced. Fraud was ruled out by having investigators bring their own Polaroid cameras containing their own film, keep them in their own possession throughout the session, and then develop the photographs on the spot. Between 1964 and 1967, Serios evidently produced over 400 images that could not be explained by ordinary means. Many trained observers, including photographers, physicists, engineers, and magicians, signed affidavits for Eisenbud testifying that they had witnessed the productions of pictures under conditions rendering normal explanations inconceivable (PP, 419; WTS, 100–10).

In paranormal photography, one does not, of course, directly see the effects being produced psychokinetically, but only the results when the film is developed. Some people, however, have evidently been able to produce directly observable motion in physical objects. One such person was Nina Kulagina, a Russian woman who had served with distinction as a senior sergeant in the defense of Leningrad in World War II. Extensive experiments with her by Soviet scientists led to a film, shown at an international parapsychological conference in Moscow in 1968, in which she evidently produced various kinds of paranormal movement, such as selectively moving one matchstick among several scattered on a table, purely by the power of concentration. Her abilities were later confirmed by both American and British scientists (Broughton PCS, 144–46).

One of the religiously most important types of expressive psi is spiritual or psychic healing. Some authors distinguish between these two forms of healing, speaking of "spiritual healing" when the direct influence of the healer's mind seems to be on the mind (or spirit) of the other person, whose changed state of mind brings about a change in her or his physical state, and speaking of "psychic healing" when the healer's mind seems to affect the body of the other person (or animal, or even plant) directly. Both of these forms of healing, in which the mind of the healer exerts influence at a distance, must be distinguished from "faith healing," in which the person's faith in the healer and/or other healing agents, such as God, angels, Jesus, Mary, or other saints, brings about the healing psychosomatically. Of course, at one time scientific orthodoxy, with its epiphenomenalism, could not even accept genuine psychosomatic healing. However, now that such self-healing is more widely accepted, many suspect that all so-called psychic and spiritual healing is really faith healing.

Experiments have been devised, accordingly, to exclude this possibility. For example, one pioneer in this area, Bernard Grad, who has documented the effects of a healer on mice and plants (Edge et al. FP, 248), reports that he chose to work with such organisms, rather than human beings, in order to exclude the likelihood that the beliefs of the individual who was healed played a role (EO, 151, 154). Experiments with human beings have excluded the faith-healing interpretation by having the healer or a group pray for the person at a distance without the person knowing about it. The results of experiments of these two types, along with other types, have been summarized and evaluated in a 1993 book by Daniel Benor (HR).

As this brief summary of types of evidence provided by parapsychologists illustrates—and it is a *very* brief survey, summarizing only a tiny portion of the available evidence (see Broughton PCS; Edge et al. FP; Murphy FB)—there is considerable reason to believe that psi, both receptive and expressive, occurs. This fact is potentially of great importance for philosophy of religion and theology. As mentioned before, however, this evidence has been almost entirely ignored, the reason for which brings us to our next topic, the scientific status of parapsychology.

PARAPSYCHOLOGY: SCIENCE OR PSEUDOSCIENCE?

In their discussions of science and religion, liberal theologians and philosophers of religion have generally ignored the various types of evidence summarized above, evidently because they have accepted the conviction, widespread in intellectual circles generally, that parapsychology is a pseudoscience. For example, some years ago I went through all the then-published issues of *Zygon: Journal of Religion and Science,* looking for references to parapsychology. In all those approximately sixty issues, I found only one such reference—and it was simply an aside, referring to parapsychology as a pseudoscience. Given that characterization of parapsychology, it would follow that there can be no scientific evidence for paranormal occurrences. The claim is often made, indeed, that evidence for such occurrences does not exist, or is so minimal that it can be safely ignored as anomalous.

We have just seen, however, that not only is the evidence for such occurrences abundant, but also that some of the evidence can be claimed, at least *prima facie*, to be scientific. Most of the claims made by intellectuals as to the lack of evidence seem to be based on ignorance. A few intellectuals, however, do know that a considerable body of evidence has been built up over the past century, since the establishment of scientific organizations dedicated to the investigation of ostensibly paranormal phenomena. Some reason is needed for rejecting this evidence. One possibility would be simply to admit prejudice, as did psychologist Donald Hebb, who wrote:

> [W]hy do we not accept ESP as a psychological fact? Rhine has offered enough evidence to have convinced us on almost any other issue. . . . Personally, I do not accept ESP for a moment, because it does not make sense. . . . I cannot see what other basis my colleagues have for rejecting it. . . . [M]y own rejection of [Rhine's] view is—in a literal sense—prejudice. (RNI, 45)

Although Hebb's admission is admirably candid, most thinkers who are aware of the evidence feel the need for some more respectable basis for justifying their wholesale rejection of it.

The most common basis is the claim that parapsychology, the discipline providing this evidence, is not really a science, but merely a pseudoscience, so that its evidence can be ignored without violating one's intellectual responsibility to take account of all relevant scientific evidence. For example, psychologist James Alcock concludes his *Parapsychology: Science or Magic?* with this summary statement:

> Parapsychology is indistinguishable from pseudo-science. . . . There is *no* evidence that would lead the cautious observer to believe that parapsychologists . . . are on the track of a real phenomenon, a real energy or power that has so far escaped the attention of those people engaged in "normal" science. (PSM, 196)

For those taking this line, the fact that the Parapsychological Association has been a member of the American Association for the Advancement of Science (AAAS) since 1969 has been an embarrassment. In 1979, for example, physicist John Wheeler launched a campaign to get the Parapsychological Association disaffiliated with an address to the AAAS titled "Drive the Pseudos Out of the Workshop of Science" (*New York Review of Books,* 13 April 1979). During the question and answer period, Wheeler groundlessly accused the best-known experimental parapsychologist, J. B. Rhine, of fraud (Broughton PCS, 75n).

Although Wheeler retracted this accusation in a "Correction" published in the AAAS magazine *Science* (13 July 1979: 144), he did not take back the characterization of parapsychology as a pseudoscience. In an article written at about the same time entitled "Parapsychology: Science or Pseudoscience?", philosopher Anthony Flew argued that because parapsychology is the latter, the Parapsychological Association should be "politely disaffiliated" from the AAAS (PSP, 529).

Some Common Reasons for the Pseudoscience Charge

The effort to discredit parapsychology and thereby its data with the claim that it is a pseudoscience requires careful examination by philosophers of religion and theologians. One reason this is so is that the very idea of a clear line of demarcation between genuine science and pseudoscience is problematic, being based less on objective criteria than on the polemical desire to reject the enterprise put in the latter category. This point is recognized even by one of the best-known critics of parapsychology, psychologist Ray Hyman, who says in response to the criteria used by Alcock:

> [W]hether it makes sense to insist on unambiguously categorizing parapsychology as science or pseudoscience is questionable. . . . The categories of both science and pseudoscience are fuzzy. . . . It looks very much like the criteria themselves were chosen in order to exclude parapsychology. (EQ, 176)

More generally, the attempt to formulate criteria for establishing a line of demarcation between science and pseudoscience—a line that would show all generally recognized sciences to be in the former category and all the distasteful ones, such as parapsychology, to be in the latter—has proved to be a failure (Laudan DSP; Grim PSO). What can be said objectively is that parapsychology is, as James McClenon calls it, a *deviant science,* in that its phenomena seem to violate some of the laws or at least metaphysical assumptions generally accepted in the scientific community (DS, 11). A deviant science, however, is still a science. To dismiss parapsychology as a pseudoscience, sociologists of science increasingly recognize, is merely propaganda.

To illustrate: One of the criteria sometimes given for considering parapsychology a pseudoscience is the fact that it does not have an agreed upon theoretical framework for discussing and testing its phenomena. As sociologist of science Marcello Truzzi points out, however, employing this criterion across the board would mean relegating most of the social and psychological sciences to the category of pseudoscience (SL, 42). More generally, to employ this criterion is to ignore the fact that in most of the currently recognized sciences, the phenomena they investigate were recognized to exist long before there was an agreed upon theory for understanding them. Indeed, the very purpose of establishing a branch of science for investigating these phenomena was to come up with such a theory. In terms of contemporary discussions in the philosophy and sociology of science, the worst thing that one can say about parapsychology in this regard is that it is not yet a "mature science" and that it has remained in a state of "immaturity" for a long time. But these points are unanimously accepted by parapsychologists themselves. Indeed, one of their complaints is that relatively little progress has been made in developing workable theories because the extreme prejudice against the field has resulted in its being grossly underfunded and understaffed.

A second reason not to accept the dismissal of parapsychology as a pseudoscience uncritically, to which philosophers of religion and theologians should especially be alert, is that this dismissal of parapsychology is often made on the basis of criteria used also to dismiss religious beliefs. This connection is especially clear in the writings of Paul Kurtz. As indicated by the full title of his previously cited book, *The Transcendental Temptation: A Critique of Religion and the Paranormal,* Kurtz believes that both paranormal and religious beliefs result from succumbing to the temptation to escape to an imagined transcendent world, rather than resting content with the worldview of modern science, which is materialistic and atheistic, telling us that "the universe possesses no purpose or meaning per se and is indifferent to human achievement and failure [and] is not divine in origin or sustenance" (TT, xi–xiv, 22, 23, 363). Very similar is the position of Alcock, who begins his examination by asking about "processes which might be capable of generating *both* religious and paranormal beliefs" and whose categorization of parapsychology as a pseudoscience follows from his contention that it, like religion, is essentially based upon magical thinking, which reigns in the human unconscious (PSM, 7, 196). For critical thinkers, Alcock says, both religious and paranormal beliefs are ruled out by the fact that the worldview vouchsafed by science is not only materialistic and atheistic but also mechanistic and deterministic, showing belief in free will to be illusory (PSM, 24, 32, 45, 195). This perceived connection between religious and paranormal beliefs, along with the related fact that some advocates of parapsychology have pointed to the support it provides for a religious view of reality, is one of the reasons given by Kurtz (TT, 339, 350, 363) and Alcock (PSM, 106; SS, 51, 53) for the conclusion that parapsychology is a pseudoscience.

That conclusion is, of course, spurious: On the one hand, any will-to-believe present in advocates of parapsychology can be matched (and often surpassed) by the will-to-disbelieve among its detractors. On the other hand, as Truzzi, following many others, has pointed out, the motives of scientists, whether they be religious or anti-religious, are irrelevant to the question of the validity of their results (SL, 48).

Another unsound reason sometimes given for an *a priori* rejection of parapsychology and its data is the allegation that there has been a long and consistent history of fraud in parapsychology. For example, in his introduction to *A Skeptic's Handbook to Parapsychology,* titled "More than a Century of Psychical Research," Paul Kurtz gives a disproportionate amount of space to people who definitely or at least quite likely used fraud (at least at times) to produce apparently paranormal effects and to the two known cases of fraud on the part of parapsychologists (William Levy and S. G. Soal). Then, in later asking whether laboratory evidence for apparent psi should be given normal or paranormal explanations, he says: "No doubt of great importance in this area is the argument from fraud" (MC, xx). Antony Flew, in his contribution to the same volume, says that the "black record of fraud" provides a good reason to doubt "what might seem to be strong new cases of psi" (PSP, 528).

This allegation of a long history of fraud, however, depends almost entirely upon (1) confusing "parapsychology" with the antics of various people who have used the claim to be "mediums" for making money, or (2) confusing parapsychology as such with the human subjects investigated by parapsychologists. To begin with the first confusion: The fact that various self-styled mediums have engaged in deception says nothing about fraud in parapsychology. The former is a sociological phenomenon, the latter is a tradition of scientific research (which sometimes involved study of these self-styled mediums to evaluate their claims, often with negative conclusions). Regarding the second confusion: The fact that some subjects being tested by parapsychologists for paranormal powers have engaged in deception does not mean that there has been fraud in parapsychology. Indeed, when fraud on the part of these subjects has been detected, this detection has usually been by parapsychologists. Some parapsychologists have, to be sure, been taken in by the fraud. But in every science there are scientists who are "taken in" by this phenomenon or that, meaning that they interpret it falsely. Their faulty interpretation of the phenomenon is corrected by other scientists. This process of making and testing alternative hypotheses is what the scientific process is all about. If there had been no false interpretations in psychical research, this fact would make it amazingly unique among the sciences.

Part and parcel of the claim that parapsychology is to be rejected because of the ubiquity of probable fraud is the allegation that the process of internal control is less rigorous in parapsychology than in "real" sciences. In response to this allegation, sociologist Marcello Truzzi says:

> [M]any critics have commented upon the long history of revealed frauds in psychical research. Yet the facts clearly show that the most rigorous critics of psi experimentation have been fellow psi researchers, and the occasional frauds discovered in modern parapsychology have almost all been discovered and called to public attention by the parapsychologists themselves and not by outside critics of the field.[3] In a very important sense, the revelations

---

3. This is true of both the Levy case and the Soal case.

of occasional fraud within parapsychology are the best evidence of the rigor and scientific objectivity present in parapsychology by those doing the policing of the discipline. It is a myth that psi researchers are a gullible lot who all trust one another unreasonably. Distrust is rife in parapsychology, and there is probably a better scientific security system operating in parapsychology than in most if not all other sciences. (SL, 45)

If fraud committed by a few practitioners of a particular science meant that all the results of that science should be dismissed out of hand, we would need to reject virtually all scientific results whatsoever. The fact of fraud in science in general, including the so-called hard sciences, is documented not only by the two essays to which Truzzi refers in support of his final sentence (St. James-Roberts DS; Weinstein FS), but much more extensively by William Broad and Nicholas Wade in *Betrayers of the Truth: Fraud and Deceit in the Halls of Science*. Besides pointing out that the fraud in the Levy case was promptly reported by J. B. Rhine (BT, 123)—more promptly, the rest of their book suggests, than is usually the case in science—Broad and Wade also support Truzzi's point about the greater scientific rigor of parapsychology in particular, saying: "Because parapsychology is widely regarded as a fringe subject not properly part of science, its practitioners have striven to be more than usually rigorous in following correct scientific procedure" (BT, 122). This point is also supported by Robert Rosenthal of Harvard, one of the world's leading theoreticians of the social sciences. In a report on the controls exercised by the various sciences to ensure the reliability of experimental reports, Rosenthal rated parapsychology higher than all the other human sciences (Broughton PCS, 323–24).

The Real Reason for the Pseudoscience Charge

The relegation of parapsychology to the category of pseudoscience on the basis of fraud in the field, motivations of its practitioners, or failing to live up to a generally valid criterion of a real science, therefore, can rather quickly be seen to be invalid. The real reason for the pseudoscience charge, which generally lies behind these various other charges, is *the conviction that parapsychology's alleged phenomena conflict with science.* For example, in an essay asking "Is Parapsychology a Science?", Kurtz says that its findings "contradict the general conceptual framework of scientific knowledge" (IPS, 510). Alcock says that parapsychological occurrences, if genuine, would imply a "relationship between consciousness and the physical world radically different from that held to be possible by contemporary science" (SS, 19). Fellow CSICOP member Christopher Scott, in addressing the question "Why Parapsychology Demands a Skeptical Response," says that "psi conflicts with the corpus of existing scientific knowledge" (WPD, 498). In his well-known 1955 essay "Science and the

Supernatural," George Price relegated the phenomena of parapsychology to the realm of the supernatural, meaning the nonexistent, primarily on the grounds that "parapsychology and modern science are incompatible" (SS, 151).

This claim is connected with the pseudoscience charge by the principle, repeated in virtually every criticism of parapsychology, that *extraordinary claims require extraordinary evidence.* Scott, for example, says:

> If psi does involve a fundamental conflict with existing science, then this is an extremely strong argument for caution; for this scientific understanding is founded on a body of experimentation and scholarly thought far more extensive than the whole corpus of parapsychology. The evidence needed to overturn such a huge volume of work would have to be extremely powerful. (WPD, 499)

Kurtz says: "[E]xtraordinary claims . . . demand extraordinary amounts of evidence. These claims are not to be dismissed out of hand, but if they are to be accepted the evidence cannot be fragmentary or questionable" (TT, 366).

The general principle invoked here is sound: We are not being unreasonable when we demand stronger evidence for claims about types of events that seem to us impossible or at least highly improbable. A problem arises, however, with Kurtz's claim that the evidence must be unquestionable or, as he puts it elsewhere, incontestable (MC, xviii), because *any* evidence can be questioned or contested by those who for some reason do not want to accept it. The same problem is raised by Alcock's version of the principle, which involves an analogy with a legal distinction:

> When one is weighing evidence in law, the distinction is made between "beyond all reasonable doubt" and "on the balance of probabilities." . . . Because psi is a concept that would probably revolutionize science. . . , most skeptics implicitly use the criterion of beyond all reasonable doubt, while accepting conclusions made on the balance of probabilities where only "normal" and noncontroversial phenomena are concerned. (SS, 24)

The problem is that, as we know from watching criminal cases on television, the distinction between a *reasonable* doubt and a merely *possible* one can be very subjective. Virtually no evidence seems beyond a reasonable doubt for jurors who have strong reasons for wanting to doubt it. By analogy, if we believe ESP, PK, and OBEs to be impossible, or at least exceedingly improbable, then any possible basis for doubting the evidence for them will likely seem a reasonable basis for doubting it.

For example, C. E. M. Hansel takes this approach in his self-styled "scientific" examination of parapsychological experiments (ESP; EP; ET). With regard to telepathy, Hansel has said: "When examining the experiments, it is not unreasonable to assume that telepathy cannot occur in view of the *a priori* arguments against it" (ET, 177). On that basis, Hansel assumes that if he can posit some possible way in which the results of a particular experiment could have been produced by fraud or error, he has undermined the evidential value of the experiment. His alternative scenarios, he says, need not be probable or even plausible: Their mere possibility is sufficient to provide a reasonable doubt that anything paranormal occurred. With regard to reported evidence for ESP, for example, Hansel says: "If the result could have arisen through a trick, the experiment must be considered unsatisfactory proof of ESP, whether or not it is finally decided that such a trick was in fact used" (ESP, 18). However, as Truzzi says,

> the burden of proof for such alternatives must rest with the proponents for such alternatives (many of which have been shown to be quite inadequate to fit the facts). The alternatives of incompetence and fraud need demonstration. Accusations of fraud without reasonable supporting evidence . . . are too frequently taken for proof. . . . Science primarily grows through continued research and not mainly through criticism based upon what might have gone wrong in experiments that we might wish to see invalidated. One can *hypothetically* explain away *any* result in science. Criticisms, just like the claims of the proponents, must also be falsifiable if we are to consider them scientific statements! (SL, 46, 39)

Even Hansel's fellow CSICOP member Ray Hyman criticizes his approach, saying that if we begin with the assumption that ESP is impossible, "then no amount of empirical evidence can change our position." Hyman continues:

> Hansel's position. . . is a dogmatism that is immune to falsification. There is no such thing as an experiment that is immune to trickery. . . . In practice, it would be impossible even to take into account all the known variables that could allow some form of deception. (EQ, 294–95)

As McClenon says, the principle that "exceptional claims require exceptional proofs" can easily be employed in such a way that "it is not possible to produce a proof for certain claims because of their extraordinary nature" (DS, 87). That is, no matter how strong the evidence, the claim can always be made that because the existence of psi would be so extraordinary, still stronger evidence is required. This strategy, McClenon points out,

> makes the "perfect" ESP experiment an impossibility. Sooner or later, the critic will ask for information that is no longer available or for a degree of experimental control and exactitude that is desirable in principle but impossible in practice. One rhetorical ploy is to demand total perfection.

It is always possible for critics to think of more rigid methodological procedures after an experiment has been conducted. (DS, 89)

This approach, in one form or another, is used by most critics of parapsychology as the main basis for the claim that no truly "scientific" or "credible" evidence for psi has been provided by parapsychologists.

This attitude, according to which the positive results claimed by some parapsychologists are thought to be in such conflict with the scientific worldview that no amount of evidence can be considered convincing, goes back to David Hume's chapter "On Miracles," which is often cited. For example, George R. Price, in the essay "Science and the Supernatural," admitted that on the basis of the empirical evidence alone (he cited J. B. Rhine's 1940 book, *Extra-Sensory Perception After Sixty Years*), he had accepted the reality of ESP. He says, however, that he changed his mind after becoming familiar with Hume's argument. The essence of this argument, as quoted by Price (SS, 148–49), runs as follows:

A miracle is a violation of the laws of nature; and as a firm and unalterable experience has established these laws, the proof against a miracle, from the very nature of the fact, is as entire as any argument from experience can possibly be. . . . no testimony is sufficient to establish a miracle, unless the testimony be of such a kind that its falsehood would be more miraculous than the fact which it endeavours to establish. . . . the knavery and folly of men are such common phenomena, that I should rather believe the most extraordinary events to arise from their concurrence, than admit of so signal a violation of the laws of nature.

On the basis of this Humean *a priori* argument, Price concluded that all evidence for ESP must arise from fraud and error.

Hume's argument, which is often cited in support of the principle that extraordinary claims require extraordinary evidence (e.g., Kurtz TT, 366), speaks of violations of "the laws of nature." In adapting Hume's argument, the claim is sometimes made that psi would be "miraculous" because it would violate particular scientific laws. At other times the "laws of nature" that would allegedly be violated are taken to be more philosophical in nature, so that the conflict is not with any particular laws of science but with the scientific worldview as such. I will look at these two forms of the argument in order.

The Claim that Parapsychology's Phenomena Conflict with Scientific Laws

One problem with the claim that ESP and PK would violate the laws of nature, or scientific laws, is that this claim often presupposes the idea that these laws are prescriptive. As both Truzzi (SL, 44) and philosopher of science Stephen Braude

(LI, 22) point out, science's laws are descriptive, not prescriptive: They are generalizations about what happens, not prescriptions as to what must happen. The understanding of laws as prescriptive is a hangover from the early modern worldview, in which the laws of nature were thought to have been imposed on the world by its omnipotent Creator. Most laws, furthermore, are stochastic or statistical in character: They describe the mass result of the behavior of many entities, not the behavior of individual entities. They at most describe how individual entities of a certain type usually behave, not how they always behave. It involves an outmoded understanding of scientific laws, accordingly, to say that they *forbid* ESP and PK.

That point aside, furthermore, the claim that ESP or PK would necessarily violate some well-established scientific law is problematic. For example, the two laws probably most often cited in this regard are mentioned in Alcock's statement that "it would seem that psi cannot occur without violating well-tested laws of physics—such as the law of conservation of matter and energy and the inverse square law of energy propagation" (SS, 20). The latter of these two laws is also cited in a statement by aforementioned psychologist Donald Hebb, quoted by Alcock (PSM, 128), in which Hebb says: "If it doesn't matter how far apart the two heads are when one brain radiates to another (in telepathy), then there is something fundamentally wrong with physics." However, as pointed out by Braude (LI, 16) and Mundle (EE, 200), there is no clear evidence that, if telepathy occurs, it does not exemplify the inverse square law. That the same is true for the law of the conservation of energy—even if we knew, which we do not, that it is always "obeyed" in ordinary physical processes—is pointed out by Braude (LI, 16) and Margenau (ESP, 221).

Furthermore, even if these and other principles turn out not to be exemplified by psi processes, this result would not "contradict the laws of physics." As Truzzi says, it "would merely limit the domain of the accepted principles to their previous areas of generalization; they would not be falsified for that limited domain" (SL, 44). Braude makes the same point, saying that the result would only imply that the phenomena lay outside the domain of physics (LI, 16, 18). Of course, many late modern thinkers think that the domain of physics is, at least in principle, coextensive with reality, so that the laws of physics are in principle sufficient to describe all processes. But to endorse that view, as Braude points out, is simply to endorse metaphysical reductionism, which is a matter of faith and philosophy, not of science itself (LI, 17, 19).

Sometimes the claim about violating laws of science is made not in terms of specific laws but in more general terms. For example, Paul Kurtz, speaking of "strange aspects of the paranormal universe of parapsychology that do equal violence to our scientific and common-sense view of the world," says:

Could "mental energy" change the state of physical entities without the interposition of physical forces? If so, this would mean a major exception to the laws of physics. (TT, 362)

This claim, however, is confused. The laws of physics have nothing to say, one way or the other, about the possibility of psychokinesis. As Braude points out, besides the fact that these laws are not prescriptive but merely descriptive, what they describe is the physical realm *in abstraction from intention and volition*. The question of the possibility of psychokinesis, accordingly, lies outside the domain of physics. As such, it poses no more threat to the laws of physics than does the influence of ordinary volition upon the body (LI, 16–19). Of course, from the perspective of materialistic physicalism, as discussed in the previous chapter, all bodily behavior is said to be describable in principle in terms of the laws of physics. But that reductionistic idea is, again, a matter of philosophy, based on faith. It is not part of the body of scientific knowledge, in the sense of being vouchsafed by any scientific evidence.

The Claim that Parapsychology's Phenomena Conflict
with the Scientific Worldview

As suggested by the previous paragraphs, the claim that some basic laws of science would be contradicted by parapsychology's phenomena, if genuine, is really a confused way of saying that they would conflict with the worldview currently dominant in the scientific community. Antony Flew has made this point clearly. Saying that "it is not easy to think of any particular named law of nature—such as Boyle's Law or Snell's Law or what have you—that would be . . . 'violated' by the occurrence of psi" (PSP, 527), Flew says that the conflict is with some principles that are "more fundamental than even the oldest and best established of named laws of nature" (C, 268). The conflict, in other words, is not with any particular results of science but with some of its basic assumptions. Some version of this claim, it turns out, is what is really meant by most of those who say that parapsychology is in conflict with science.

The crudest version of this claim is based on the view that science is materialistic. Alcock, for example, makes this claim in an essay with the title "Parapsychology as a 'Spiritual' Science" (which alludes to statements from some advocates, especially biologist John Randall [PNL, 241], characterizing parapsychology approvingly in these terms). Alcock says:

> To call parapsychology a "spiritual science," as a few parapsychologists have done, would appear to be a contradiction in terms: How can a science of the spirit exist, given that science is by its very nature materialistic? If one believes in the reality of the paranormal, then one must either ultimately change the basic foundations of science or accept that paranormal phenomena lie beyond science, in either case overthrowing materialism. (PSM, 562)

Parapsychology's sin against the scientific perspective, Alcock makes clear, is the fact that its data, if genuine, would seem to imply a dualistic rather than a materialistic view of human beings.

> It is the axiom of mind-body dualism, with its obvious anti-materialistic ramifications, that not only links parapsychology with religion but keeps parapsychology outside of science. (PSM, 558)

Alcock made the same point in another essay, "Parapsychology: Science of the Anomalous or Search for the Soul?"

> [I]t is hard to escape the conclusion that the concept of paranormality implicitly involves *mind-body dualism*, the idea that mental processes cannot be reduced to physical processes and that the mind, or part of it, is non-physical in nature. . . . The influence of dualistic thinking creates a deep schism between parapsychology and modern science. (SS, 18, 19)

The claim that science as such, even *modern* science as such, is materialistic is, of course, indefensible, whether one is speaking historically or normatively. With regard to the historical question, Alcock describes "the path of science" as "a path laid down upon the foundation of materialism" (PSM, 562). As we saw in Chapter 5, however, modern science was originally based precisely upon the dualistic and religious perspective that Alcock considers inherently unscientific. And Alcock knows this, pointing out that "the founders of modern science were steeped in religious beliefs." He seems to think that this does not invalidate his point, however, because of an essential difference between those founders of modern science and present-day parapsychologists: The former, Alcock claims, carried out their research "despite religious needs and beliefs," whereas the latter carry out their (psychical) research "because of such needs" (PSM, 555). Even if that distinction were historically true (which, as we saw in Chapter 5, it is not), it would provide no support for Alcock's claim that modern science was "laid down upon the foundation of materialism."

Let us look now at the normative question of whether, in spite of its past history, science is now inherently materialistic, so that no work implying a nonmaterialistic worldview can be considered scientific. In support of this claim, Alcock (PSM, 558) quotes the following statement from M. A. Thalbourne:

> Whether we like it or not, and despite the best efforts of an Eccles or a Popper. . . , the dominant mode of thinking among present-day scientists is that of Central-State Materialism. Parapsychologists alone constitute a professional group where Dualism is still the most popular assumption. (CFP, 13)

Whatever the validity of this argument, Alcock's quotation of it reveals that he knows that some very prominent neurophysiologists, such as John Eccles, and some very prominent philosophers of science, such as Karl Popper, have rejected materialism in favor of dualism. Thalbourne's statement about currently fashionable philosophical views, furthermore, cannot validly be used by Alcock to draw conclusions about the nature of science itself, no more than could one in the late seventeenth century, on the basis of majority vote, have properly said that science necessarily implies a dualistic-supernaturalistic worldview.

Parapsychology and Basic Limiting Principles

Most of those who argue that parapsychological phenomena contradict the philosophical worldview presupposed and supported by science do not identify this worldview, as least as explicitly as does Alcock, with materialism. Rather, the appeal is usually to what C. D. Broad called "basic limiting principles" (RPPR, Ch. 1). Although Broad's original account contained thirteen such principles, they can be stated with fair adequacy in terms of the following four points:

1. There can be no causation and (therefore) no perception at a distance— either at a temporal distance, which rules out precognition and retrocognition, or at a spatial distance, which rules out telepathy and clairvoyance.
2. Besides the fact that all influence on the mind is mediated by the brain (which the first principle implies), all of the mind's influence upon the world beyond the brain is mediated by the brain.
3. Minds cannot experience apart from brains, which rules out survival of death apart from a supernatural act.
4. An efficient cause cannot come after its effect(s), which means that there can be no retrocausation and (therefore) no precognition.

These principles have had a strange history in the discussion: Broad's own point was that, although these principles are usually accepted nowadays in both science and ordinary life, the evidence provided by parapsychology suggests that they, or at least most of them, are *not* universally valid. Most critics of parapsychology, however, do not mention this point, simply citing the conflict of these principles with claims for psi as good reason to suspect the latter to be false. For example, Paul Kurtz, after summarizing these principles, says:

> These general principles have been built up from a mass of observations and should not be abandoned until there is an overabundant degree of evidence that would make their rejection more plausible than their acceptance. (TT, 50)

These principles, thus characterized, are central in the case against parapsychology made by Kurtz (and therefore, to a great extent, by CSICOP in general). This centrality is seen in a statement by Kurtz only partly quoted earlier:

> [E]xtraordinary claims that violate our basic limiting principles demand extraordinary amounts of evidence. These claims are not to be dismissed out of hand, but if they are accepted the evidence cannot be fragmentary or questionable. (TT, 366)

Thus employed, the basic limiting principles virtually entail, from Kurtz's perspective, that we could never rationally accept any claim for psi, because the evidence will always be "fragmentary" and considered "questionable" by sufficiently determined critics. McClenon, summarizing with approval the conclusions of some other sociologists of science, says that "a scientist's belief in the 'reality' of a phenomenon affects his or her evaluation of research into the phenomenon" (DS, 43).

To give a second example: Antony Flew, who had Broad's principles in mind in speaking of assumptions of science more fundamental than its particular laws, is extremely confident of their universal validity. Having said that "psi phenomena are in effect defined in terms of the violation of certain 'basic limiting principles,'" he characterizes the latter as "principles that constitute a framework for all our thinking about and investigation of human affairs, and principles that are continually being verified by our discoveries" (PSP, 527). More strongly still, he refers to them as "nomological propositions which are constantly tested and—it would appear—never found wanting in the daily living of scientists, technologists, and plain men." The fact that psi phenomena would violate such fundamental principles, says Flew, provides an "overwhelmingly good reason for dismissing the alleged phenomena . . . as just plumb impossible" (C, 268).[4]

One response to this way of declaring parapsychology in conflict with science would be to say, with various thinkers, that it is wrong to define science in terms of any substantive principles. For example, Truzzi, in response to Kurtz's claim that "the chief claims of parapsychology contravene the basic principles of both science and common sense," says that "science's *basic* principles are methodological and not substantive" (SL, 40). Truzzi's statement echoes William James, who said: "Science means, first of all, a certain dispassionate method [not] a certain set of results that one should pin one's faith upon and hug forever" (WJPR, 41). However, although this kind of answer is largely correct,

---

4. To be more precise: Flew's phrase "just plumb impossible" is used in relation to extrasensory perception and psychokinesis, which he takes to be *contingently* impossible, in distinction from (true) precognition, which he (rightly) considers *logically* impossible.

it would not completely fit with the position I have taken in this book, according to which scientists do rightly presuppose scientific naturalism and thereby at least one substantive principle—namely, that there are no interruptions of the most basic causal principles. From the viewpoint of this analysis of the "most fundamental principles" of science, the question is whether Broad's "basic limiting principles" exceed the requirements of scientific naturalism as such.

In approaching this question, we should recall that Broad himself, one of the most circumspect and respected philosophers of the twentieth century, did not believe these principles to be required by either science or common sense. He believed, instead, that the various forms of psi vouchsafed by parapsychology "call for very radical changes in a number of our basic limiting principles" (RPPR, 22).

In the second place, the characterization of these principles by both Kurtz and Flew as inductive generalizations "built up from a mass of observations" and as "never found wanting in . . . daily living" is simply false. Polls show, for example, that a large percent of the population in America believe that they have had at least one telepathic or clairvoyant experience, which shows that they do not accept the first and second principles. With regard to the third principle, which implies that existence apart from a biological body never occurs, no one has had experiences supporting this principle (no finite set of experiences can prove a negative), and polls show, again, that many people believe that they have had experiences, such as out-of-body experiences and the perception of apparitions, that imply that such existence does sometimes occur. At least most of Broad's basic limiting principles do not, therefore, have behind them the weight of universal human experience.

In the third place, far from being inductive generalizations from experience, most of Broad's principles are *deductions* from the late modern worldview, with its sensationism, mechanism, and materialism (or, to be more precise, epiphenomenalism: Broad's phrasing of the principles, while attributing no independent capacity to the mind either to receive or exert influence, did distinguish the mind from the brain). And, as we saw in Chapter 6, this worldview, far from being implicitly presupposed in all scientific practice, is *contradicted* by the presuppositions of all human practice, including scientific practice. The fact that psi phenomena would contradict various principles of this worldview, then, emphatically does *not* put parapsychology in conflict with any necessary assumptions of science. It may turn out, in fact, that it is precisely by contradicting many of the basic limiting principles that parapsychology is in harmony with various presuppositions of scientific practice.

Parapsychology and Influence at a Distance

Although these points may be generally granted, one of these principles has been so widely considered part and parcel of modern science that it needs special attention. This is the idea, contained in the first and second principles, that there can be no causal influence to or from a mind except that which travels by means of a chain of contiguous causes. It is primarily because belief in psi phenomena violates this principle that it is widely assumed to involve "magical thinking" incompatible with the basic assumptions of the scientific worldview. For example, in his *Parapsychology: Science or Magic?*, Alcock says:

> The practitioner in magic believes that he can control objects or events by magical behaviour, although there is no objective causal link between the behaviour and the object or event. (PSM, 8)

Later, having said that the "parallels between psi and magic should not be forgotten," Alcock (PSM, 193) quotes the following passage from James Frazer's *The Golden Bough:*

> [B]elief in the sympathetic influence exerted on each other by persons or things at a distance is the essence of magic. Whatever doubts science may entertain as to the possibility of action at a distance, magic has none; faith in telepathy is one of its first principles. (GB, 25)

In his essay "Science and Supernaturalism," George Price, having said that "[t]he essence of magic is animism," indicated that we should be receptive to novel claims only when we can "imagine a detailed mechanistic explanation" (SS, 152–53). Flew likewise, in characterizing parapsychology as a pseudoscience, indicates that the decisive objection to its alleged phenomena is the lack of a "conceivable mechanism" (PSP, 532). All these statements, especially those by Price and Flew, support McClenon's sociological observation that, "When we reject an anomalistic claim on the basis of its not having a suitable explanation, we indicate that it lacks a 'mechanistic' explanation that does not conflict with present physical theory" (DS, 54). This, in fact, was the issue behind Donald Hebb's rejection of ESP on the basis of "prejudice." He would have found the evidence for ESP convincing, he suggested, if he could have made "some guess as to the mechanics of the disputed process" (RNI, 45).

An example of an *explicit* use of the basic limiting principle in question against parapsychology is provided by philosopher Jane Duran. The relevant principle, as formulated by Broad (RPPR, 9), is that "any event that is said to cause another event (the second event being referred to as an 'effect') must be related to the effect through some causal chain." Although Broad believed the evidence for telepathy and clairvoyance to be strong enough for us to reject this principle, Duran, by contrast, says that

> the absence of a specifiable and recognizably causal chain seems to con-stitute a difficult, if not insurmountable, objection to our giving a coherent

account of what it means to make such a claim. As long, at least, as our ordinary notions of causality remain intact, there seem to be strong philosophical reasons for concluding that telepathy [and] clairvoyance. . . are not possible. (PD, 202)

For Duran, in other words, the equation of causal influence with causation through a chain of contiguous events is so complete that, besides considering telepathy and clairvoyance to be impossible, she suggests that the very concepts may be meaningless.

Why has it seemed virtually self-evident to most modern thinkers that causal influence between noncontiguous things or events must be mediated through a chain of contiguous things or events? The primary factor has been the mechanistic-materialistic view of nature upon which late modern thought has been based. The deep conviction within the intellectual community that influence at a distance is unnatural, therefore impossible, represents one of the most powerful effects of the double transition through which the late modern worldview was created: the seventeenth-century transition to a dualistic, supernaturalistic worldview embodying a materialistic, mechanistic account of the physical world, followed by the nineteenth-century transition to a materialistic, mechanistic view of reality as a whole. After the first transition, events involving influence at a distance were impossible except with supernatural assistance. After the second transition, with its rejection of supernaturalism, such events came to be considered, in Flew's words, "just plumb impossible." It is essential to realize, however, that this conclusion is due not simply to the acceptance of naturalism but to the fact that this naturalism retained the mechanistic philosophy of nature and sensationist theory of perception, which had originally been adopted to support a dualistic and supernaturalistic worldview. Within a wider naturalism, such events may not seem so self-evidently impossible.

Parapsychology and Repeatable Experiments

Before we explore this possibility, however, we need to consider one more reason for the charge that parapsychology is a pseudoscience. As McClenon points out, although parapsychologists rest their case on empirical evidence, their critics rely primarily upon "a priori philosophical arguments" (DS, 82). One of the main arguments of these critics, however, has at least the appearance of being empirical. Second in importance only to the *a priori* conviction that psi events are impossible has been the demand that, before parapsychology can be accepted as a legitimate science, it must produce a repeatable experiment, analogous to repeatable experiments in physics and chemistry. Critics have used the continuing absence of such an experiment to argue that we lack sufficient evidence rationally to believe that psi exists. On this basis, they say that parapsychology does not deserve to be considered a science with a distinctive subject matter. For example, Flew, comparing parapsychology today with its condition several decades earlier, says:

> [T]here is still no repeatable experiment to demonstrate the reality of any putative psi phenomenon. . . . [T]here is no repeatable demonstration that it [parapsychology] does in truth have its own peculiar and genuine data to investigate. (PSP, 520)

Kurtz agrees, saying:

> A basic criterion in the experimental sciences is that there must be a replicable experiment that can be repeated in any laboratory by independent or neutral observers. This condition has not been satisfied in the parapsychological laboratory so that the very existence of psi-phenomena is still in question. (TT, 366)

Kurtz's demand, that an experiment be repeated by "independent" or "neutral" scientists, is probably a demand that cannot be fulfilled by definition. As Truzzi points out, "*some* parapsychologists began experiments as critics and were converted to psi by what they have reported to be evidence" (SL, 43). Once such people become converted, however, they are no longer, from the viewpoint of determined critics, "neutral" or independent." They are now (gullible) "believers." Edmund Gurney stated this position ironically long ago in describing the typical attitude of scientists toward evidence for an extraordinary fact:

> The fact is so improbable that extremely good evidence is needed to make us believe it; and *this* evidence is not good, for how can you trust people who believe in such absurdities? (TQ, I: 264)

In spite of Gurney's having pointed out the circularity of the critics' demand, it is still repeated. For example, Alcock says—without irony—that fraud must always be suspected, "Since replication by independent, impartial researchers seems not to be possible in parapsychology" (PSM, 138).

In any case, the reason why this repeatability of all successful experiments is thought to be essential is the modern belief in universal laws of nature to which all events are subject. Galileo, for example, said that Nature acts through "immutable laws which she never transgressed" (McClenon DS, 66). A central point in John Stuart Mill's philosophy of science was the "Axiom of the Uniformity of the Course of Nature" (SL, 326–27). One feature of the transition to materialism from dualism is the fact that this principle of uniformity is now thought to apply not only to (physical) nature but to all events whatsoever, including all distinctively human events. Critics on this basis assume that, if ESP or PK were to occur, it would be as lawlike as phenomena produced by the interactions of atoms and molecules, so that if ESP or PK were ever manifested under experimental conditions, the experiment should be as repeatable as ordinary experiments in physics and chemistry. Flew brings out this connection in an essay titled "Parapsychology Revisited: Laws, Miracles, and Repeatability," arguing that a natural science presupposes the idea of a "law of nature" which in turn requires the idea of repeatability (PR, 263). Flew then argues that this criterion of *natural* science is relevant to *historical* inquiry involving reports of extraordinary phenomena, including reports made by present-day parapsychologists. Because their phenomena are not repeatable in the strong sense—according to which "certain things will necessarily and repeatably happen, given the appropriate and stated conditions"—they "cannot be subsumed under any natural law." We have good reason, therefore, for doubting "that paranormal phenomena have ever occurred" (PR, 268).

Repeatability is evidently not as central to science, however, as these critics allege. In a chapter called "The Limits of Replication," Broad and Wade report that very little attempt to replicate experiments occurs in any of the sciences. "The notion of replication, in the sense of repeating an experiment in order to test its validity," they say, "is a myth, a theoretical construct dreamed up by the philosophers and sociologists of science" (BT, 77). With reference to the claim that absence of replicability is unique to parapsychology, Truzzi (SL, 43) says: "[A]bsence of replicability is present for significant claims in many other accepted sciences (especially in psychology and sociology but also in such fields as astronomy)." McClenon supports this point, saying:

> [T]here should be no need for a completely repeatable experiment to make any field legitimate. If psychology and sociology were stripped of all their "nonrepeatable" experiments, they would be restricted to only their most simplistic and absurdly commonsensical formulations. (DS, 12)

The demand for highly repeatable experiments, McClenon pointedly adds, is "selectively applied to deviant sciences like parapsychology" (DS, 12).

The response of the critics to the fact that there are few if any repeatable experiments in sociology and orthodox psychology is that such experiments are not as necessary when orthodoxy is not being challenged. "Some sciences may be exempt from the replicability criterion," says Kurtz, "but this is the case only if their findings do not contradict the general conceptual framework of scientific knowledge, as parapsychology seems to do" (IPS, 510). Alcock agrees: "On this basis, repeatability is, in general, less important in psychology than in parapsychology" (SS, 27). These critics confirm, therefore, McClenon's point that the demand for high repeatability is selectively applied. And they should explain why psi or paranormal capacities, if real, should manifest more uniformity than do abilities considered normal, especially given the fact, as discussed below, that there is good reason to expect them to manifest even less uniformity.

Another response by defenders of the scientific status of parapsychology is to point out that repeatability is a matter of degree. McClenon says:

> It should be recognized that there are degrees of replicability. Some forms of inquiry produce higher levels of replicability than others. . . . As is found in all other statistically oriented sciences, analysis of long-term lines of research reveals various trends demonstrating a degree of replicability. (DS, 12, 13)

Truzzi agrees, saying that

> replicability is . . . a matter of degree, and many experiments in parapsychology have been replicated with some consistency by different experimenters (to say nothing about replications within the same experimental set or replications with a single subject). (SL, 43)

K. Ramakrishna Rao also concurs, saying:

> Replication is not an all or none phenomenon. It admits of degrees. Psychological phenomena are not replicable to the same degree as physical phenomena are. Admittedly parapsychological phenomena are even less replicable than most psychological phenomena. This is not to say, however, that psi phenomena are not replicable. (C, 108)

An important question for the philosophy of science would be whether there are good reasons why the various sciences reveal various degrees of replicability, and why parapsychology in particular should have even less replicability than psychology dealing with more normal dimensions of experience.

The critics, in any case, will have none of this. Not only do they apply the demand for replicability selectively to parapsychology. They demand that parapsychology's experimental results be even *more* repeatable than those in sociology and orthodox psychology. This tactic is illustrated by Ray Hyman. Having agreed that repetition is "not the only hallmark of science," because there are unique events in sciences such as astronomy and archeology, Hyman says:

> [F]or a new discovery to become part of science, it has to have both rational consistency within the accepted framework and repeatability under specifiable conditions. And *if the discovery is weak on one of these criteria it must be doubly strong on the other.* The problem with the claims for paranormal phenomena . . . is that they are weak on both of these criteria. Since, by their very nature, the claims are inconsistent with the accepted framework, the need for reliable, reproducible, and robust phenomena in this area is especially critical. (EQ, 211; emphasis added)

However, what if there were good reason to believe that parapsychology's phenomena are *not highly repeatable precisely because they are the type of human experience that is most inconsistent with the currently accepted orthodoxy?* If so, the demand by Hyman and other critics—that the less orthodox a phenomenon is the more repeatable it must be—would be unreasonable. It would be a "Catch-22," a demand that by definition could not be met. We can explore this question in terms of the recognition by both parapsychologists and its critics that if psi exists, it is an extremely "elusive" phenomenon.

The Elusive Nature of Psi

Christopher Scott is one critic who has commented upon psi's elusive nature, which prevents replication upon demand. Near the beginning of an essay entitled "Why Parapsychology Demands a Skeptical Response," Scott says: "Parapsychologists can reasonably blame nature for the nonrepeatability of psi. It is not their fault if results refuse to come at the will of the investigator" (WPD, 498). This is an objective comment, which manifests an apparent intention to be fair. At the end of the essay, however, Scott concludes that "sensible people should always maintain an attitude of skepticism toward the paranormal, at least unless and until a paranormal phenomenon is found that can be demonstrated at will" (WPD, 501). We have here a clear example of a Catch-22. On the one hand, it is agreed that if psi exists, it is an elusive phenomenon, meaning that it "refuse[s] to come at the will of the investigator." It is recognized, furthermore, that this refusal is rooted in "nature," not in any temporary failure of parapsychologists. On the other hand, however, we are told that, if we are sensible, we will not believe that psi exists until it can be "demonstrated at will." Scott's demand is self-contradictory, because it implies that we should believe in psi, understood to be an inherently elusive phenomenon, only if it is proved to exist by a repeatable experiment, which would show that it is *not* inherently elusive! Scott's appearance of fairness, therefore, is only apparent.

It is Ray Hyman who has especially cast himself in the role of the fair and responsible critic of parapsychology. He says, for example:

> [M]y task as a responsible critic is more complicated when I am confronted by the mounting evidence . . . that is emerging from parapsychological laboratories in ever greater numbers. I think it is fair to say that the standard criticisms that are still leveled against parapsychology by casual critics just cannot be justifiably maintained by anyone who seriously undertakes to examine carefully the reports that are currently available in the several parapsychological journals. (EQ, 234)

And, in a review of the *Handbook of Parapsychology* edited by Benjamin Wolman, he says that

> the critic will be surprised by both the quantity and quality of the evidence presented in the chapters on experimental results. The obvious and usual charges of misuse of statistics, selection of data, inadequate experimental controls, fraud, or other common artifacts do not appear sufficient to account for the entire array of findings. (EQ, 170)

The fact that most of the criticisms leveled against parapsychology are unfair is repeatedly stressed by Hyman. Here are some representative statements:

> [Parapsychologists] are being rejected by most scientists on grounds that have nothing to do with what they are doing now. Most scientists don't even know what is going on in parapsychology. . . . I find it dismaying that most of the criticism of current parapsychological research is uninformed and misrepresents what is actually taking place. . . . [W]hen a critic does take the time to examine the literature in depth, he or she invariably concludes that the methodology or statistics cannot be faulted. These critics . . . typically find that the experimental reports, taken at face value, make an irrefutable case, by current scientific standards, for the existence of the phenomena they call *psi*. Then they switch tactics and try to challenge the evidence on the basis of the possibilities for deception on the part of the experimenters, the subjects, or both. Such tactics explicitly or implicitly concede the case for parapsychology has passed muster on ordinary scientific grounds. . . . The case made by the parapsychologists is much stronger and more sophisticated than my fellow skeptics typically acknowledge. (EQ, 9, 11, 234–35)

Such statements certainly suggest that Hyman means to be a fair and responsible critic. Indeed, other critics might wonder whether, with such statements, Hyman has "given away the store." They might even wonder if he is really a "believer" in skeptical clothing. Hyman, however, has merely cleared away what he considers irrelevant distractions in order to focus on what he considers the decisive problem, which is pointed to by the title put on his collection of essays, *The Elusive Quarry.*

Hyman's introduction of this issue, like Christopher Scott's, describes the problem faced by parapsychologists in an objective, even sympathetic, manner:

> My own criticisms . . . are of things that parapsychologists don't have control over. The way I see it, parapsychologists are doing their darnedest to do the best kind of science. . . . The problem is that they don't have a good

phenomenon by the hand. If they do have one, it is a very elusive thing. It is very weak. It is very sporadic. (EQ, 9)

After recognizing this point, however, Hyman then employs it polemically, saying that parapsychology should not be recognized until it has performed the impossible. Having said again that most of the reasons given by critics for dismissing parapsychology are bogus, he then adds:

> [T]here are also good reasons for orthodox scientists to be skeptical of the claims of their unorthodox colleagues. . . . [T]he reasons, I believe, are sufficient to justify withholding any attention to the claims for the paranormal on the part of orthodox science. . . . [T]he most important reason is the elusive and erratic nature of the phenomena—if there be such. The psychic researchers must somehow catch and tame their quarry before they can place a legitimate claim upon the time of their fellow scientists. (EQ, 206)

This demand that parapsychologists "catch and tame their quarry" means that they must be able to specify the *sufficient conditions* for producing it repeatably on demand. This criterion is so essential, Hyman claims, that all experiments apparently demonstrating psi, no matter how flawless, can be ignored unless their results can be reproduced anytime, anywhere, by any researchers:

> [A] single experiment or a series of experiments that seemingly demonstrates some effect or relationship has no force if subsequent investigators in independent laboratories cannot reliably obtain this same effect or relationship. It does not matter if we cannot retrospectively find loopholes or flaws in the original experiments. We do not have a candidate for a scientific phenomenon if we cannot specify some sufficient conditions for observing the effect. (EQ, 210)

As Hyman reveals elsewhere, this demand for strong repeatability is based on the assumption of the uniformity or lawlikeness of nature. The general criteria for assessing all scientific claims, he holds, include "lawfulness." Given this criterion, a series of experiments yielding "significant departures from chance" are insignificant unless they yield "consistent and lawful patterns" (EQ, 145). Hyman's assumption, in other words, seems to be that, to modify Hegel, the regular is the real and the real is the regular, and that this necessary assumption of science will forever prevent parapsychology from being included.

> [W]hat if, as I believe, a field of inquiry cannot gain general acceptance as science unless it deals with lawful phenomena that are sufficiently robust to withstand scrutiny under varied conditions? If a group of inves-

tigators choose to investigate a phenomenon that does not have this required degree of robustness, then no matter how competent and conscientious they may be, they will not achieve general recognition as scientists. This failure, however, is beyond their control. (EQ, 176)

In this passage, Hyman seems to be making an objective comment about the psychology and sociology of science, leaving open the possibility that the exclusion of parapsychology is unfair. As we saw above, however, Hyman endorses, even encourages, this attitude, saying that parapsychology has no "legitimate claim" to the status of a science until it can, by specifying the sufficient conditions for ESP or PK, produce a universally repeatable experiment. Until that occurs, he contends, scientists should not even accept the existence of psi as a phenomenon requiring an explanation:

> Only when the parapsychologists settle upon a standardized paradigm [yet another demand not enforced in sociology or, Hyman's own field, psychology], tidy up the procedures [also not generally enforced in the social and psychological sciences], demonstrate that the results follow certain laws under specified conditions [also selectively enforced against parapsychology], and that these results can be duplicated in independent laboratories [also selectively enforced], will we have something that needs "explaining." (EQ, 294)

In spite of his pose of greater fairness, Hyman's stance, as this statement shows, is not really much different from that of other critics. His position, in effect, is that until we know virtually everything about psi, so that we can specify sufficient conditions for producing it at will, we should assume that it does not exist and thereby deny funding to support more intense investigation. McClenon has commented upon this strategy:

> The demand for a replicable experiment is actually an aspect of the demand for a solution to the puzzle that psi presents. A totally replicable experiment would require recognition of all the factors associated with the occurrence of psi. . . . But anomalies, by definition, are phenomena that lack such solutions. When anomalies are granted high ontological status, they generate research programs using scientific resources to seek answers to the questions they present. The demand for a replicable experiment from a deviant science is thus a strategy to deny scientific resources for investigating the rejected anomaly. (DS, 90–91)

The rejection of parapsychology on the basis of an unfulfilled demand for highly repeatable experiments, as the above discussion shows, reflects in a twofold way the assumption that the orthodox worldview is essentially correct. First, the idea that the real and the regular are equivalent reflects the belief in a deterministic worldview in which all events exemplify universal laws. Of course, as most of the critics know full well, there are few if any highly repeatable experiments, in the sense required of parapsychology, in the psychology of humans or other mammals.[5] It is at this point that the assumption that orthodoxy is correct plays its second role, justifying the selective application of the strict repeatability demand: Psi must be "doubly strong" in terms of the criterion of repeatability, Hyman says, because it fails to exemplify "rational consistency within the accepted framework" (EQ, 211). Given the fact that Hyman has agreed that the evidence for psi is otherwise good and that all the other criticisms of parapsychology are invalid, we can see that *the only reason for rejecting the evidence of parapsychologists for the reality of psi is the assumption that the "accepted framework" is correct.* Although it is probably through the analysis of Hyman's writings that it becomes most evident, this assumption is really at the root of the other critics as well. Psi is assumed to be nonexistent and parapsychology therefore not to be a genuine science most fundamentally because psi is incompatible with the materialistic, mechanistic, sensationistic worldview that has been the "accepted framework" in the scientific community for about the past 150 years.[6]

It must be admitted that this would be a powerful reason *if* this worldview provided a self-consistent framework that is actually adequate, as often claimed, to all well-evidenced phenomena except psi. As we have seen in the previous chapter, however, that claim is far from the truth. The sensationistic, materialistic, deterministic worldview cannot explain the unity of our conscious experience. It cannot even explain how consciousness can exist. It cannot explain how purposes can influence the body—how, for example, we can intentionally wiggle our fingers. It cannot explain the freedom that we inevitably presuppose. It cannot explain why we inevitably presuppose the reality of mathematical and logical truths and moral and aesthetic norms. It cannot, as we will see in the following chapter, explain how our world could have been created through an

---

5. C. D. Broad has pointed out that Pavlov's famous experiment with the salivating dog was far less repeatable than it has usually been portrayed. For example, "the presence of a stranger, or even of the experimenter himself, completely upsets the reaction of the animal" (RPPR, 106).

6. An important book on parapsychology of which I became aware only after my manuscript was completed is Dean I. Radin's *The Conscious Universe: The Scientific Truth of Psychic Phenomena.* Written by one of the leading experimentalists in the field, the book will be especially helpful to those interested in the evidence for psi phenomena gathered from controlled laboratory experiments. I especially recommend the book for its meta-analysis of parapsychological experiments, of which Radin is perhaps the leading practitioner. Also to be highly recommended is the chapter called "A Field Guide to Skepticism," which can be used to supplement my discussion of the "pseudoscience" issue, especially with regard to more recent developments. Tragically, the book has also turned out to illustrate the point about the extreme bias that still exists with regard to parapsychology: Its publication led to the dismissal of Radin from his position at the University of Nevada, Las Vegas.

evolutionary process. And this is to name only a few of its inabilities. Given the almost complete inadequacy of this worldview, the fact that psi is not consistent with it provides no reason whatsoever for being suspicious of psi.

PSI IN THE FRAMEWORK OF PANEXPERIENTIALISM

Reflecting on the principle that "exceptional claims require exceptional proofs," McClenon has said: "The degree to which psi is considered 'exceptional' is the fundamental issue" (DS, 103). The previous discussion illustrates the truth of this observation. Given the principle that exceptional claims require *proportionately* exceptional evidence, the view that psi is a *complete* exception to ordinary processes will inevitably lead to the demand that evidence for it meet standards of perfection that would be impossible for reports of any alleged phenomenon, let alone one as elusive as psi. The reality of psi, accordingly, is not likely to be widely accepted in the intellectual community apart from the transition to a worldview making psi seem less exceptional. As Flew as said, it would help shake up present notions of what is possible and impossible "if the supposedly impossible could be naturalized by the excogitation of a fresh theory simultaneously explaining both these recalcitrant occurrences and a whole lot else of undisputed authenticity" (C, 269). What we need, in other words, is a framework that includes psi along with well-established scientific facts and hard-core common sense.

Most modern believers in psi, realizing that materialism could not allow for it, have adopted a dualistic philosophy. The idea that we and other animals have minds that are different in kind from the physical stuff to which the laws of physics and chemistry apply, it is thought, makes psi seem less improbable. As we saw in the previous chapter, however, ontological dualism created an insoluble mind-body problem, which contributed greatly to the move to materialism. As we have seen in the present chapter, furthermore, the association of parapsychology with dualism has been one of the main reasons for rejecting it. To be sure, the materialistic philosophy accepted by most of these critics has even more difficulties. But these difficulties do not change the fact that dualism is unintelligible. We cannot justify the acceptance of one absurdity by pointing out that we have rejected an even greater one. The attempt to get the scientific and philosophical communities to return to dualism, accordingly, is not a promising approach for getting the evidence for psi generally accepted.

The formal thesis of this book is that the wider naturalism based on Whitehead's philosophy provides a helpful framework for overcoming the modern conflicts between science and religion. The present chapter applies this thesis to parapsychology, but in this case what needs to be overcome is the conflict of this science not with religion but with the remainder of the sciences—that is, with the worldview generally presupposed by their present practitioners. The thesis here is that Whiteheadian panexperientialism can provide the basis for what has always been the dream of the parapsychological community: a worldview that, while accommodating all the facts of everyday experience and normal science, also accounts for the possibility of, as well as the infrequency of, paranormal events. As we saw in the previous chapter, this form of panexperientialism affirms the main point of dualism, which is that the mind is not identical with the brain or some aspect thereof. But it does this without the problematic point of dualism, which is the ontological difference between the mind and the entities of which the brain is composed.

Panexperientialism, of course, has been even less popular in scientific and philosophical circles than dualism. In terms of the pragmatic issue, accordingly, one might argue that having parapsychology associated with panexperientialism would provide even less hope for overcoming the prejudice that has prevented an open-minded assessment of its evidence. However, the difference in the reasons for the unpopularity of dualism and panexperientialism is crucial. Dualism has been widely adopted and defended for over three centuries and found wanting after having been extensively analyzed. Panexperientialism, by contrast, has been rejected without being explored. It has been unpopular for the same reason as has parapsychology: prejudice. The two prejudices have, in fact, the same root: the continuing paradigmatic power of the early modern adoption of the mechanistic view of nature, which defined all contrary views as "occult" and "superstitious." Thanks to many developments, a few of which were mentioned at the end of Chapter 4, there is good reason to believe that this mechanistic view is losing its hold, so that open-minded consideration of alternatives to both dualism and materialism will become increasingly possible. If this occurs, panexperientialism should, because of its superiority in terms of adequacy and self-consistency, become the preferred view.

Our first concern, in any case, should be truth, not pragmatic concern with success, and our best test of the probable truth of any comprehensive theory is still its capacity to deal with all the facts of experience in a self-consistent way. Parapsychology has demonstrated that psi belongs to the "facts of experience" to which a philosophy striving for adequacy must do justice. One of the strengths of Whiteheadian panexperientialism is that it shows psi not to be so exceptional as to seem impossible or even highly improbable.

The Possibility of Receptive Psi or Extrasensory Perception

In discussing how panexperientialism makes psi possible, I will begin with receptive psi, or extrasensory perception, although some of this discussion will also be applicable to expressive psi, to be discussed below. The first relevant point about panexperientialism is the very fact that it describes the basic units of the world as *experiences*. Part and parcel of the seventeenth-century claim that influence at a distance is inconceivable was the view that the ultimate units of nature were bits of matter with no inside in which "occult" (simply meaning *hidden*) powers, such as the power to exert or receive influence at a distance, could be residing. As Mary Hesse's study of the concept of action at a distance in physics shows, that concept lost repute after the introduction of the idea of matter as devoid of any inside (FF, 118, 125, 291). Given such an account of the ultimate units of nature, it seemed self-evident that causation could only be exerted by contact. Given the idea, by contrast, that every basic unit is an "occasion of experience," it is not self-evident that causal influence can occur only between contiguous things, because it seems at least thinkable that one experience could receive influence from another distant experience. (This is probably why for most people, telepathy—the direct influence of one mind on another—seems more intelligible than those kinds of paranormal events thought to involve insentient matter.) In Whitehead's account, furthermore, each occasion of experience is a subject for itself before it becomes an object for others. Every actual entity, in other words, has an "inside" before it has an "outside." It may be that all such units do have powers, hidden from outside observers, to exert and receive influence at a distance.

Whitehead's philosophy, indeed, suggests that they do. With regard to *receiving* such influence, this power is the capacity for prehension. As we saw in the previous chapter, each occasion of experience begins with prehensions of previous occasions. The new point to add here is that these prior occasions need not be contiguous, whether spatially or temporally, with the prehending occasion. Although it is likely to be constituted *primarily* out of its prehensions of contiguous ones, it will also be partially constituted out of its prehensions of occasions that are not spatially and temporally contiguous.

With regard to *exerting* influence at a distance, the hidden power involves the distinction between the physical and the mental pole of an occasion of experience. In its physical pole, which is constituted out of its prehensions of prior occasions, the occasion is related to the past actual world. In its mental pole, it is related to the realm of possibilities, which subsist in what Whitehead calls the "primordial nature of God." In the occasion's mental pole, therefore, it is related not to time but to eternity. With this point in mind, Whitehead says that the mental pole is "out of time," in contrast to the physical pole, which is "in time" (PR, 248).

This distinction between the physical and mental poles provides the basis for a distinction between two ways in which an occasion can be prehended by subsequent occasions: If it is prehended in terms of its *physical* pole, we have a *pure* physical prehension. If it is prehended in terms of its *mental* pole, the transaction is called a *hybrid* physical prehension. (In the latter case, the prehension is still called *physical* because its object is another actual entity, not a mere possibility.) It is in these hybrid physical prehensions, in which an experience prehends a previous one in respect to the dimension of it that is out of time, that Whitehead saw influence at a distance as most likely. Writing in

1929, he said, "provided that physical science maintains its denial of 'action at a distance,'[7] the safer guess is that [pure physical prehension] is practically negligible except for contiguous occasions" (PR, 308). Whitehead saw no reason to suppose, however, that *hybrid* physical prehensions should be limited to contiguous occasions. In fact, he said, the denial of this limitation would be the more natural supposition. Instances of telepathy, he added, provide some empirical support for this suggestion.

Given the fact that causation is simply the reverse side of physical prehension (PR, 236), the point of the prior paragraph can be translated from the language of "prehension" into that of "causation." Insofar as the occasion exerts influence in terms of data in its physical pole, we can speak of *pure physical causation.* Insofar as it exerts causation in terms of data in its mental pole, we can speak of *hybrid physical causation.* An occasion of experience's capacity to exert hybrid physical causation on subsequent events may be its "hidden" power to exert causal influence at a distance.

The Rarity of Conscious Extrasensory Perception

The next question to ask about receptive psi is why, if it is possible, it seems to occur so seldom. In approaching this issue, we must begin by distinguishing between extrasensory *perception*, understood as direct *conscious* perception of finite actualities beyond one's own body, and extrasensory *prehension*, which may or may not become conscious, of such actualities.[8] As indicated earlier, extrasensory prehension is, by hypothesis, occurring all the time: One is always prehending the entire past world, not simply that which is contiguous with one's mind. If that is so, however, one is obviously not consciously aware of most of it. In fact, most people, most of the time, are not consciously aware of *any* of the world beyond their own bodies except that which they have perceived through their senses. If we are indeed prehending the world extrasensorily all the time, at least most of these prehensions do not rise to consciousness.

The first step in explaining why extrasensory perception is so rare, accordingly, is to point out that (by hypothesis) extrasensory *prehension* is *not* rare: What is rare is only for some of those prehensions to rise to the level of conscious awareness. With regard to this issue, we need to add an element not brought out explicitly in the discussion of consciousness in Chapter 6. The "stuff" of which the mind is composed is not consciousness, but experience, which may or may not rise to the level of consciousness. If consciousness does

---

7. Whether it has, given recent discussions of "nonlocality," is debatable.

8. Specifying that the actualities are *finite* distinguishes this type of nonsensory prehension from that in which God is the object.

occur, it occurs only in the final phase of an occasion of experience, being the "subjective form" of an "intellectual prehension," meaning one that contrasts what *is* with what *is not*. This means that consciousness, rather than existing prior to the prehension of other things, must be evoked into existence by prehensions with sufficient data and strength to do so.

The main point relevant to the present discussion is that we can be conscious of only a very few of the virtually infinite number of things that we are prehending at any moment. There is, as it were, a struggle among our numerous prehensions to survive into the final phase of an occasion of experience, where they will be illuminated by consciousness. Only those objects that have been prehended with the greatest intensity will survive to this phase; the rest are blocked from becoming conscious. Memories and sensory perceptions tend to win out over extrasensory perceptions. Of course, even most memories are subliminal. In any moment, we are not consciously aware of most of the things that we have lived through. Also, even with regard to sensory perception, much of it is subliminal, below the level of that which is lit up by consciousness. Nevertheless, it remains true that, of those prehensions that do rise to consciousness, memories and sensory perceptions are by far the most numerous. The second step in the explanation of the rarity of (conscious) extrasensory perception, therefore, is to ask why this is the case. Why do memories and sensory perceptions normally block out potential extrasensory perceptions from consciousness, rather than the other way around?

One factor involves the distinction between pure and hybrid physical causation. Pure physical causation is, at least usually, far more powerful than hybrid physical causation. Insofar as causation involves a transfer of creative energy, most of this transfer occurs at the level of pure physical causation. Hybrid physical causation, by contrast, is extremely weak, at least most of the time. Only at the level of the higher animals, and especially at the level of human experience, does hybrid physical causation become very significant, and even then it is quite weak, at least usually, in comparison with pure physical causation.

This point is relevant to the question of why sensory perception is more likely to become conscious than extrasensory: The latter, being prehension of noncontiguous things, is hybrid physical prehension. This means that the causal influence to which it is responding is hybrid physical causation. This causal influence will in general be very weak, in comparison with the causal influence exerted by those prehensions that result in sensory perceptions, because those latter prehensions, arriving through chains of contiguous events, are based on chains of *pure* physical causation. Those prehensions that are capable of resulting in sensory perceptions, accordingly, will generally be much more intense and thereby much more likely to make it through to the conscious level of experience. And in so doing, they will tend to block out those prehensions that might otherwise become (conscious) extrasensory perceptions.

The Possibility of Expressive Psi or Psychokinesis

The next question is whether Whiteheadian panexperientialism can make expressive psi, or psychokinesis, seem possible. Much of the groundwork for understanding this possibility has already been laid in the prior discussion, with its explanation of the possibility of influence at a distance, especially through hybrid physical causation. Also, the fact that causation and physical prehension are simply two sides of the same relation implies that, if extrasensory prehension is possible, so is psychokinesis. That is, in telepathy, one mind has prehended another mind directly. This means that, from the other end of the relation, the one mind has causally influenced the other at a distance. The one mind has, in other words, directly exerted influence on the world beyond itself without using its body to mediate this influence. And that is expressive psi or psychokinesis.

To be sure, the direct influence of one mind upon another has traditionally been called "thought transference" or "telepathic influence," while the term "psychokinesis" has usually been defined as the influence of a psyche or mind *on matter*. Although that definition originally presupposed a dualism between mind and matter, which panexperientialism denies, we can still use the term "matter" to refer to aggregational societies of low-level occasions of experience, such as rocks and animal bodies. It is precisely this reconception of the nature of "matter," however, that removes the earlier idea that "telepathic influence" (on other minds) is different in kind from "psychokinetic influence" (on matter). Just as this reconception of the nature of matter overcomes the older view, based on dualism, that clairvoyance is more difficult to explain than telepathy, by showing them both to involve experiences of distant experiences, it also shows psychokinetic influence on matter, no less than telepathic influence between human minds, to involve the influence of experience on noncontiguous experience. Just as thought transference from one mind to another involves extrasensory prehension by the latter mind, so the psychokinetic influence on bacteria or matchsticks involves a prehension of the psychokinetic agent's mind by the bacteria or the molecules in the matchsticks. To accept the possibility of extrasensory perception within the panexperientialist context, therefore, is to accept the possibility of psychokinesis, at least in principle.

This acceptance, however, might mean accepting at most the possibility of very slight influences, perhaps the kind that shows up only in the statistical analysis of large numbers of experiments. From a philosophical viewpoint, to be sure, the crucial barrier has been crossed with the acceptance of *any* direct influence of the mind on the world beyond itself, especially the "physical" world. From a psychological viewpoint, however, the large-scale effects reported in "poltergeist" cases and even in some experimental situations, such as paranormal photography, are more likely to make the mind boggle. Is it really plausible, even if the mind has a little power to produce psychokinetic effects, that it has the power directly, say, to lift heavy tables? Epiphenomenalism and materialism have taught, of course, that the mind as such—that is, as distinct from the brain—has no power whatsoever to exert causation on other things, not even upon its own body. Most of us do not accept this idea. Modern thought has led, however, to a widespread acceptance, at least implicitly, of a viewpoint near the epiphenomenalist end of the spectrum of possible views, which we can dub "semi-epiphenomenalism." That is, although people cannot accept the view that the mind has *no* power in relation to its body, they are likely to assume that its power is minimal, being limited to the power to influence the motor-muscular system. For those who have explicitly accepted this view, it is even surprising to learn of the "placebo effect" (according to which, say, pills with no medicinal value may be almost as effective as the indicated prescription if the patients *believe* that they are receiving the real thing). The effect of one's mental state on ulcers, cancer, and AIDS is also surprising. The appearance of stigmata in the hands of people who have long focused on the cross of Jesus is even more surprising. All these phenomena can often be incorporated within a semi-epiphenomenalist view, however, because one can assume the existence of previously unknown paths of contiguous influences, beyond the motor-muscular system, from the brain to the rest of the body. The power of the mind required to produce these phenomena, accordingly, can be assumed to be no greater than that required for moving one's arms and legs. To keep within this semi-epiphenomenalist viewpoint while accepting psychokinesis, however, would require that the psychokinetic effects all be very small. The problem is that, evidently, they are not (for examples, see my PPS, Ch. 2).

The special question raised for our panexperientialist naturalism, therefore, is whether it can account for the ability of human minds, at least *some* human minds, to produce large-scale effects psychokinetically. Closely related is the question as to why it seems to be only human minds that have these extraordinary powers. That is, although there is evidence that other animals can produce some psychokinetic effects, the power to produce large-scale effects seems to be limited to the human psyche. Why should the human psyche be so distinctive in this respect?

One step in explaining the psychokinetic powers of (at least some) human minds was provided in the account of compound individuals in Chapter 6. According to this account, there is a hierarchy of compound individuals, in which the higher-level, more inclusive ones have more power, meaning the twofold power of self-determination and of causal influence upon others. Panexperientialism runs quite counter, accordingly, to the reductionistic viewpoint of materialism and epiphenomenalism, according to which all power is lodged in the lowest level of nature, which entails that all vertical causation must run upward. Panexperientialism does not endorse the *complete* opposite of epiphenomenalism, which would be "hypophenomenalism"—the idealistic view according to which all vertical causation runs downward, so that what we call "matter" would be simply a nonefficacious by-product of mind. Panexperientialism holds, instead, to the middle way, according to which causation runs both up-wards and downwards (as well as horizontally). But it does say that the mind has more power, not less, than lower-level individuals.

The idea that the mind has more power than its brain cells does not mean that, in the interaction between mind and brain, the mind exerts more power on the brain than the brain exerts back on the mind. After all, if the principle of the conservation of creative energy applies in this interchange, at least roughly, we would expect parity. The mind at any one moment is a single actual entity, how-ever, while the normal human brain is composed of over 100 billion brain cells. If the interaction between mind and brain involves a fair exchange, therefore, it would seem that the mind would be billions of times more powerful than a single brain cell. Reductionism, from this point of view, is not even close to the truth.

Although this antireductionist perspective provides a necessary basis for approaching a philosophical understanding of psychokinesis, it does not help explain why the minds of human beings alone, at least among creatures on this planet, seem to have strong psychokinetic power. That humans would have more than mice or even dogs is not surprising. But the genetic difference between humans and other primates is slight. Why should their minds apparently be so different with regard to expressive psi? A likely answer is provided by John Cobb: The distinctiveness of human beings involves "surplus psychic energy," meaning energy beyond that needed for the well-being of the body. The thresh-old dividing humans from other primates occurred, suggests Cobb, when "the surplus psychic energy became sufficient in quantity to enable the psychic life to become its own end rather than primarily a means to the survival and health of the body" (SCE, 39). This surplus psychic energy is the precondition for religion and all those other activities in which the mind treats itself as an end in itself, not simply as the director of its bodily activities. It also might help explain, although Cobb himself did not mention this issue, the human mind's distinctive psychokinetic powers.

Cobb's account of the distinctiveness of the human psyche can also help explain the fact that psychokinetic power, to the extent that it is present in the human mind, usually seems to be largely, if not totally, beyond conscious con-trol. In asking what is distinctive about human existence, Cobb agrees with the standard evolutionary accounts insofar as they make symbolism central. He does not agree, however, with the focus on the practical advantages for survival pro-vided by symbolic language. Rather, suggests Cobb (who was influenced by the Jungian historian of consciousness Erich Neumann), the evolutionary develop-ment constituting human distinctiveness consisted of "the greatly increased unconscious psychic activity organizing the whole of experience for its own sake" (SCE, 39). The much greater supply of surplus psychic energy, in other words, exists primarily in what we call the unconscious portion of the mind. The unconscious psychic activity of symbolization did, to be sure, result in "a new and incomparably richer mode of consciousness" (SCE, 41). But the great increase in surplus psychic energy occurred primarily in the unconscious and remains there to this day, being used for symbolizing activity that is, as revealed in dreams, largely autonomous from conscious symbolizing activity. If we assume that this same surplus psychic energy lies behind the power to produce expressive psi, Cobb's view would explain why this energy or power would be, at least for most people most of the time, beyond conscious control. This suggestion, that most of the power of the psyche is in a level below consciousness, fits with the account of consciousness provided earlier, according to which it arises only in the last phase of an occasion of experience, after most of the occasion's activity has already occurred.

If the power that can produce psychokinetic effects, at least of the large-scale variety, is in the portion of the mind that is below consciousness, how is it possible that occasionally—either now and then in one person or somewhat regularly in an occasional person—conscious effort is able to produce such effects? An answer suggested by panexperientialism is that the conscious mentality of one occasion of experience, while quite weak in itself, can in principle activate and direct the unconscious power of succeeding occasions of experience to bring about extraordinary effects. The power to produce such effects intentionally, therefore, would depend upon a special attunement between the conscious and unconscious portions of the psyche. This suggestion is in harmony with the fact that the power to produce such effects is generally correlated with the power to become conscious of extrasensory prehensions. It also coheres with the fact that this twofold power often develops as a side effect of spiritual disciplines that serve—even if their purpose is not so described—to bring the conscious and unconscious dimensions of the mind into attunement.

Given this discussion of how expressive psi or psychokinesis in general is possible, it may be helpful to look at levitation, a type of psychokinesis that, from our ordinary experience, seems so bizarre. In levitation, a physical thing, perhaps the body of the psychokinetic agent, seems unaffected by the Earth's gravitational pull. Because in our experience the force of gravity seems so strong and ever-present, the idea that a mind could counteract it can understandably seem incredible, so that stories of saints floating during prayer or Jesus walking on water would have to be regarded as supernatural interventions or pious inventions. Given the power of the psyche plus the idea of a hierarchy of compound individuals, however, the possibility of levitation need not seem so remote. Although the gravitational force seems very strong, this appearance is due to the very large number of particles involved. The force of gravitational attraction on each subatomic particle is actually extremely weak, being 10^{43} times weaker than the electromagnetic force. A psyche could (perhaps unconsciously) neutralize the gravitational force on its body, accordingly, merely by inducing some level of compound individuals in its body—perhaps the cells, or macromolecules, or atoms—to exert a tiny bit of counteractive force upon the events at the subatomic level.

The Special Case of Apparent Precognition

On the basis of Whitehead's panexperientialist worldview, I have argued, we can understand in principle how both receptive and expressive psi can occur. There is, however, one kind of putative psi that cannot be accepted from this standpoint. This is precognition, which means, if taken literally, direct noninferential knowledge of an event before it occurs. This knowing is usually taken to be based on perception, which would require, to coin a term, *preprehension*—the prehension of an event prior to its existence. Because prehension is the reverse side of causation, preprehension would imply "backward causation"—the influence of the future event back upon the present. All of this is nonsense from a Whiteheadian viewpoint. As Whitehead says, there can be no prehension of a future event because "no future individual occasion is in existence" (AI, 193). And an event cannot be "known" before it exists because the existence of every individual event involves an element of self-determination in the moment, which is in principle not knowable in advance. To put the point in terms of causation, there can be no backward causation from the future on the present because in the present the future occasions do not yet exist: They cannot exert causation until they have passed from the subjective to the objective mode of existence, and they have not even begun the subjective mode. In still other terms: They will exert causation on the basis of what they will *be;* what they will be will be determined by what they *become;* and this process of becoming involves partial *self-determination in the moment.* The idea that they could exert causal influence before they have decided just what they are to be is nonsense. True precognition, accordingly, is impossible. Of the four basic limiting principles of C. D. Broad given earlier, in other words, I agree with the fourth, which says that "an efficient cause cannot come after its effects(s), which means that there can be no retrocausation and (therefore) no precognition."

The idea that apparent precognition is different in kind from the other phenomena discussed by parapsychologists is not unique to a Whiteheadian perspective. Some advocates of parapsychology, in arguing that it requires the adoption of a new worldview, have lifted up precognition as especially impossible to account for within a materialistic framework. Some critics of parapsychology make the reverse point—that precognition especially shows the impossibility of the alleged phenomena of parapsychology. Flew, for example, believes that ESP and PK would involve violations of our best supported ideas of "contingent impossibility" (PSP, 529). But the obstacle to accepting precognition, he says, is different in kind: "Because causes necessarily and always bring about their effects, it must be irredeemably self-contradictory to suggest that the (later) fulfillments might cause the (earlier) anticipations" (PSP, 533). The idea of precognition, in other words, is *logically* impossible.

Several advocates of parapsychology, furthermore, agree with this point, saying that experiences that are sometimes interpreted as true precognition must be given some alternative explanation (Braude LI, 256–77; Eisenbud PF; Mundle DCP; Roll PP; Tanagras PE). Several possible "real time" explanations have been given. For example, we could say of a so-called precognitive experience that it involved the perception not of the *actual* future but merely the *probable* future, given the present trajectory of events (which is, in fact, how many "fortune tellers" present their predictions, telling the person that by taking appropriate action an undesirable "foreseen" event can be prevented). With this interpretation, no violation of the reality of time would be involved, because probabilities are present realities. A second kind of interpretation of apparently precognitive experiences involves the notion of a dream or waking vision that brings to consciousness an inference based on information acquired by means of unconscious clairvoyance. For example, a woman may have a dream about a rockslide that destroys a school building the night before the occurrence of such an event. She may believe that the actual rockslide is what triggered her dream, whereas in reality the dream was the result of her unconscious awareness, learned clairvoyantly the day before, that tremors had made the mountain unstable. Yet a third possibility, sometimes called the "psychokinetic interpretation," is that instead of the future event causing a precognitive dream about it, the dream helps brings about the future event. In any case, it is not necessary for our present purposes to discuss all the possible interpretations, but only to note that real time interpretations exist (thirteen of which I have discussed in PP, 270–75). The importance of this fact, of course, is that parapsychology's data can thereby be seen to be not inconsistent with the perspective of the Whiteheadian worldview, with its insistence on the ultimate reality of asymmetrical, irreversible time (as discussed in Griffin PAP and PUST), and thereby, more generally, with our hardcore commonsense assumptions about time and causation.[9]

---

9. It might be thought that my denial of true precognition on the grounds of impossibility is no different from the rejection by other philosophers of ESP and PK on the grounds of impossibility, of which I am critical. In reply, four points are relevant. First, rather than criticizing other philosophers for using their judgments of possibility in assessing claims for paranormal occurrences, I only suggest that when claims of contingent impossibility are involved, philosophers should allow their judgments to be reversed by overwhelming evidence to the contrary. The second point is that empirical evidence is *not* relevant in the same way in relation to matters of logical or analytic impossibility. The third point is that it is one thing to insist that all data must be given a nonparanormal interpretation, quite another to offer an alternative paranormal interpretation of some of the data. Fourth, none of the data come with "precognition" written on them. Indeed, I have already conceded too much by referring to "apparent precognition," as if precognition were somehow the *privileged* interpretation, which I am "rejecting." Given our normal understanding of causation, according to which it goes from the present to the future, the "real time" explanations should be considered privileged, which those who speak of true precognition are rejecting.

THE RELIGIOUS IMPORTANCE OF PARAPSYCHOLOGY

At the outset of this chapter, I cited with approval both the suggestion of J. B. Rhine and others that parapsychology could be considered "religion's basic science" and the prediction of William James that once religion becomes allied with a "radical empiricism"—one based on nonsensory perception in general and psychical research in particular—a "new era of religion" will begin. The distinctive feature of this new, postmodern era, I have suggested, will be a spirituality based upon a form of religious belief that is liberal in basic worldview and method—rejecting supernaturalism and authoritarianism—but conservative in content, providing the basis for a robust religious life. It is now time to fill out that suggestion, indicating concretely some ways in which parapsychology can help overcome the thinness of modern liberal theology.

Negative and Positive Importance

This thinness, I have argued, has resulted from the acceptance of some distinctively modern assumptions, especially the sensationist doctrine of perception, the mechanistic doctrine of matter, and—since the latter part of the nineteenth century—a materialistic view of reality as a whole. The first two doctrines—sensationism and the mechanistic doctrine of nature—were first accepted in large part to support a supernaturalistic view of the universe as a whole. This supernaturalism lies behind the authoritarian method and many of the substantive doctrines of conservative-to-fundamentalist theologies to this day. *Negatively*, parapsychology's greatest theological significance is the fact that it undermines this supernaturalism, showing that the kinds of events traditionally classified as supernatural "miracles," besides happening in all religious-cultural traditions, can be understood as products of natural, if somewhat extraordinary, powers. Those interested in defending theologies based on the idea of supernatural interventions have been right to feel threatened by parapsychology.

Modern liberal theology developed its thinness when the acceptance of sensationism and the mechanistic view of nature was combined with the rejection of supernaturalism and dualism in favor of a more or less completely materialistic worldview with no room for divine influence, life after death, and most of the other beliefs that have been associated with robust religion. *Positively*, parapsychology's greatest theological significance is that it provides empirical evidence that serves to undermine the late modern assumptions that have ruled out a significantly religious worldview. Those who despise religion of every type—not just supernaturalistic religion—have been right to feel threatened by parapsychology. They have rightly seen that, even if philosophers of religion and theologians have seldom taken advantage of this fact, parapsychol-

ogy supports, whether intentionally or only inadvertently, various religious presuppositions that the late modern worldview had ruled out. It is on this positive
significance of parapsychology for philosophy of religion and theology that I
will concentrate.

Religious Experience

Perhaps the most obvious implication is that, by supporting the reality of
nonsensory perception, parapsychology supports the possibility of genuine religious experience, in the theistic sense of a direct experience of a personal God.
The obviousness of this implication, however, should not lead us to underesti-
mate its importance. Late modern thought has been based upon the assumption
that genuine religious experience, in this sense, is not possible. Most theories of
religion, such as those of Hume, Feuerbach, Marx, Comte, Freud, and Durkheim,
have been created to explain why, given our modern understanding that no one
really has religious experiences, religion should be so ubiquitous, having evidently been central to every known human society. Parapsychology provides the
basis for a more straightforward explanation—that there is a Holy Reality and
that people always and everywhere have been religious because they sometimes
directly perceive this Holy Reality by means of the same capacity for nonsensory
perception that sometime results in telepathic experiences. Parapsychology provides support for an even stronger contention. By providing evidence that we are
directly prehending other minds all the time, so that those exceptional experiences classified as telepathic experiences are exceptional only by virtue of rising
to consciousness, it suggests, by analogy, the view entailed by Whitehead's
philosophy: that we are directly prehending God all the time, so that those
exceptional experiences classified as "religious experiences" or "experiences of
the holy" are exceptional only by virtue of rising to consciousness.

The importance of this point can be illustrated by the theological reflections
of Gordon Kaufman on the origin of ideas of God. Agreeing with Kant not only
that concepts without percepts are empty but also that percepts are exclusively
sensory, he asks to what the word "God" refers. "Certainly not to anything we
directly experience," he answers. "Whatever else this term might signify, it does
not seem to name anything experientially perceived" (IFM, 415). Kaufman's
conclusion is that the idea of God is entirely an "imaginative construction." In
one sense, he rightly says, all our symbolic ideas, including "tree" as well as
"God," are products of the imagination (IFM, 39). However, in another sense, he
maintains, the idea of "God" is different in kind:

> God is simply not the sort of reality that is available to direct observation
> or experience. . . . The idea of God, thus, should not be regarded as
> epistemically similar to the idea of a perceptual object (for example, a table

or a person or a mountain); it is not based on direct human perceptions of God. Rather it is constructed imaginatively in the mind. (IFM, 323)

In saying that the idea of God is constructed imaginatively, therefore, Kaufman is saying that it is not constructed, in the slightest, out of any kind of experience of a really existing Holy Reality.

On the basis of Whitehead's alternative to the Kantian epistemology, supported by empirical evidence from parapsychology, we can think of the idea of God as indeed "epistemically similar" to ideas of (other) "perceptual objects," especially other persons, in the sense of "other minds," which we can know not only inferentially, on the basis of our sensory percepts of their bodily behavior, but also directly, telepathically, as Whitehead suggests (SMW, 150). We can, therefore, think of various ideas of God as more or less accurate responses to (nonsensory) perceptual experiences of an actuality distinct from all the finite actualities comprising the world (see my RE and RWS, Ch. 2).

Normative Experience

Besides providing support for the reality of religious experience, parapsychology also supports the reality of moral, aesthetic, and other normative experiences. As I have argued, we cannot help presupposing in practice the reality and importance of Truth, Beauty, and Goodness, in the sense of the reality of cognitive, aesthetic, and moral ideals. However, late modern thought, primarily through the sensationist theory of perception, has made it impossible to understand how we could experience such ideals. Because cognitive, aesthetic, and moral ideals, not being physical objects, are obviously not the kinds of objects that can excite our sensory organs, the assumption has been that, even if they somehow existed in the nature of things, we would have no way to perceive them. By undermining the sensationist doctrine of perception, parapsychology has undermined the primary basis for that assumption.

It would be difficult to exaggerate the importance of this fact. Late modern thought, including much of what has been called "postmodern" thought, has affirmed, or at least implied, an ultimately nihilistic view of life, according to which moral and aesthetic judgments are said to be incapable of being either true or false, because they are not rooted in experiences of any realities to which these judgments could possibly correspond. As Martin Heidegger (WN) put it in explicating the meaning of Nietzsche's saying that "God is Dead," one no longer believes in a "supersensuous" realm, meaning a realm of transhistorical norms or values knowable through some means other than sensory perception.

Whereas types of postmodern thought have also arisen from Whitehead's philosophy, they are, in distinction from the deconstructive types of postmodernism associated with Heideggerian thought, constructive types of postmodern think-ing. One of the crucial differences is the fact that Whitehead reaffirms the old idea of a supersensuous realm, which includes normative ideals. For Whitehead, this realm exists as the content of the "primordial nature of God," which is God's primordial envisagement of the possible forms of existence with an appetite for them to be realized, each in due season. The divine appetite for Truth, Beauty, and Goodness to be realized in human existence lies behind our more or less explicit awareness of the existence of cognitive, aesthetic, and moral ideals. In Whitehead's words:

> There are experiences of ideals—of ideals entertained, of ideals aimed at, of ideals achieved, of ideals defaced. This is the experience of the deity of the universe. The intertwining of success and failure in respect to this final experience is essential. We thereby experience a relationship to a universe other than ourselves. We are essentially measuring ourselves in respect to what we are not. . . . The universe is thus understood as including a source of ideals. The effective aspect of this source is deity as immanent in the present experience. (MT, 103)

Implicit in Whitehead's final sentence is his "ontological principle," according to which only *actual* entities can act, which means that nonactual things, such as normative ideals, cannot exert any agency on their own, which would mean that they could not affect human experience. Whitehead agrees, accordingly, with the widespread view—illustrated by Willem Drees in Chapter 3 and the Deweyan naturalists in Chapter 4—that a nontheistic worldview implies the nonefficacy of normative ideals.[10] Whitehead affirms the efficacy of such ide-als for human experience partly by regarding them as actively entertained by the primordial mind of the universe, partly by regarding human experience as capable, by virtue of its nonsensory mode of perception, of prehending this primordial mind. Parapsychology supports the Whiteheadian affirmation of the experience of normative ideals by supporting telepathy between human minds, which provides an analogy for the direct perception of a divine mind by hu-man minds.

A theology based on Whitehead's philosophy and parapsychology's evi-dence, therefore, can, unlike Kaufman (IFM, 328, 364), resist the assumption of late modern thought that all our values are entirely the products of human imagination, rather than being to some extent rooted in a nonsensory perception of normative ideals in the nature of things.

---

10. The view that we could not possibly be influenced by ideal or "abstract" entities was discussed in relation to mathematical objects in notes 5 and 6 of Ch. 3. Philosophers who have endorsed this view in relation to moral norms include Gilbert Harman (NM, 9–10) and John Mackie (E, 38–39).

The Reality of God

The other presupposition involved in this explanation of religious and normative experiences—the existence of a divine mind that can be prehended by human minds, so that it can influence them—is also supported by parapsychology. Belief in God as a personal being distinct from the totality of finite things has been closely connected with an interactionist view of the mind-body relation, because the latter provides an analogy for the former. The move from dualistic interactionism to materialistic epiphenomenalism and identism has, by destroying that analogy, been accompanied by the move to a nontheistic view of the universe: If we cannot think of the human mind as exerting downward causation upon the body, we have no basis for affirming downward causation from a divine mind to the world. Kaufman's nontheistic view of the universe, for example, is in line with his assumption of mind-brain identism. Having used the term "God" to refer to the trajectory in the evolutionary process that led to human beings, Kaufman says:

> Bringing "God" into our considerations here does not commit us to the existence of some additional *being* (either in the world or beyond the world), from which these evolutionary forces proceed, any more than speaking of selfhood (for example) commits us to an additional "something" alongside the body, which brings about our bodily movements. (IFM, 349)

Whitehead's philosophy, as we have seen, allows for a form of interactionism without dualism. The mind or psyche is again affirmed as an actuality distinct from the brain with its own power, through which it can exert downward causation upon the body. By documenting the reality of psychokinesis, which evidently cannot be explained in terms of powers of the brain, parapsychology supports the distinctness and power of the mind. It thereby provides analogical support for the idea of downward causation from a mind of the universe.

Divine Action

This idea of divine action in the world is, of course, of utmost importance in theistic religions. Besides presupposing the reality of religious experience in general and the communication of the divine concern for various values (such as truth and justice), these religious traditions have affirmed the reality of divine revelation, inspiration, providence, incarnation, and answer to prayer, all of which become vacuous without the notion of divine action in history, as illustrated by the positions of Drees and Kaufman. The inability to speak of divine causal influence in the world in any straightforward manner has been responsible for much of the increasing vacuity of modern liberal theology. One of the ways in which parapsychology proves itself to be "religion's basic science" is by documenting the reality of both telepathy and psychokinesis, thereby providing analogical support for the idea of divine action in history.

These theistic religions, of course, do not limit divine action to human history. They also speak of God's action in "nature," especially in creating it. In our day, of course, a doctrine of the world as divine creation would include not only the origin of our universe but also ongoing influence in the evolutionary process over billions of years. Modern liberal theology, given its presuppositions, has been especially unable to reaffirm the biblical conviction of the world as God's creation. Rudolf Bultmann's program to "demythologize" the Bible, for example, entailed reinterpreting the biblical message so that it no longer contained any cosmological affirmations (see my APC, Ch. 3). As we will see in the following chapter, Whitehead's philosophy provides the basis for a theistic evolutionism. An essential part of this basis is his panexperientialism, which allows all actualities, analogously to human minds, to prehend God, thereby explaining how God can internally affect the world at all levels, even the inorganic level. The fact that parapsychology provides evidence not only for the telepathic influence of one human mind upon another, but also for the psychokinetic influence of mind upon "matter," provides analogical support for this idea of God's influence in the physical world.

Spiritual Discipline

The theological importance of parapsychology's validation of psychokinesis and thereby of the extraordinary power of the human soul, however, extends beyond the support that this notion gives for the idea of downward causation from God to the world. The evident fact that the human soul has such power is important in its own right. For example, one reason for the decline of spiritual discipline in the modern world has surely been the idea that the human mind, being a virtually impotent by-product of the brain, has neither the capacity nor the need for spiritual discipline. By contrast, just as the evidence for psychogenic illness shows that the soul has power that, if informed by unhealthy attitudes, can be destructive to one's own body, the reality of psychokinetic effects shows that the soul has power that, if informed by such attitudes, can be directly dangerous to the world beyond one's own body. As such, it needs self-discipline if it is to be a power for good rather than a destructive force. Carl Jung reports, for example, that it was his awareness of his own psychokinetic power that, by making him realize that his psyche did not belong to him alone, led him to devote his life to the psyche (MDR, 192).

The fact that most of the power to produce psychokinetic effects is evidently, as illustrated most extremely by "poltergeist children," in the unconscious portion of the psyche, beyond the control of the conscious will, shows even more clearly the need for spiritual discipline. Of course, most people do not produce poltergeist or other conspicuous psychokinetic effects. The evidence that even ordinary people can produce psychokinetic effects that are statistically detectable, however, suggests that poltergeist outbreaks are merely extreme manifestations of a type of power that, for good or for ill, all of us are radiating from our souls all the time. Spiritual discipline would seem to be important, accordingly, not only for ourselves considered individually, but also for the common good.

Spiritual Healing

Parapsychology has provided evidence not only for psychokinetic effects in general, but also for psychical or spiritual healing in particular. At one time, the church's ministry of healing was central to its existence, not only because the capacity to heal, even if only occasionally, was part of its drawing power, but also because it provided evidence for the reality of spiritual power. In churches informed by modern liberal theologies, however, this ministry has generally been absent, reflecting the fact that these theologies have discouraged, if only by silence, belief in the power of prayer to heal. In the meantime, a wide readership is being attracted for books citing evidence for this power—books written by physicians, not theologians or ministers (Benor HR; Benson BRR; Dossey HW; Siegel LMM).

The theological perspective of the present book, taken in conjunction with the evidence from parapsychologists and now some physicians, provides a postmodern basis for again taking seriously the power of prayer for healing. Of course, this kind of prayer would not be based on belief in the power of God to heal illness unilaterally. That belief, although resulting in prayer that is sometimes practically efficacious, increases the theoretical problem of evil, as it seems to presuppose a deity that, needing to be cajoled to use its healing powers, is less concerned about healing than we are. And, given the fact that prayer is only sometimes followed by healing, that belief suggests divine arbitrariness.

The kind of prayer suggested by the present book would exemplify its naturalistic theism, according to which divine action, instead of interrupting or canceling out the causal powers of the creatures, works with these causal powers, seeking to transform them to bring about the best results possible in each concrete situation. According to this understanding of prayer, which has recently been beautifully articulated by Marjorie Suchocki (IGP), all things come about through literal cooperation between divine and creaturely power. Given this naturalistic theism, the world's evils provide no occasion for doubting the perfect love of God for all creatures suggested by the scriptures of most traditions. From this perspective, we can assume that the divine healing energies are always being exercised to the maximum—that God is already, prior to our prayers, doing everything possible to bring about healing of every type in each situation. The purpose of our prayers for our own healing, therefore, would be to try to align ourselves with the divine aims for us, so that our power of self-determination would be working with, not against, these divine aims. The purpose of our intercessory prayers for the healing of others would be to seek to align the direct causal influence of our souls upon others with the healing energy of God for them. The assumption behind such prayer would be that, in some situations, the obstacles to healing are such that they can be overcome if and only if our prayers reinforce the healing possibility provided by the divine aims. This perspective, combined with the empirical evidence for the efficacy of intercessory prayer provided by parapsychology, could provide the basis for a recovery of the church of its healing ministry.

The evidence for the reality of psychical and spiritual healing is also religiously important for Christianity in terms of the reliability of the New Testament, especially the gospels, and also for understanding Jesus and the early Jesus movement. Jesus is presented in the gospels as one who had extraordinary healing powers. For a long time, modern liberal historians assumed that all the healing stories in the gospels had to be relegated to the category of myth. More recently, given wider awareness of the reality of psychogenic healing, some New Testament scholars have decided that those stories that can be interpreted in these terms may contain a kernel of truth. The wider awareness of the reality of psychic and spiritual healing will enable scholars in the future to assume that a wider range of the stories may be based upon kinds of healing that actually occurred through the influence of Jesus (and later, some of his followers, especially, evidently, Peter and Paul).

Life After Death

Although there are several other ways in which parapsychology is religiously important, I will bring this discussion to a close by referring to its relevance to the question of life after death. Most modern liberal theologians, having accepted the dominant view of the time, would find it very difficult, perhaps somewhat indecent, to reopen this topic. For these theologians, the idea of a literal life after death, including the idea that Jesus in some literal sense appeared to his disciples after death, is assumed to belong to the mythological form taken by Christian faith in bygone ages. Such theologians, while rejecting the "Greek" and "dualistic" idea of the "immortality of the soul," may praise the "biblical" and "wholistic" idea of the "resurrection of the body," but they then reduce this notion to its "symbolic value" or "existential meaning," assuring us that a salvation in a life after death is not really the point of Christian hope and that, with maturity, the need for such a hope disappears. "Yet," says John Cobb in an essay on "The Resurrection of the Soul," the question of what, if anything, happens after we die has not disappeared from the range of human concerns. It has simply moved out of professional theology into other hands. Our sophisticated equivocations on this topic have contributed to our general irrelevance to the religious interests of our contemporaries. (RS, 13)

Cobb, of course, is referring to liberal theologians. Conservative-to-fundamentalist theologians have continued to speak of a salvation in a life beyond this one, with the result that their theology is regarded as highly relevant by a great number of people. However, that theology, including and especially its doctrine of salvation in a life after bodily death, is based upon the method of authority, which liberal theologians have (rightly) rejected. Liberal theologians will be able to speak of life after death only if they come to see that it can be affirmed on the basis of reason and empirical evidence. Being able to do so will involve the movement from modern to postmodern liberal theology.

Within modern liberal theology, reason and experience seemed to conspire against belief in life after death. Reason seemed to say that the mind is simply the brain or an epiphenomenal by-product thereof, so that life after death would be unthinkable. All the relevant experience, from our dependence on our physical sensory organs for knowledge to the ways in which memories and personalities become disoriented through brain disease or damage, seemed to point in the same direction. Of course, all sorts of stories from times past, and even some from today, seemed to suggest the reality of life after death. But, given the fact that modern science and philosophy had so completely ruled out its possibility, one could safely assume that all such stories were to be dismissed as superstition, wishful thinking, fabrications, delusions, and the like.

However, given the rejection of modern mechanism and sensationism, as we saw in Chapter 6, "reason" comes to a quite different conclusion about the relation of the mind to its brain and accompanying sensory organs. The mind is distinct from the brain, having power of its own to exercise self-determination and causal efficacy upon others. It thereby has one of the necessary conditions for surviving apart from the brain. Also, sensory perception, rather than being our only or even our basic mode of perception, is a derivative mode. The idea that a discarnate mind might be able to perceive (prehend) other realities is, therefore, not unthinkable. Given this verdict of reason, the relevant empirical evidence becomes expanded. In the first place, the parapsychological evidence for extrasensory perception, by supporting the position of James and Whitehead that we have a nonsensory mode of perception, provides empirical evidence for the possibility of perceptual experience in a discarnate state. The evidence for expressive psi, furthermore, suggests that, if we found ourselves in a postcarnate state, we might be able not only to perceive, but also to express ourselves, acting so as to influence other beings.

This empirical evidence for the *possibility* of existing beyond the body is greatly strengthened by credible evidence that some souls *actually* do thus exist for brief periods of time. These "out-of-body experiences" (OBEs) have been studied with increasing intensity in recent decades by members of the medical profession, especially in relation to "near-death experiences." Of course, given the continued prevalence of the late modern view of human beings, the dominant assumption in the medical profession is that all such reported experiences are, when not dismissable as lies, to be interpreted as fantasies, induced by drugs, lack of oxygen, fear of death, and the like. An open-minded study of the evidence, however, shows that none of the proffered "intrasomatic" interpretations of OBEs can do justice to the facts. Especially suggestive of the truth of the "extrasomatic" interpretation of at least some OBEs is the recent documentation, mentioned above, of corroborated veridical perceptions during the time the person was behaviorally unconscious. Beyond these brief comments, I can here only refer the interested reader to my extensive discussion elsewhere (PPS, Ch. 8) of the relative merits of the intrasomatic and extrasomatic interpretations of OBEs.

On the basis of Whitehead's postmodern philosophy, especially combined with the evidence from ESP, PK, and OBEs suggestive of the possibility of discarnate existence, one has a basis for examining the actual evidence for life after death. And when one does so openmindedly, it turns out to be surprisingly good. Here, at the end of an already long chapter, I cannot even begin to summarize this evidence, any discussion of which, to be worthwhile, must be quite lengthy. I have, in my recent book-length study of parapsychology, devoted chapters to the evidence from mediumistic messages, cases of the possession type, cases of the reincarnation type, and apparitions (PPS, Chs. 4–7). Although much of the evidence is susceptible to a nonsurvivalist interpretation, at least if one allows for "superpsi" (very powerful receptive and expressive psi), which I do, some of it does not seem to be.

Also, if one seeks to give a nonsurvivalist interpretation of all the evidence (as I originally did), one faces a dilemma: On the one hand, the various kinds of evidence for life after death, especially when considered together, are so overwhelming that they virtually require belief in life after death, *unless* one allows for superpsi of the most extraordinary sorts. On the other hand, if one does allow for such superpsi in order to give a nonsurvivalist interpretation of the evidence, then one has undermined the main reason for not believing in life after death—the assumption that the mind or soul is not the kind of entity that *could* survive apart from the body. To put the issue in terms of "antecedent probability": If the mind or soul can do all the things it would have to do if the superpsi alternative to the survivalist interpretation of the evidence is true, then it would be the kind of reality that one would also expect to be able to exist apart from the body. My own conclusion, in any case, is that some of the evidence cannot plausibly be explained in terms of superpsi on the part of the living, and that all of the evidence,[11] when taken cumulatively, points strongly in favor of the reality of postcarnate existence. For the reasons supporting this conclusion, I can only refer the reader to my aforementioned discussion.

If accepted, this conclusion, of course, has major implications. It would not necessarily mean returning to traditional views of heaven and hell. It would also not necessarily mean returning to a form of spirituality in which the importance of the present life is understood primarily as a preparation for a future life. Indeed, within the Whiteheadian worldview advocated here, neither of these consequences would follow. Belief in life after death would take the form of the anticipation of a continued journey, in relation to the Holy Reality of the universe, with fellow travelers. And, far from distracting us from the enormous social, political, economic, and ecological issues facing us, this anticipation would empower us to deal with them more wholeheartedly by freeing us from paralyzing anxieties related to death and the sense of ultimate meaninglessness. However, the implications of belief in life after death, like the kinds of evidence for it, are far too extensive to be discussed at the end of this chapter. I will only state my conviction that the incorporation of belief in life as a journey that continues beyond bodily death is probably an essential ingredient if there are to be religious communities combining a vital spirituality with a liberal approach to religious belief. I suspect, in other words, that James's prediction—of a new era in religion following upon the association of religion with radical empiricism—is more likely to prove true if this radical empiricism includes parapsychology taken as providing evidence for belief in life after death. Life as a continuing journey beyond death can be understood as a further development in the evolutionary process brought about by God's creative influence in the world— which is the topic of the next chapter.

---

11. "The chief danger to philosophy," said Whitehead, "is narrowness in the selection of evidence," adding that this narrowness arises from the idiosyncracies and timidities not only "of particular authors, of particular social groups, [and] of particular schools of thought" but also "of particular epochs in the history of civilization" (PR 337). In the late modern epoch, this narrowness in the selection of evidence has probably been most extreme with regard to phenomena suggestive of life after death and paranormal occurrences more generally.

8

CREATION AND EVOLUTION

INTRODUCTION

At the heart of the conflict between religion and scientific naturalism since the latter part of the nineteenth century has been the issue of creation and evolution. As Robert Wesson points out, "Darwinism became the banner of those who would overthrow what they saw as an irrational, superstitious view of human origins.... The theory of evolution became the focus of the confrontation of science and religion" (BNS, 20). This centrality of evolution in the conflict has remained to this day. As Holmes Rolston says in his *Science and Religion,* "evolutionary theory... is that particular science most troubling to religion" (SR, 102). There is good reason for the fact that this theory has been so troubling. On the one hand, the belief that our world was created by a purposive deity lies at the foundation of Christianity and the other biblically based religions (as well as most other religions). On the other hand, the evolutionary theory of origins that has been adopted by the scientific community—the present version of which is neo-Darwinism, although it is often still simply called "Darwinism"—not only rejects the supernaturalistic idea that the various forms of life were created *ex nihilo,* but also excludes *any* kind of divine activity in the evolution of the various forms of life. Although, as Neal Gillespie points out in *Charles Darwin and the Problem of Creation,* the theory as Darwin himself formulated it involved a deistic creation of the universe and even of the first organic beings (CD, 130–31), most of Darwin's followers have believed the spirit of Darwinian scientific naturalism to imply a completely nontheistic account of the universe. "Science" has, therefore, seemed to stand in opposition to belief in a personal, purposive Creator. Michael Denton is probably at least close to the truth when he says that "the decline in religious belief can probably be attributed more to the propagation and advocacy by the intellectual and scientific community of the Darwinian version of evolution than to any other single factor" (E, 66).

Agreement on the central role played by Darwinian evolutionary theory in the genesis of late modern atheism can be found in innumerable authors. For example, having said that the most important event in the "demystification of life" was the Darwinian theory of evolution, which "showed how the wonderful capabilities of living things could evolve through natural selection with no out-side plan or guidance," physicist Steven Weinberg adds: "The demystification of life has had a far greater effect on religious sensibilities than has any discovery of physical science" (DFT, 246). Historian of science William Provine, counter-ing the claim that modern science in general and evolutionary theory in particu-lar are not in conflict with religion, says: "Most [evolutionary biologists] are atheists, and many have been driven there by their understanding of the evolu-tionary process and other science" (PE, 68).

Besides being central, the debate about creation and evolution has also been the most heated of the controversies involving apparent conflicts between religion and science. It has been so heated not only because the stakes are so high, but also because two or more of the possible sources of conflict discussed in Chapter 1 are often involved. In the most publicized form of the conflict, that between Darwinian evolutionists and "scientific creationists," *all four* sources are involved: On the one hand, scientific creationists not only advocate ontologi-cal supernaturalism, according to which God can interrupt the ordinary causal processes and create things *ex nihilo* (which results in conflict of the first kind). They also, on the basis of an epistemic supernaturalism according to which the Bible gives an inerrant account of the creation of our world, are led to declare the Earth to be only a few thousand years old and to make innumerable related affirmations about matters of fact that conflict with well-established ideas (thereby creating conflict of the second kind). On the other hand, Darwinian evolutionists not only go beyond the denial of supernaturalism to reject all forms of theistic influence whatsoever and many other beliefs presupposed by religion, including freedom, life after death, and an ultimate meaning to life (thereby creating conflict of the third kind). They also, on the basis of this extremely restrictive naturalism, rather than on the basis of any evidence, make declarations about matters of fact that contradict basic religious beliefs, such as the declaration that the world was not created by an intelligent, purposive Creator, that life in gen-eral and human life in general were in no sense intended, and that our intuitions about morality and ultimate meaning do not reflect anything objective in the universe (thereby creating conflict of the fourth kind). The way in which Dar-winists typically create these conflicts of the third and fourth kinds was illus-trated by Provine's statement quoted in Chapter 2.

Given the total opposition between Darwinian evolutionists and scientific creationists on these matters of ultimate concern, it is no wonder that the antagonism between the two camps has been so extreme. Creationists see evolutionism, along with the scientific naturalism that it both reflects and supports, as a threat to everything good and holy. The belief in "evolutionary naturalism" is often portrayed as the primary source of the widespread moral relativism and sense of meaninglessness in modern society. Evolutionists such as Provine, who are clear-thinking enough to see the implications of Darwinism and brave enough to state them forthrightly, show that the charges of the creationists are not wholly without foundation. The creationists, in any case, sometimes use every possible means to stop the spread of evolutionism—from personal attacks on Darwin, to the exploitation of differences of opinion among evolutionists (sometimes quoted out of context) for the purpose of casting doubt on evolution as such, to the use of the courts to block the teaching of evolutionary science or at least to get "equal time" for creation science. Darwinists, likewise seeing their sense of what is good and holy under attack, sometimes find it difficult to distinguish valid criticisms of the presently dominant theory of evolution from wholesale attacks on evolution and scientific naturalism as such. The dogmatism that typically characterizes the chief spokespersons for the currently dominant worldview is intensified. As Wesson suggests, "Biblical fundamentalism in the United States may be the chief reason that Darwinist fundamentalism is especially strong in this country" (BNS, 20). The reverse may also be true.

However, although the conflict between Darwinism and scientific creationism has been especially intense, the disappearance of scientific creationists would not mean an end to the conflict between science and religion around the issue of evolution. As Provine rightly says, "the conflict between science and religion exists [not] only in the naive, literalist minds of the [scientific] creationists." Rather, the conflict is much more fundamental: Scientific evolutionism as usually understood stands in conflict with any significantly religious view of the world (PE, 70). Harmony between science and religion will only be possible, accordingly, if our culture settles on a position that transcends both supernaturalistic creationism and Darwinian evolutionism.

Fortunately for the prospect of an integral worldview that is satisfactory for both religious and scientific purposes, there is good reason to believe that both of these worldviews are false. In earlier parts of this book, I have given my reasons for rejecting anti-evolutionary creationism. In the middle sections of the present chapter, I provide various reasons for suspecting the falsity of Darwinian evolutionism. In the final two sections, I will show how the Whiteheadian form of theistic naturalism developed in previous chapters provides resources for developing a version of theistic evolutionism that can do justice both to the facts that count in favor of evolution and those that count against the neo-Darwinian theory of it. It is first necessary, of course, to see exactly what this theory is.

Various Dimensions of "Darwinian Evolutionism"

The need, more precisely, is to look at the range of doctrines to which the terms "Darwinism" and "Darwinian evolutionism"—and sometimes simply "evolutionism," understood as the Darwinian doctrine thereof—are applied. Becoming clear about this range of doctrines is especially important because problems are sometimes created simply by the fact that these terms are used in different ways. For example, due to the fact that the terms "evolution" and "evolutionism" are often employed for a wide range of ideas, evidence for (or against) evolution in one sense is often taken to constitute evidence for (or against) evolution in one or more other senses, even though the one may not really entail the other(s). If an examination of the evidence is to be used not simply as propaganda either to attack the idea of evolution as such or to defend the currently dominant theory of it, but instead to see the variety of considerations to which an adequate theory of origins must today do justice, it is necessary to sort out the various dimensions of that comprehensive theory known as "Darwinian evolutionism." Only in this way can we distinguish those dimensions of it, if any, that are well supported from those, if any, that are not. Discussing fourteen dimensions of Darwinian evolutionism, I will distinguish these various dimensions by using subscripted numerals after "Darwinian evolutionism" and, where possible, by employing an identifying adjective or phrase.

Before beginning, it is necessary to discuss the terminological problem of whether to speak of "Darwinian" or "neo-Darwinian" evolution. The currently dominant theory is neo-Darwinism, which involves not only a merger of original Darwinism with Mendelian genetics but also some other modifications. In spite of the fact that some of these modifications are very important, most neo-Darwinists refer to their position as "Darwinism," seeing it as true to the *spirit* of Darwin's own position despite more or less drastic departures from the letter. In spite of the fact that this practice often creates confusion, I will follow it, while indicating which dimensions involve departures from Darwin's own position. With these clarification, I now offer an analysis of "Darwinian evolutionism" in terms of fourteen dimensions.[1]

Darwinian Evolutionism₁

Darwinian Evolutionism$_1$

Microevolution (Darwinian evolutionism$_1$) involves minor genetic and sometimes phenotypical changes within a species (as when a type of bacteria evolves resistance to an antibiotic), or even the transformation of members of one species into a new species in one technical sense of that term. That is, if the term "species" is

---

1. From the perspective of advocates of particular types of neo-Darwinism, some doctrines not mentioned here may seem vitally important. My focus, however, is on points that distinguish neo-Darwinism in general from other positions.

defined as a group of interbreeding organisms, then offspring who are reproductively isolated constitute a new species. Evolutionism even in this limited sense contradicts the idea that all species are *absolutely* fixed. The Darwinian view of this microevolution is that it occurs entirely in terms of natural selection operating on random variations. In any case, Darwinian evolutionism$_1$, sometimes called Darwin's *special* theory of evolution, is now uncontroversial. For example, although the subtitle of Michael Denton's book refers to evolution as "a theory in crisis," he says: "The validation of Darwin's special theory has been one of the major achievements of twentieth-century biology" (E, 86). This validation, he adds, "has inevitably had the effect of enormously enhancing the credibility of his general theory of evolution." To this *general* theory, the (neo-)Darwinian theory of macroevolution—which involves at least the second through the tenth dimensions—we now turn.

Darwinian Evolutionism$_2$

The idea of *macroevolution* is most fundamentally the idea that *all present species of living things have in some way descended from previous species over a very long period of time* (Darwinian evolutionism$_2$). Although as the phrase "in some way" indicates, there is nothing distinctively Darwinian about evolutionism$_2$, Darwin himself said that this doctrine, through which the idea of the *separate creation* of each species was replaced by *descent with modification,* was the main issue, with his distinctive doctrine of natural selection being of secondary importance (Gillespie CD, 130). When evolutionists distinguish the *fact* of evolution from the (Darwinian) *theory* of evolution, it is usually evolutionism$_2$ to which they refer. For example, Stephen Jay Gould has said: "What's not in doubt is the fact of evolution, which is merely the fact of genealogical connection and descent with modification" (GP, 8). Creationists, of course, do not agree that evolutionism$_2$ is a fact. In any case, although evolutionism in this sense is not distinctively Darwinian, it is an essential part of evolutionism in the Darwinian sense. It rules out not only biblicistic creationism, according to which all the species were created at virtually the same time only a few thousand years ago, but also "progressive creationism"—the view, discussed in relation to Alvin Plantinga and Phillip Johnson in Chapter 3, that accepts the current consensus on the dating for the rise of bacteria, eukaryotic cells, multicellular life, mammals, and so on, but maintains that each species was created *ex nihilo.*

Darwinian Evolutionism$_3$

Also not distinctive of Darwinism, but equally essential to it, is the stipulation that, whatever the explanations for macroevolution be, they must be entirely *naturalistic*, at least in the minimal sense of not involving miraculous, supernatural interruptions of the normal causal processes (Darwinian evolutionism₃). This stipulation, which is about the nature of scientific explanations in general, reflects the fact that Darwinism was part of, and soon became the spearhead of, the movement toward *scientific naturalism,* in which all ideas about supernatural interventions would be excluded from the positive content of science. The fact that Darwin's theory clearly embodied this belief was one of the reasons for its rapid endorsement by other members of the movement for scientific naturalism. As George Romanes said at the time:

> [A]ll minds with any instincts of science in their composition have grown to distrust, on merely antecedent grounds, any explanation which embodies a miraculous element. . . . Now, it must be obvious to any mind which has adopted this attitude of thought, that the scientific theory of natural descent is recommended by an overwhelming weight of antecedent presumption. (Lovejoy AOE, 374)

Although I have placed this stipulation, that the scientific explanation of the origin of the species must be naturalistic, after evolutionism₂, this stipulation is really the presupposition of that earlier doctrine, because what "descent with modification" rules out is the separate creation of each species, which would involve supernatural interventions. Whereas I will be arguing that most of the further specifications of "naturalism" discussed below should be rejected, I hold that evolutionism₃, and thereby evolutionism₂, should be accepted.

Darwinian Evolutionism₄

Closely related to the rejection of supernaturalism is the Darwinian stipulation that naturalism must involve *uniformitarianism,* which means that only causal factors operating in the present can be employed (Darwinian evolutionism₄). In Darwin's own mind, this stipulation, which has been one of the most contested dimensions of Darwinian evolutionism, involved two dimensions: ontological uniformitarianism and geological uniformitarianism. *Ontological* uniformitarianism, which rules out (among other things) supernatural divine interventions, has been contested, of course, by supernaturalists. *Geological* uniformitarianism, which rules out occasional catastrophes, has been contested not only by biblical creationists, primarily because of the catastrophic flood reported in the biblical book of Genesis, but also by recent paleontologists who have said that the record does suggest that several catastrophic extinctions have occurred. In Darwin's mind and that of many of his followers in earlier times, these two meanings of uniformitarianism were closely linked, partly because many of those arguing for catastrophes were creationists. In recent times, however, these two issues have become unlinked, as the evidence for catastrophic extinctions has become undeniable and the realization has grown that these catastrophes could have had purely natural causes (such as meteorites and ice ages). Darwinian evolutionism, accordingly, no longer necessarily involves geological uniformitarianism, so evolutionism$_4$ is limited to *ontological* uniformitarianism. Evolutionism in this sense poses no problem for a harmony of science and religion, given a theology based on naturalistic theism.

Darwinian Evolutionism$_5$

A further specification of neo-Darwinism, however, is that biological evolution occurs *without any theistic guidance whatsoever* (Darwinian evolutionism$_5$). Evolutionism in this nontheistic sense rules out not only supernatural interventions, in which the ordinary causal processes of finite beings and events are interrupted, but also *any* kind of directing influence by a purposive power transcending these finite beings and events. Darwinian evolutionism$_5$, in other words, rules out naturalistic as well as supernaturalistic theism. And, as Phillip Johnson sees, this *metaphysical* exclusion of ongoing divine activity is not simply an optional interpretation of the *scientific* doctrine of Darwinism, which seeks to show how all more complex forms of life have come about solely through random variations and natural selection. Rather, "the metaphysical statement is . . . the essential foundation for the scientific claim" (DT, 168). The "scientific doctrine," in other words, does not involve merely a *methodological* atheism, according to which scientists exclude God from their scientific theories without thereby implying that God is not real or at least not presently influential in the world. Rather, the Darwinian theory of macroevolution, according to which it involves nothing but natural selection operating on random variations, is intended as an explanation of how the world got the way it is on the assumption that there has been no divine influence. This is the meaning behind Richard Dawkins' oft-quoted statement that "Darwin made it possible to be an intellectually fulfilled atheist" (BW, 6). That is, although it was logically possible to be an atheist before Darwin, says Dawkins, one would have had no alternative explanation for all those things in the world that, by appearing to be designed, had hitherto been seen as evidence for the existence of a designing deity. What Darwin provided, claims Dawkins, is this alternative explanation.

This dimension of Darwinism evidently accounted for its rapid acceptance in certain circles from the outset. Darwin's theory was enthusiastically welcomed, Brooke says, by those who wanted a science fully autonomous from theology: Whereas this ideal had already been achieved in most other branches, Darwin's theory effected the complete divorce of historical biology from theology (SR, 263, 271). The idea that Darwin's theory has removed the need to posit any divine influence in the formation of our world is widely shared in the scientific community. For example, Heinz Pagels said that "the evidence of life on earth, which promoted the compelling 'argument from design' for a Creator, can be accounted for by evolution" (DR, 158). Steven Weinberg has said that Darwin "showed how the wonderful capabilities of living things could evolve through natural selection with no outside plan or guidance" (DFT, 246).

Darwinian Evolutionism$_6$

The exclusion of theism of every type is based in part on a further dimension of Darwinian evolutionism, its *positivism,* according to which all causes of evolution must be at least potentially verifiable through sensory observations (Darwinian evolutionism$_6$). Darwin's theory of evolution by means of natural selection was part and parcel of a movement for a completely positivistic science, in the sense of one that excludes all explanatory factors not empirically observable—with *empirically* equated with *sensorily.* The doctrine of evolution in this positivistic sense rules out all explanations in terms of divine causation, because such explanations would be neither verifiable nor falsifiable through sensory observations. But Darwinian evolutionism$_6$ goes even further, also ruling out all causal explanations employing any other kinds of forces not sensorily perceptible, such as Alister Hardy's suggestion that unconscious telepathic influence might explain certain features of the evolution of form (LS, 253–61).

The passion with which positivism in this sense is still thought by many to be essential to truly scientific explanations can be illustrated by the response to Rupert Sheldrake's proposal of "a new science of life" based on "the hypothesis of formative causation." This hypothesis, which develops the suggestion made by Hardy in different language, proposes that morphogenesis may be partly explainable in terms of "morphic fields," which, although similar to other fields in some respects, are not physical, as least in the usual sense of the term, because influence is exerted at a distance (NSL). Although morphogenesis is generally acknowledged to be mysterious, thus far inexplicable in conventional terms, the editor of *Nature* magazine angrily denounced Sheldrake's book as "pseudo-science" and as an "infuriating tract," calling it "the best candidate for burning there has been for a long time." Sheldrake's major sin was said to be an interest in "finding a place for magic within scientific discussion."[2]

This insistence on explanation in terms of nothing but causal factors that are observable through the senses, at least in principle, is virtually identical with the insistence on exclusively physical or material causes, in that only such causes

---

2. This editorial, along with responses to it and other responses to Sheldrake's *A New Science of Life*, are contained in the second edition of that book.

are in principle detectable through the physical senses. Therefore, Darwinian evolutionism$_6$, which I have called positivistic evolutionism, can equally well be called *materialistic* evolutionism. Gillespie, in speaking of "the implicit metaphysics of positivism" as "the belief that all events are part of an inviolable web of natural, even material, causation" (CD, 148), reflects the equation by Victorian naturalists of *natural* causation with *material* causation, which implied the equation of *scientific naturalism* with *scientific materialism.*

Darwin's theory was evidently embraced so rapidly by so many fellow scientists not because Darwin had convinced them that natural selection was the sole, or at least the primary, factor, which he did not, but simply because his theory, in being fully naturalistic in this materialistic sense, fit into the program of Victorian naturalism by suggesting that the existence of life and even human culture could be understood in completely materialistic terms. For example, just three pages before the most famous line in John Tyndall's "Belfast Address"— "We claim, and we shall wrest from theology, the entire domain of cosmological theory" (FS, 530)—he said: "The strength of the doctrine of evolution consists, not in an experimental demonstration . . . but in its general harmony with scientific thought" (FS, 527). Tyndall's statement, which indicates that the crucial issue is not empirical evidence but materialistic metaphysics, coheres with Gillespie's description of "the basic positivist assumption that when sufficient natural or physical causes were not known they must nonetheless be assumed to exist to the exclusion of other causes" (CD, 115). (Most of what I know about Tyndall's thought I have learned from Stephen Kim's excellent study [JT].)

Darwinian Evolutionism$_7$

As Gillespie's reference to *sufficient* physical causes indicates, another ideal of Darwinian scientific naturalism, closely related to its materialism, is *predictive determinism.* Indeed, Gillespie argues that "materialism" to Darwin primarily meant the doctrine that the world is, without exception, a deterministic system of causes and effects (CD, 138–41). In any case, Darwin accepted the idea, which had been growing since the eighteenth century, that science requires predictability, which in turn requires determinism, which in turn requires the exclusion of all teleology or final causation in the sense of purposive causation. As Frederick Gregory says, "Scientific explanation had come more and more to mean mechanical explanation, so much so that even reference to 'naturalistic' explanation could be intended to connote the exclusion of final cause" (IDE, 370). As Gillespie puts it: "The essence of the positive science was predictability: caprice had no place in its cosmos" (CD, 152).

Darwin's acceptance of this ideal meant that no exception could be made for human beings, which implied that, in Gillespie's words, "free will . . . fell under his ban" because it would introduce "an element of caprice" (CD, 139).

Indeed, materialism for Darwin meant primarily that human thoughts and decisions, like those of all other animals, are determined by the brain (CD, 139). Determinism, in other words, leads to ontological reductionism, according to which all vertical causation goes upward, so that every "whole" is determined by its parts: The whole as such exerts no self-determined causation back upon its parts. Darwin's acceptance of the idea that any properly scientific theory had to be deterministic was also one of his reasons for excluding any theistic influence in the evolutionary process, for such influence, Gillespie points out, would likewise introduce an element of "caprice" (CD, 120). Darwinian evolutionary theory, accordingly, is completely reductionistic and deterministic (although some Darwinists today reject this feature because of quantum indeterminacy).

Darwinian Evolutionism$_8$

Now, having summarized its metaphysical background, we can turn to the strictly scientific side of Darwinian evolutionism, which in most general terms is the doctrine that *all macroevolution is to be understood entirely in terms of the processes involved in microevolution* (Darwinian evolutionism$_8$). This insistence follows from Darwinian evolutionism$_4$, or ontological uniformitarianism, according to which the only causes that can be posited are those that are presently operating, combined with Darwinian evolutionism$_6$, which restricts these to causes observable through the senses. These dicta imply, in the words of Alexander Mebane, "that lifeforms in the past must have been molded *only* by the forces the breeders of plants and animals observe to be molding them *now*" (DCM, 4).

Darwinian evolutionism$_8$ is at the heart of the controversies between neo-Darwinists and other evolutionists, as the latter do not believe that macroevolution can in principle be explained in terms of the factors sufficient for microevolution, whereas the former do. For an example of Darwinian confidence, we can cite Douglas Futuyma, who declares that "the known mechanisms of evolution [provide] both a sufficient and a necessary explanation for the diversity of life" (EB, 449). Richard Dawkins, even more confidently, says that Darwinism "has no difficulty in explaining every tiny detail" (BW, 302). Some more candid or at least more circumspect Darwinists, however, admit that this has not been *shown*. For example, Walter Bock, who believes that macroevolution can *in principle* be explained in terms of microevolution, says: "One of the major failures of the [neo-Darwinian] synthetic theory has been to provide a detailed and coherent explanation of macroevolution based on the known principles of microevolution" (SE, 20). In any case, we will now look separately at these principles, usually listed as two: random variations and natural selection.

Darwinian Evolutionism₉

As usually characterized, Darwinism is the doctrine that all subsequent species of life, both living and extinct, have come about through evolutionary descent from the first form of life *solely* through natural selection operating upon random variations, understood primarily in terms of random genetic mutations[3] (Darwinian evolutionism₉). Actually, this is the neo-Darwinism worked out in the 1930s and 1940s, sometimes called the "modern synthetic theory" or the "evolutionary synthesis," which synthesizes original Darwinism with Mendelian genetics and makes some other modifications. For example, Darwin himself could speak of random "variations" but did not know about Mendelian "genes" and certainly not about DNA. He also did not limit the causes of evolution to random variations and natural selection, but also allowed for other factors, including the (later forbidden) inheritance of acquired characteristics, insisting only that natural selection was the *primary* factor (Cronin AP, 82–83).

Another complication is that much that passed for "Darwinian evolutionism" from the 1860s to the 1930s disagreed with both Darwin himself and his astute theological critics that—in the words of one such critic, Charles Hodge—"the essential aspect of evolutionary theory was natural selection" (Gregory IDE, 378-79). These non-Darwinian forms of evolutionism, many of which were formulated by reconcilers of theism and evolution, involved various types of purposive causes, which Darwin himself had emphatically excluded (Gregory, IDE, 379, 383; Provine PE, 60). As Provine has suggested, the so-called evolutionary synthesis would be better termed the "evolutionary constriction," because it was

> not so much a synthesis as it was a vast cut-down of variables considered important in the evolutionary process. . . . What was new in this conception of evolution was not the individual variables . . . but the idea that evolution depended on relatively so few of them. (PE, 61)

The new agreement, in particular, was that "purposive forces played no role at all." This constriction, which for the first time limited the variables solely to random variations and natural selection, thereby excluding all purposive forces, rendered untenable, Provine stresses, the idea that evolution could be compatible with theistic religion (PE, 62). This constriction brought Darwinism back to what Charles Hodge called it in 1874: In a book entitled *What is Darwinism?*, Hodge said that it is evolution based on natural selection resulting from *unintel-*

---

3. Although some authors write as if genetic mutations were the only source of random variations, such variations can also result from sexual recombination.

ligent causes (Gregory IDE, 376). The central fact about Darwinian evolutionism₉, accordingly, is that it is *nonpurposive* in every possible sense, reflecting neither the purposes of a universal creator nor those of the local creatures.

The main element in Darwinism that had opened the way for the addition of purposive influences was the idea of random (or chance) variations. To understand this Darwinian concept, it is first important to understand what it does *not* mean. Many writers have been confused by the language into assuming that Darwinism, by saying that mutations happen "randomly" or "by chance," thereby implies that evolution is not fully determined. Holmes Rolston, for example, assumes that mutations are random in the sense of being "without necessary and sufficient causal conditions," thereby completely contingent (SR, 104, 91). This view, if true, would mean that Darwin's theory fully embodied the "caprice" he regarded as incompatible with science. But, as Gillespie points out, in speaking of "chance," Darwin often meant, as he himself said, only "our ignorance of the cause" (CD, 55). Indeed, Darwin explicitly said that "the variations of each creature are determined by fixed and immutable laws" (VAP, 248–49). "At no time," Gillespie adds, "did he see the world as the result of fortuity in a metaphysical sense" (CD, 56). Later Darwinists have often been explicit about this point. In an essay titled "Chance and Creativity in Evolution," Theodosius Dobzhansky denied that there is any "principle of spontaneity inherent in living nature." To say that mutations are "chance events" is not to say that they are uncaused, he adds, but only that we are ignorant of the causes (CCE, 313–14, 329). Dawkins says: "Mutations are caused by definite physical events; they don't just spontaneously happen" (BW, 306). Niles Eldredge likewise says: "Mutations have definitive, deterministic causes" (RD, 133).

What Darwinists do mean by calling mutations random is, in the first place, that *they are not biased in favor of the adaptation of the organism to its environment*. In the words of Eldredge, "mutations are random with respect to the needs of the organisms in which they occur" (RD, 133). As Stephen Jay Gould puts it, genetic variation is "not preferentially directed towards advantageous features" (MM, 325). We can call randomness in this sense "randomness$_{na}$" (for "not advantageous" or "not adaptational"). This doctrine rules out, among other things, the idea that the organism might, in terms of its desire or need to adapt to its environment, influence its genes to mutate in a particular direction. There can be, in other words, no need-induced mutations.

This doctrine that mutations are random$_{na}$ is important to Darwinists for several reasons: The idea that the organism's purposes could influence evolution would contradict the ideal of making biology a purely mechanistic, deterministic science. Also, the idea that purposes could give a bias to genetic mechanisms seems impossible to most Darwinists. (Dawkins, for example, says that "nobody has ever come close to suggesting any means by which this bias could come about" [BW, 312].) And, perhaps most important, the idea that variation is somehow directed toward adaptation would reduce the importance of the central

Darwinian conception, natural selection. The "essence of Darwinism," says Gould, "is the creativity of natural selection" (ESD, 44). That is, natural selection creates new organs and species by scrutinizing all variations, the overwhelming majority of which are deleterious, selecting out the very rare one that just happens to give the organism an edge over its rivals in the competition for survival. However, if variations themselves were directed towards adaptation, says Gould, natural selection would be unnecessary or would at least play a less central role (ESD, 44; HT, 138). Dawkins, expressing Darwinian orthodoxy emphatically, says: "It is selection, and only selection, that directs evolution in directions that are non-random with respect to advantage" (BW, 312).

The desire to regard natural selection as virtually the sole creator of all living forms leads to the tendency, in fact, to insist that mutations are random in an even stronger sense. Gould, for example, sometimes says that variation is random in the sense of being "undirected," of "aris[ing] in all directions," of having "no determined orientation" (PT, 79; HT, 138, 334). Randomness in this sense, which we can call "randomness$_{eps}$" (for "every possible sense"), says that besides not being directed towards adaptation to the immediate environment, mutations are also random in every other possible sense of the term (except, perhaps, in the sense of not being fully determined by antecedent causes). This doctrine rules out, for example, the idea that mutations might be biased toward the production of beauty (even if this beauty conferred no selectionist advantage). This stronger meaning of randomness is intended, of course, to rule out the idea of any form of theistic directivity, which, besides violating the ban on nonmaterial and purposive causes, would reduce the role of natural selection. Gillespie, expressing Darwin's perspective, says that "the assumption of an active and directing deity made redundant nonsense of natural selection" (CD, 210). The idea that variations are random$_{eps}$, by contrast, supports the idea that Gould calls the "cardinal tenet" of Darwinism: "that selection is *the* creative force in evolution" (HT, 138). This tenet is simply the reverse side of what Gould elsewhere calls "the central Darwinian notion," which is that evolution is completely "undirected" (PT, 38).

One problem with all this, however, is that it is pure speculation. Neither Darwin nor anyone else has had any way of knowing that all mutations are random$_{na}$, let alone random$_{eps}$. Darwin was clear that he did not know the cause of mutations, saying that "variations in the domestic and wild conditions are due to unknown causes." And he asked: If God does not, then "what the devil determines each particular variation?" (Gillespie CD, 120). The admission that the ultimate causes of variability were unknown opened the door to advocates of an evolution influenced by purposes, creaturely and/or divine (CD, 121). The architects of the "evolutionary constriction" known as neo-Darwinism meant to close that door. But, of course, they no more than Darwin himself knew the ultimate causes of genetic mutations. We do know that some mutations are caused by cosmic rays; but we do not thereby know that *all* mutations are due to these or analogous causes. Many neo-Darwinians, nevertheless, express great confidence in the truth of this speculation—a confidence that, in light of the number of confidently held ideas that have in the past turned out to be false, is somewhat awe-inspiring. For example, Jacques Monod, arguing that random mutations "constitute the *only* possible source of modifications in the genetic text," so that "chance *alone* is at the source of every innovation, of all creation in the biosphere," added in an oft-quoted statement (CN, 112–13):

> Pure chance, absolutely free but blind, at the very root of the stupendous edifice of evolution: this central concept of modern biology is no longer one among other possible or even conceivable hypotheses. It is today the *sole* conceivable hypothesis. . . . And nothing warrants the supposition—or the hope—that on this score our position is likely ever to be revised.

In any case, Darwinian evolutionism$_9$, by insisting that all evolution is due to natural selection acting on random variations, is fundamentally the doctrine that our world, at least since the rise of life, has been created by impersonal, nonpurposive forces.

The other most controversial factor excluded by the evolutionary constriction known as neo-Darwinism is "the inheritance of acquired characteristics." Although many neo-Darwinists are embarrassed by this fact, Darwin himself accepted this factor, as had Lamarck before him and many others. But, although it was only one of many factors in Lamarck's theory (the idea of need-induced mutations was more central [Gould PT, 77]), it was the main element in the neo-Lamarckism that did battle with Darwinism around the turn of the century. Becoming known as "Lamarckian inheritance," it became the chief heresy, so that evidence for it has seldom been entertained dispassionately. Neo-Darwinists have devoted much more energy to attacking the idea than seriously testing whether there might be an element of truth in it, and they typically portray the idea as thoroughly disproved. Helena Cronin, for example, says: "Everybody knows that the theory was eventually abandoned because acquired characteristics turned out not to be inherited" (AP, 39). But, of course, not "everybody" does know this: Cronin herself remarks on the "puzzling" fact that "Lamarckism has had such an enduring attraction" (AP, 37). The inheritance of acquired characteristics, in any case, is excluded by neo-Darwinism, even if not by the more pluralistic theory of Darwin himself.

Darwinian Evolutionism$_{10}$

Another dimension, which follows from many of the previous dimensions, is that *macroevolution proceeds gradually, through a step-by-step process comprised of tiny steps* (Darwinian evolutionism$_{10}$). As Darwin famously said:

Natural selection acts only by the preservation and accumulation of small inherited modifications, each profitable to the preserved being. . . . [N]atural selection [will] banish the belief of the continued creation of new organic beings, or of any great and sudden modification of their structure. (OS, 100)

Richard Dawkins has recently reaffirmed this approach. Asking how living things, which are "too improbable and too beautifully 'designed' to have come into existence by chance," did come into existence, Dawkins replies:

The answer, Darwin's answer, is by gradual, step-by-step transformations from simple beginnings. . . . Each successive change in the gradual evolutionary process was simple enough, *relative to its predecessor,* to have come into existence by chance. (BW, 43)

This dimension of Darwin's proposal was doubly radical. On the one hand, it went against the traditional typological view, according to which minute changes would result in incoherent, unviable organisms. According to Georges Cuvier (1769–1832), who articulated this typological position most fully, the principle of the correlation and interdependence of parts rendered the evolution of one species into another (Darwinian evolutionism$_2$) improbable. As Brooke summarizes Cuvier's position (which was articulated in opposition to Lamarck, Darwin's predecessor in advocating gradual transformism): "There simply could not be a gradual accumulation of variation in any one part, unless all could change in concert. And that, for Cuvier, was simply too fanciful" (SR, 246). In addition to challenging this traditional theoretical idea, Darwin also had to take on the empirical evidence: The fossil record simply did not support the idea of gradual evolution, because it revealed almost nothing but well-defined types, with few if any intermediate varieties. Darwin handled this problem, if with some discomfort, by claiming that the geological record must be "imperfect to an extreme degree" (OS, 438).

Darwin was warned against this gradualism by several fellow scientists,[4] including his advocate Thomas Huxley. In response to Darwin's acceptance of the dictum "nature does not make jumps" (OS, 181, 191, 256, 435), Huxley wrote: "You have loaded yourself with an unnecessary difficulty in adopting *natura non facit saltum* so unreservedly" (LL, II: 176). Huxley, however, evidently did not understand that, given the various philosophical principles behind Darwin's theory, the difficulty was not "unnecessary." As Wesson says, "Darwin

---

4. In summarizing the five major reviews of the first edition of Darwin's *Origin* by paleontologists, Niles Eldredge says (RD, 67): "Each of them expresses concern that Darwin does not acknowledge the by then well-known generalization that *once a species appears in the fossil record, it tends to persist with little appreciable change throughout the remainder of its existence.*"

insisted on gradualism as the essence of naturalism and the repudiation of divine intervention" (BNS, 38). Gillespie points out that Darwin considered any suggestion of evolution *per saltum* (by jumps) to be a disguised appeal to miraculous creation (CD, 82). The reason why Darwin had to deny jumps is summarized by Howard Gruber: "[N]ature makes no jumps, but God does. . . . [N]ature makes no jumps, therefore if something is found in the world that appears suddenly, its origins must be supernatural" (DM, 125–26). Darwin, recognizing that his whole theory was here at stake, said:

> If it could be demonstrated that any complex organ existed, which could not possibly have been formed by numerous, successive, slight modifications, my theory would absolutely break down. (OS, 171)

Given his principles, Darwin could not "save" his theory by allowing divine insertions here and there to explain the apparent jumps in the record. In response to Charles Lyell's belief that an exception to ontological uniformitarianism *was* required to account for the human mind, Darwin, as mentioned in Chapter 2, wrote:

> If I were convinced that I required such additions to the theory of natural selection, I would reject it as rubbish. . . . I would give nothing for the theory of Natural Selection, if it requires miraculous additions at any one stage of descent. (DLL, II: 6–7)

Having quoted this passage, Dawkins adds:

> This is no petty matter. In Darwin's view, the whole *point* of the theory of evolution by natural selection was that it provided a *non*-miraculous account of the existence of complex adaptations. For Darwin, any evolution that had to be helped over the jumps by God was not evolution at all. . . . In the light of this, it is easy to see why Darwin constantly reiterated the *gradualness* of evolution. (BW, 249)

As we will see in the next section, it is Darwinian evolutionism$_{10}$ that still creates the greatest problems for neo-Darwinism.

Darwinian Evolutionism$_{11}$

A complete explanation of why Darwinian evolutionism implies gradualism involves yet another philosophical principle behind Darwinism, its *nominalism* (Darwinian evolutionism$_{11}$). This philosophical principle is a rejection of the position often called "Platonic realism," according to which forms, archetypes, or ideas—from the Greek *eidos,* meaning form or shape—are really real, being somehow inherent in the nature of things. The traditional view has been that these essences or forms, not being able to exist on their own, "subsist" in the mind of God. Nominalism—from the Latin *nomen,* meaning name—is the doctrine that the names for these forms are *merely* names, not pointing to entities that really exist (or even "subsist") in any sense. Whereas realism with respect to forms, whether in a more Platonic or a more Aristotelian framework, was pervasive throughout the Middle Ages, nominalism has been increasingly prevalent in modern times. In fact, some writers (Weaver IHC) date the beginning of modernity in the fourteenth century with the rise of nominalism (as an aspect of that extremely voluntaristic form of supernaturalism discussed in Chapter 4). Darwinism, in any case, is fully nominalistic, rejecting the realism about forms upon which the typological approach of Cuvier and other traditionalists, such as Linnaeus, was based. For example, Ernst Mayr, a prominent Darwinian theoretician, has said:

> I agree with those who claim that the essentialist philosophies of Plato and Aristotle are incompatible with evolutionary thinking. . . . For the typologist, the type (eidos) is real and the variation an illusion, while for the populationists (evolutionists) the type (average) is an abstraction and only the variation is real. (PSE, 4)[5]

This nominalism is reflected in one of Darwin's oft-quoted statements, in which he contrasts his position with the traditional view that the various taxonomic groups reflect forms in the mind of a creator:

> [T]he Natural System is founded on descent with modification. . . . [T]he characters which naturalists consider as showing true affinity between any two or more species, are those which have been inherited from a common parent, all true classification being genealogical. . . . [C]ommunity of descent is the hidden bond which naturalists have been unconsciously seeking, and not some unknown plan of creation. (OS, 391)

This nominalism provides a link between the gradualism of Darwinism and its rejection of *any* type of theistic influence on the evolutionary process, not simply supernatural interruptions of this process. Darwinists from Darwin himself to Dawkins have usually assumed that the rejection of gradualism in favor of any version of saltationism would necessarily imply supernaturalism. However, if we assumed the real existence of forms or archetypes in the nature of

---

5. Mayr's statement is especially significant in light of the fact that he, as Eldredge points out (RD, 29), has been one of the few prominent neo-Darwinists to regard species as important to evolution.

things, we could suppose that they might serve as "final causes" or "attractors," through which the jump from one coherent type to another could be rendered somewhat intelligible. It is difficult, however, to think of forms as simply existing in the void, as it were, and even more difficult to think of them as exerting any influence, even of an attractive sort. These intuitions lay behind the medieval doctrine that the forms subsist in God, who can give them both a home and efficacy. This medieval God was, to be sure, endowed with supernatural powers, having created the world *ex nihilo*. But the forms could equally well be given a home and efficacy—albeit not such *sudden* efficacy!—by a soul of the universe. A realism about forms and thereby a more saltationist evolutionism, in other words, could be consistent with a *naturalistic* theism. This move is impossible for Darwinism, however, because of its commitment to principles that require the exclusion of all theisms, whether supernaturalistic or not. This exclusion necessitates that Darwinian evolutionism be nominalistic and thereby gradualistic.

Darwinian Evolutionism$_{12}$

If Darwinism rules out theistic influence, naturalistic as well as supernaturalistic, it might seem to follow, as writers such as Dawkins, Gould, and Provine indicate, that it is completely atheistic (Darwinian evolutionism$_{12}$). This is arguably true, in the sense that the principles of Darwinism imply atheism, so that the "spirit of Darwinism" can be said to be atheistic. It is not true, however, that Darwin himself was a complete atheist. Rather, he endorsed, if somewhat waveringly, what is now usually called *deism*. Darwin wrote, for example, of "the laws impressed on matter by the Creator" (OS, 449), saying that "some few organic beings were originally created, which were endowed with a high power of generation, and with the capacity for some slight inheritable variability" (NS, 224). Elsewhere, rejecting the idea of a special creation of each species, Darwin quotes with approval the statement by a "celebrated author and divine" that "it is just as noble a conception of the Deity to believe that He created a few original forms capable of self-development into other and needful forms" (OS, 443). Some interpreters, to be sure, have suggested that Darwin's deism was a passing phase, that in his later years he became fully atheistic or at least agnostic. The implausibility of this contention, however, has been shown by Gillespie (CD, 140–45). For example, in a personal letter in 1881, Darwin wrote that the universe cannot be conceived to be the result of chance, "that is, without design or purpose." And, after reading William Graham's *The Creed of Science* (1881), which defended both evolution and theism, Darwin told the author that this book expressed his "inward conviction . . . that the Universe is not the result of chance." In spite of these and many more statements cited by Gillespie (CD, 142), some interpreters, evidently not wanting to believe that Darwin could have held a view that they consider so superstitious,[6] suggest that none of these statements should be taken literally: Some should be interpreted metaphorically, the rest as self-protective deception. However, as Gillespie points out, this view would require us to believe that Darwin had deceived his best friends and even himself for over twenty years (CD, 134). As indicated by a chapter title in a recent biography by Anthony Desmond and James Moore (D), Darwin was "Never an Atheist."

What *is* arguably true, however, is that Darwin *should* have been an atheist. In this sense, those of his followers who try to portray him as an atheist are right: Given his fundamental philosophical principles, self-consistency would have led to the complete atheism characteristic of later Darwinists. One argument for this conclusion involves the principles involved in Darwin's understanding of the nature of science, especially naturalism, uniformitarianism, and positivistic materialism. Darwin was part of the movement, as we saw, for a scientific naturalism, which meant a separation of science from the supernaturalistic framework within which it had operated from the time of Descartes, Boyle, and Newton. Darwin only achieved, however, a separation of *historical biology* from that framework. Given his published view that the universe, including the Earth, was designed and created by a supernatural intelligence, he left cosmology, physics, chemistry, and early geology within that framework. Indeed, even early historical biology, insofar as it deals with the origin of the first form or forms of life, was not separated from the old framework. Those today who seek to understand the origin of life and even the universe itself in completely nontheistic terms, perhaps by means of some kind of prebiotic natural selection, can be said to be carrying out the naturalistic spirit of Darwinism. A similar conclusion follows from the principle of uniformitarianism, understood ontologically. Darwin protested against Lyell's violation of this principle, and thereby naturalism, with regard to the origin of the human mind. But Darwin himself violated uniformitarianism and thereby naturalism with regard to the origin of the universe, positing at that time a divine agency absolutely different

---

6. A recent example of portraying Darwin as an atheist, while knowing better, is provided in *Darwin's Dangerous Idea* by Daniel Dennett. "Darwin's dangerous idea," Dennett tells us, "is that Design can emerge from mere Order via an algorithmic process that makes no use of pre-existing Mind" (DDI, 83). Placing Darwin in opposition to John Locke, who thought design inconceivable without Mind (83), Dennett says that Darwin's "new and wonderful way of thinking" completely overturned Locke's "Mind-first way" (65–66). The idea "that evolution is a mindless, purposeless, algorithmic process," Dennett says, "permitted Darwin to overthrow Locke's Mind-first vision" (320). The basic rhetoric of the Dennett's book, therefore, depends on the idea that Darwin was really an atheist. However, in a few passages (67, 149–50, 164, 180), Dennett mentions the fact that Darwin himself believed in divine design as the basis of the laws of nature behind the evolutionary process. A more honest title, therefore, would have been "Neo-Darwinism's Dangerous Idea." (Whereas Dennett, disliking deism, plays down its role in Darwin's thought, Michael Corey, who loves it, emphasizes the deistic roots of Darwinism, provocatively calling his book on deistic evolution *Back to Darwin*.)

from anything occurring in the world today. The contention that Darwin should have been an atheist follows also from the principle of positivistic materialism, according to which science is to appeal only to causes that can in principle be verified by means of the physical senses. In speaking of an intelligent First Cause of the universe, Darwin obviously violated this positivistic ideal. Unless he had modified some of his philosophical principles, Darwin could have achieved a self-consistent position only by moving on to complete atheism.

A second argument for the view that a consistent Darwinism is atheistic involves the problem of evil combined with Darwin's determinism. Regarding evolution and evil, Gavin de Beer says:

> Darwin did two things: he showed that evolution was a fact contradicting scriptural legends of creation and that its cause, natural selection, was automatic with no room for divine guidance or design. Furthermore, if there had been a design, it must have been very maleficent to cause all the suffering and pain that befell animals and men. (E, 23)

This statement is misleading insofar as it implies that Darwin himself, partly on the basis of the problem of evil, was led to reject belief in a designing Creator altogether. It is true, however, that this problem led him close to that view, or alternatively to the view that the design was "very maleficent." The world's evil bothered Darwin intensely. In addition to the kinds of evil that have made it difficult for many others to maintain belief in Omnipotent Goodness, Darwin's profession forced upon his attention what he described as "the clumsy, wasteful, blundering, low, and horribly cruel works of nature" (Gillespie CD, 126). As Darwin famously wrote to Asa Gray:

> I cannot persuade myself that a beneficent and omnipotent God would have designedly created the Ichneumonidae with the express intention of their feeding within the living bodies of Caterpillars, or that a cat should play with mice. (MLD, I: 194)

Darwin's solution—his theodicy—involved deism plus natural selection. That is, God's activity was removed from any direct activity with the often cruel and grotesque contrivances of nature. God was directly responsible only for the general structure of the universe, the laws of nature, and the first form or forms of life, with all the details left to random variations and natural selection. On this basis, Darwin said, the contrivances that are "abhorrent to our ideas of fitness" are not morally intolerable:

> We need not marvel at the sting of the bee . . . causing the bee's own death, . . . at the instinctive hatred of the queen bee for her own fertile

daughters; at the ichneumonidae feeding within the living bodies of cater-pillars; or at other such cases. (OS, 436)

This argument is problematic, however, when it is taken in conjunction with the principle of determinism. As we saw, Darwin was convinced that science requires complete predictability and thereby a completely deterministic world. Pushing all of God's creativity activity back to the beginning of things does nothing, therefore, to save the divine goodness: The intelligent First Cause, if truly intelligent, would have known all the consequences that were to follow from the first forms of life interacting on the basis of the laws of nature that had been "impressed on matter by the Creator."[7] As Dov Ospovat said: "An omniscient creator could of course have known beforehand every detail of his evolving universe. But if so, why work through the wasteful and irrational process of chance variations?" (DDT, 276 n.55). Indeed, Darwin himself said at the close of his 1868 study: "An omniscient Creator must have foreseen every consequence which results from the laws imposed by Him" (VAP, II: 515).

From this perspective, we can agree with those who imply that, in rejecting a Divine Creator on the basis of the problem of evil, they are reflecting Darwin's own position. That is, without the belief that the move from theism to deism mitigated the problem of evil, this problem, having led many other sensitive souls to atheism, would probably have done the same for Darwin. In the light of Darwin's own determinism, however, the mitigation was illusory. The fact that Darwin himself did not adopt an atheistic stance, accordingly, does not contravene the conclusion that atheism is implied by the spirit of Darwinism. Of course, the move from deism to atheism has severe consequences, one of which is a completely meaningless universe. But better a world with *no* meaning, many believe, than a world with a *horrible* meaning. Darwin arguably would have agreed. Being sympathetic to that perspective myself, I can only be grateful that I happened onto a philosophical perspective that allows a third alternative. In any case, we can, in agreement with most Darwinists, include atheism among the dimensions of Darwinian evolutionism.

Darwinian Evolutionism$_{13}$

Adding that dimension implies yet another dimension (part of which was just introduced) that was not part of Darwin's own worldview: the idea that the

---

7. This point is, of course, in tension with the contingency of the evolutionary process emphasized by Gould (WL), insofar as this view is attributed to Darwin himself. It is likely that true contingency, which would have been necessary to alleviate the problem of evil, remained in tension with divine foreknowledge in Darwin's mind, as it has in the minds of many other thinkers.

evolutionary process is not only *meaningless* but also *amoral*, in the sense that it provides no moral norms (Darwinian evolutionism$_{13}$). With regard to the first of these two issues: We have seen Provine's statement that evolutionary biology implies that the universe has no ultimate meaning (PE, 70). Gould agrees, saying that, because there is no meaning in nature, we have to create our own (ESD, 13; PT, 83; HT, 93). With regard to the amoral character of the evolutionary process: Provine says that the scientific worldview implies that "there are no inherent moral or ethical laws" (PE, 65). Gould again agrees, saying that "there is no 'natural law' waiting to be discovered 'out there'" (ISJ, 118). This impli-cation is spelled out more fully by Helena Cronin in a discussion of how recent neo-Darwinists have dealt with the problem of altruism:

> [P]erhaps the very notion that there is an objective moral code at all is only an illusion, a belief built into us by natural selection. . . . [I]t could be a belief that corresponds to nothing "out there." It could be no more than a reinforcer, just another of natural selection's tricks for oiling the machin-ery of altruism. (AP, 353)

As should be clear from these statements, to say that Darwinian evolutionism provides no moral norms is not merely to say that one part of the universe—the evolutionary process—provides no basis for morality, which could leave open the possibility that this basis could be sought elsewhere. According to Darwin-ism, at least in its fully atheistic form, the "evolutionary process" encompasses the whole universe: There is nothing beyond it that could provide what it does not. Accordingly, Darwinian evolutionism$_{13}$ entails that, insofar as we have the twofold sense that our lives are finally meaningful and that some ways of living are morally better than others from an ultimate perspective, this sense is illusory. Although these nihilistic beliefs were no part of the Darwinism of Darwin him-self, they are implications of contemporary neo-Darwinism.

Darwinian Evolutionism$_{14}$

A final dimension of Darwinian evolutionism is that it is nonprogressive (Dar-winian evolutionism$_{14}$). There is said to be no general trend behind or within the macroevolutionary process to produce organisms that are "higher" or "better" or "more valuable" than those that came earlier. As a result of the felt need of many "to distinguish the consequences of neo-Darwinian natural selection from the older *progressivist* theories," says Matthew Nitecki, the editor of a recent book entitled *Evolutionary Progress*, "the concept of progress has been all but banned from evolutionary biology" (EP, viii).

Gould is one of the neo-Darwinists who has especially rejected the idea of evolutionary progress, calling it "noxious" (RIP, 319). Referring to Darwin's own reminder to himself never to speak of "higher" or "lower," Gould says (ESD, 36): "If an amoeba is as well adapted to its environment as we are to ours, who is to say that we are higher creatures?" Darwin's criterion of adaptation, Gould concedes, is "improved fitness," but this means, he contends, only "better designed for an immediate, local environment," not improvement in any "cosmic sense" (ESD, 45). Gould makes clear that it is the distinctively Darwinian notion of evolution that lies behind this denial:

> The essence of Darwinism lies in its claim that natural selection creates the fit. Variation is ubiquitous and random in direction. It supplies the raw material only. Natural selection directs the course of evolutionary change. . . . Evolution by natural selection . . . is not progressive in any cosmic sense. (ESD, 44, 45)

Darwinian evolutionism rules out progress, in other words, because it portrays macroevolution as proceeding solely in terms of random$_{eps}$ variations and natural selection, which together could provide no basis for saying that some of the later products are in any sense higher than some of the earlier ones. The sole criterion is adaptability for survival in the local environment, and the extant early products, such as bacteria and amoebae, are, if anything, superior to all the later products in this respect.

The reason why Darwinism must deny evolutionary progress, so that local adaptation becomes by default the sole criterion of success, is brought out clearly by Provine. In his essay on "Progress in Evolution and Meaning in Life," he points out that although some neo-Darwinists, while denying the idea of a cosmic purpose, have tried to hold on to an idea of progress, this attempt was self-contradictory: "The difficult trick was to have the progress without the purpose. . . . The problem is that there is no ultimate basis in the evolutionary process from which to judge true progress" (PE, 63).

In an essay asking, "Can 'Progress' be Defined as a Biological Concept?", Francisco Ayala makes the same point, that a criterion for progress depends upon a standard not provided by Darwinian evolutionism. The first part of Ayala's argument clearly brings out the twofold nature of the idea of progress:

> The concept of progress contains two elements: one *descriptive,* that directional change has occurred; the other *axiological* (=evaluative), that the change represents betterment or improvement. The notion of progress requires that a value judgment be made about what is better and what is worse, or what is higher and what is lower, according to some axiological standard. (CP, 78)

The second part of Ayala's argument is that, because the notion of progress is axiological, "it cannot be a strictly scientific term," because "value judgments are not part and parcel of scientific discourse" (CP, 81).

Unlike many of his fellow neo-Darwinists, however, Ayala does believe that, if one carefully qualifies the term, speaking only of *particular* and *net* progress, not general and uniform progress (CP, 80–81), one can come up with a meaningful standard of progress, such as "the ability of an organism to obtain and process information about the environment" (CP, 90). Given this standard,

> animals are more advanced that plants, vertebrates are more advanced than invertebrates; mammals are more advanced than reptiles, which are more advanced than fish. The most advanced organism by this criterion is doubtless the human species. (CP, 92)

However, Ayala is quick to add, "there is nothing in the evolutionary process which makes the criterion of progress I have just followed best or more objective than others" (CP, 95). What he means, of course, is that there is nothing in the evolutionary process *as interpreted through Darwinian spectacles* that makes this criterion more objective than any others.

As Provine points out, this dimension of neo-Darwinism is closely related to the previous one: The denial that a cosmic purpose is being manifested in the evolutionary process not only implies that the process is meaningless, but also entails that it can provide no axiological norms for declaring progress to have occurred. The fact that Darwin himself was not fully atheistic would lead us to expect that he would not have denied progress in such a thoroughgoing way. The truth, in fact, is that progress was central to Darwin's own theory. Of course, many neo-Darwinists, anxious to show, as Robert Richards says, that their own views are sanctioned by those of the master, claim that Darwin himself did not think of evolution as progressive (MF, 146). But, as commentators more concerned with historical than neo-Darwinian correctness have shown, these claims involve misreadings. In a chapter on "Natural Selection and 'Natural Improvement,'" Ospovat shows that "Darwin never seriously doubted that progress has been the general rule in the history of life" (DDT, 212). Richards agrees (MF, 146), thereby rejecting Gould's claim, cited above, that for Darwin improvement meant only "better designed for an immediate, local environment." Indeed, Richards says, far from there being a conflict in Darwin's mind between progress and the theory of natural selection, "Darwin crafted natural selection as an instrument to manufacture biological progress and moral perfection" (MF, 131). That this was not merely Darwin's early view, later to be discarded, is shown by the fact that in the final paragraph of *The Origin of Species*, Darwin said: "Thus, from the war of nature . . . the most exalted object which we are capable of conceiving, namely, the production of the higher animals, directly follows." Ospovat, citing the passage from Gould cited above, says: "Stephen Gould has made much of Darwin's vow never to use the terms 'higher' and 'lower.' But Darwin consistently refused to adhere to his rule" (DDT, 227).

Ospovat, agreeing with John Greene's characterization of Darwin as an "evolutionary deist," shows that this conception of the universe lay behind Darwin's belief in progress (DDT, 72). In Richards' words, Darwin's belief in evolutionary progress was "a direct consequence of Darwin's . . . regarding natural selection to be a secondary cause responsive to the primary cause of divine wisdom" (MF, 142). Because he believed that the evolutionary process reflected divine wisdom and purpose, Darwin could believe that there was an objective standard in terms of which to speak of progress from lower to higher forms of life. Although Darwin did, on the basis of his theory of natural selection, craft a distinctive conception of highness, which he called "competitive highness," he also said, Ospovat points out, that this kind of highness will in the long run result in an organization that is "higher in every sense of the term" (DDT, 228). In his autobiography, Darwin included the existence of human beings, with their distinctive capacities, as a reason for believing in divine purpose, saying that it is impossible to conceive "this immense and wonderful universe, including man with his capacity of looking far backwards and far into futurity, as the result of blind chance or necessity" (A, 92). Although Darwin did give up his early theological view that the *details* of the world reflected a divine plan, he accepted until the end the belief that results of a general nature were preordained by the "general laws" imposed by the Creator. Therefore, although the fact that the details were left to chance meant that humans as we know them were not intended (at least insofar as Darwin ignored the fact that his predictive determinism left no room for "chance" in the metaphysical sense), Darwin could still believe, as Ospovat shows, that beings with moral and intellectual qualities were intended (DDT, 72–73, 226).

We see, accordingly, that it is not Darwin's biological theory as such that rules out the possibility of speaking of evolutionary progress, only this theory when placed within the context of a completely nontheistic framework. As stated above, nevertheless, neo-Darwinists are right insofar as they believe that the *spirit* of Darwinism, given the general ideals about science that Darwin accepted, should lead to a completely atheistic view of evolution. We can, therefore, add the denial of progress to the various dimensions of Darwinian evolutionism.

PHILOSOPHICAL PROBLEMS OF DARWINIAN EVOLUTIONISM

Having clarified what Darwinian evolutionism is, we are in position to examine the various reasons for considering it inadequate. These reasons are of two basic sorts: Just as Darwinian evolutionism itself is a mixture of general philosophical convictions and a more strictly scientific theory, so the reasons to question its adequacy can be divided into more philosophical and more strictly scientific issues. I will in this section discuss the more philosophical reasons to question

Too much like Phillip Johnson

contemporary Darwinism, saving the more scientific reasons for the next section. Because many of the points involve problems inherent in neo-Darwinism, as distinct from Darwin's own position, much of the subsequent discussion refers specifically to neo-Darwinism and neo-Darwinists.

A significant part of the argument in favor of Darwinism by its advocates from the outset, as we have seen, was philosophical. The argument, roughly, has been that, because we need a materialistic theory of evolution and Darwinism is the only or at least the best materialistic theory, our "antecedent presumption," as Romanes put it, should be that it is true. This type of argument can be turned against neo-Darwinism, to wit: If materialism has proved inadequate for other issues, then our antecedent presumption should be that a fully materialistic theory of evolution will probably be inadequate. We saw, for example, that materialism is inadequate for the mind-body relation. This inadequacy cannot, in fact, be considered external to the problem of evolution. As Wesson says: "Evolutionary theory cannot be deemed complete unless it can give an accounting of the mind and the behaviors it has created" (BNS, 278). This inadequacy with regard to the mind-body relation is, furthermore, simply the most obvious part of the fact that materialism provides an inadequate framework for science in general. In Wesson's words:

> The faith that all things can be attributed to analyzable material causation is, in the end, only a faith like more candid faiths. The contention that reality consists of only material particles and their modes of interaction is not even a clear-cut theory. . . . But are the laws of nature not real? Are mathematical theorems real? Are patterns real? Are thought and consciousness? It is paradoxical to deny their essentiality, for science could not exist without them. (BNS, 4)

In saying that it is paradoxical for scientists to deny the reality of those non-physical things called "laws," "patterns," and "conscious thoughts," because science itself presupposes their reality, Wesson is making the same point I have made about "hard-core commonsense notions," that it is self-contradictory to deny in theory what we inevitably presuppose in practice.

This argument from the inevitable presuppositions of practice counts against several of the philosophical doctrines on which Darwinism is based. Besides materialism, there is its epistemic corollary, *positivism,* according to which science should not affirm the reality of anything not knowable through sensory perception. The most obvious reason for rejecting this principle is that science necessarily presupposes the reality of logical and mathematical norms, which are not objects of sense-perception. Close behind in obviousness is the fact that scientists necessarily presuppose the reality and importance of that (nonmaterial) ideal known as "truth." Also, in appealing to norms such as simplicity and elegance, scientists presuppose the reality and importance of that ideal traditionally called "beauty." This hard-core commonsense argument against positivism is also undergirded, as we saw in Chapter 7, by the evidence for extrasensory perception provided by parapsychology.

Closely related is the conviction, which neo-Darwinists have felt constrained to maintain, that the evolutionary process is entirely *amoral,* providing no basis for moral norms. The difficulties that they have experienced in trying to hold consistently to this implication of neo-Darwinism suggest that a hard-core commonsense notion is at issue here as well. Given the traditional trinity of normative values—Truth, Beauty, and Goodness—we would expect Goodness to have the same status in the nature of things as the other two. If we presuppose the reality of Truth and Beauty as normative ideals, and if we can understand why we do so by recognizing that we have a nonsensory capacity for apprehending them, then we can suppose that the same is true for moral norms. The fact that we cannot help presupposing the objectivity of moral norms—that some ways of being and acting are inherently better than other ways—provides another anteced- ent presumption against the positivism upon which Darwinism is based.

This problem is usually discussed in relation to the issue of "altruism," which is introduced by Helena Cronin in these words:

> Man's inhumanity to man may indeed make countless thousands mourn. But it is man's *humanity* that gives Darwinians pause. . . . Human morality . . . presents an obvious challenge to Darwinian theory. (AP, 325)

The problem is that the theory of natural selection entails that all habits must provide a survival advantage for the organisms (or their genes), and it is hard to see how altruism, which involves self-sacrificial behavior, can be so explained. Most neo-Darwinists who have tackled this problem have tried to explain human morality solely in terms of what Cronin calls an illusion built into us by natural selection (AP, 353). However, Darwin himself took morality with utmost seri- ousness, saying that "of all the differences between man and the lower animals, the moral sense or conscience is by far the most important" (DM, I: 70). Given his emphasis on the continuity between humans and other animals, however, Darwin did not regard the difference as absolute. Rather, he regarded the human moral sense as a development out of "social feelings," constituting an incipient moral sense, found in lower animals (Cronin AP, 325–26, 349–53).

The other founder of Darwinism, Alfred Russel Wallace, was even further from contemporary neo-Darwinists, arguing that human morality can in no way be explained in terms of natural selection. Darwin, says Joel Schwartz, was

> aware that his whole concept of "evolution by natural selection" was en- dangered by Wallace's insistence that natural selection was not the only factor in the evolution of man. If one essential part of the theory was denied, the entire theory was called into question. (DW, 288)

[handwritten annotation at top of page: "What G is rejecting is not neo-Darwinism but the conflation of evol. with positivism, monism-reductionism & materialism."]

[handwritten annotation below header: "G needs to recognize that an a priori naturalism is abstracted from value consideration. It's [?] [?] to identify the abstraction with the [?] altogether."]

Cronin, who quotes Schwartz's statement, surprisingly believes that there was no such threat:

> By the time that *Descent* was published, it was very widely agreed—even by non-scientists—that evolution (either wholly or partly by natural selection) could make claim both to our bodies and to some of our mental attributes, even if not to our moral sense. So the Darwinian case for human descent was—rightly—not seen as hingeing upon the issue of morality. Even less was this one point seen as a test case for the "entire theory" of natural selection—again, rightly. (AP, 365)

To allow that some other factor could help account for human morality, however, would be to give up the most distinctive claim of Darwinian evolutionism, the claim that *all* the products of the evolutionary process can be accounted for *solely* in terms of natural selection of random variations. It is evidently Cronin's allowance for an element of pluralism here—which fits with her statement elsewhere that not *all* characteristics must be adaptive (AP, 86–87)—that permits her to embrace Darwinism so wholeheartedly.[8]

In any case, if one is considering neo-Darwinism without loopholes, then the issue must be faced squarely: If we cannot believe that human morality can be explained in terms of selection of random variations, then we have a good reason to suspect that an extra-Darwinian factor, which lies behind the moral sense in humans and perhaps some other animals, has been operative throughout the evolutionary process.

A similar point can be made with regard to the aesthetic sense. Again, there is an important distinction to be made between Darwin himself and most later Darwinists. For Darwin himself, as Cronin brings out, it was extremely important to insist that animals have "a taste for the beautiful," which, he rightly said, did not depend upon intellectual development (Cronin AP, 167). This notion was part and parcel of his theory of *sexual* selection, through which he accounted for many features of the living world, saying, for example, that "a great number of male animals . . . have been rendered beautiful for beauty's sake" (OS, 186). Realizing that the idea would be criticized, he said:

> I fully admit that it is an astonishing fact that the females of many birds and some mammals would be endowed with sufficient taste for what has apparently been effected through sexual selection; and this is even more astonishing in the case of reptiles, fish, and insects. (DM, II: 400)

---

8. This same allowance perhaps explains the fact that Cronin (AP, 378–79) quotes Dawkins' statement about our capacity for cultivating pure altruism (to be cited later) without commenting upon its contradiction with the remainder of his reductionistic program.

Darwin was right, Cronin points out, in suspecting that this idea, which fit with his portrayal of continuity between humans and other animals, would not be popular (AP, 178–79). Whereas Darwin distinguished between sexual selection, based upon aesthetic taste, and natural selection, based on purely utilitarian factors, most later Darwinists have articulated a purely utilitarian natural selectionism (AP, 155–58, 163, 183–90). Even those who today give sexual selection a central role do not do so in terms of Darwin's "anthropomorphic" attribution of an aesthetic sense to other animals—partly, surely, because they cannot attribute an aesthetic sense for the beautiful in a realistic sense even to human beings. The positivist denial of any nonsensory intuitions of nonphysical ideals, such as beauty, rules out an objectivist interpretation of aesthetic experience no less than of moral and religious experience.

Insofar as we cannot accept a purely subjectivist interpretation of all aesthetic experience as purely arbitrary preferences, we have still another reason to reject Darwinian evolutionism, with its positivism. It would seem, in fact, that Darwin's endorsement of positivism was in conflict with his own belief in a sense for the beautiful, perhaps because he did not grasp the philosophical fine point that the sense for the beautiful *as such,* as distinct from the perception of beautiful *things,* involves nonsensory perception. In any case, a theory of evolution that, by rejecting positivism, allows a genuine aesthetic sense to play a role would arguably be in harmony with Darwin's own intuitions.

Perhaps the dimension of Darwinism most clearly contradicted by our hard-core common sense is its *predictive determinism.* Even many materialistic philosophers, as we saw in Chapter 6, admit that we inevitably presuppose the reality of freedom. Although, given their own materialistic, deterministic theories, they could not draw the conclusion that we must really have this freedom, the denial of this conclusion is irrational: If we inevitably presuppose the reality of freedom in practice, including the practice of developing a philosophical or scientific theory, we violate the law of noncontradiction if the content of that theory denies the reality of freedom. A theory of evolution cannot be adequate, therefore, unless it portrays the evolutionary process as producing human beings with genuine freedom. And human freedom, as we saw in Chapter 6, cannot be made intelligible without positing some degree of freedom, in the sense of spontaneity or self-determination, all the way down.

Again, such a view would arguably be in accord with at least one side of Darwin's own Darwinism. His theory of sexual selection, Cronin points out, involved the notion that the animals' "mental capacity sufficed for the exertion of a choice" (AP, 175). The fact that sexual selection involved "will" or "choice" was one reason, in fact, that Darwin distinguished it from natural selection (AP, 231). And this was one reason that many of his more mechanistic followers did not follow him here (AP, 245). For example, Karl Groos in *The Play of Animals*

(1898) gave an account in which he reduced sexual selection to natural selection by ousting the notion of choice, explaining his rationale thus: "The selective principle involved [in Darwin's theory of sexual selection] is not the mechanical law of survival of the fittest, but rather the will of a living, feeling being capable of making a choice" (PA, 230). Only a mechanical principle, in other words, was considered acceptable by Groos. The mechanists were right, and Darwin was (evidently) wrong, to think that the idea of choice is compatible with ontological determinism, which predictive determinism must presuppose. Accordingly, insofar as we believe that the mechanists are wrong and Darwin is right about animals having the capacity, to one degree or another, to exercise choice, we have another reason to reject determinism as an ideal to which a scientific theory of evolution must conform.

Closely related to Darwinism's predictive determinism is its *nominalism,* which assumes the unreality of forms. The connection between the two is that nominalism, by positing the unreality of forms, implies the unreality of formal-final causation (that is, self-determination oriented toward the realization of particular forms). Nominalism thereby supports the all-sufficiency of efficient causation and thereby determinism. We have already seen reasons for rejecting nominalism with respect to mathematical, logical, noetic, aesthetic, and moral forms. And if such forms exist in the nature of things, there is reason to accept the equal reality of the geometrical, temporal, and other types of forms that seem to be manifested in the various species of living things. Indeed, the well-marked character of the prebiotic world, which has been revealed by physics and chemistry, provides good reason to suspect the living world to be equally oriented around distinct forms of existence. The nominalism of Darwinism provides, therefore, yet another reason for an antecedent presumption against its adequacy.

Closely related to nominalism is *atheism.* These two doctrines have, in fact, been mutually supportive: On the one hand, the conviction that there are no forms inherent in the nature of things removes one of the main reasons for affirming the existence of a Divine Reality, namely, the belief that there must be an omnipresent experience in which the various types of forms could subsist. On the other hand, the conviction that there is no such Divine Reality provides a reason for the nominalist denial that the various types of forms immanent in actual things have any reality transcending those particular things.[9] Because of this connection, the realization that nominalism is problematic should lead to a reconsideration of neo-Darwinism's postulate that a completely nontheistic theory

---

9. This mutual support does not mean, of course, that the two doctrines imply each other. On the one hand, the original nominalists were theists. On the other hand, some nontheists have, if problematically, affirmed the objective existence and efficacy of mathematical, logical, moral, and aesthetic forms. The fact remains, however, that this affirmation *is* problematic and that realism about forms can only be made intelligible in a theistic framework.

of evolution can be adequate. In this reconsideration, furthermore, there is no need to be hampered by the assumption that theism means returning to super-natural interventions and an insoluble problem of evil.

In this reconsideration, the question of the antecedent probability that a wholly nontheistic theory of biological evolution will prove adequate should be considered in relation to both prebiotic and postbiotic (or cultural) evolution. With regard to the prebiotic evolution: There has been a definite swing among cosmologists back to the idea that the origin of our universe is inexplicable apart from a cosmic intelligence. For example, astronomer Fred Hoyle, who long preferred the steady-state view of the universe because it seemed less suggestive of theism than the big bang theory, has more recently said that "a supercalculating intellect" must lie behind the basic laws of physics (TU, 12). With regard to human cultural evolution: Insofar as we can judge from the available evidence, human beings of every time and place have been religious. The simplest expla-nation for this fact, materialist and sensationist dogmas aside, is that human beings always and everywhere have been aware, more or less dimly, of a Divine or Holy Reality—which they have, naturally enough, imagined and conceived in terms of their culturally conditioned images, beliefs, and linguistic systems (Griffin RE). If nonsensory prehensions of a Divine Reality have influenced cultural evolution, then, on the basis of nondualism and ontological uniformitarianism, we should suppose that the same is true, to some degree, of evolution in general. The atheism presupposed by the neo-Darwinian theory of biological evolution, therefore, provides another reason for an antecedent presumption against its adequacy.

Closely related to this atheism, as we have seen, is the conviction by neo-Darwinists that they must, given their atheism, maintain that the evolutionary process is completely *meaningless* and *devoid of progress*. Insofar as we find this twofold conclusion of neo-Darwinism counter-intuitive, we have yet another reason for suspecting its premises.

In sum, the fact that neo-Darwinism is a materialistic, positivistic, deter-ministic, nominalistic, atheistic theory of evolution not only gives us no anteced-ent reason to suppose it, on its more scientific side, to be true, but it also gives us very strong antecedent reasons to suspect it probably to be false.

EMPIRICAL DIFFICULTIES OF NEO-DARWINISM AS A SCIENTIFIC THEORY

This *a priori* suspicion, furthermore, is supported by a range of empirical and conceptual difficulties faced by the more strictly scientific side of neo-Darwin-ism as a theory of macroevolution, namely, Darwinian evolutionism$_{8-10}$. With regard to the widespread belief that this theory "was all but proved one hundred years ago and that all subsequent biological research . . . has provided ever-in-creasing evidence for Darwinian ideas," Denton says: "Nothing could be further from the truth" (E, 77). Denton, Wesson, and others have devoted entire books simply to discussing various scientific factors that count against the adequacy of the contention that macroevolution can be fully explained in terms of natural selection operating on random variations.[10] In this relatively brief discussion of empirical issues, I can summarize only two of the major issues: evidence for need-induced inheritable variations and the clash between gradualism and the fossil record.

Evidence for Need-Induced Inheritable Variations

Darwinian evolutionism$_9$, in stipulating that all inheritable variations are random$_{na}$, rules out the possibility that some variations may be evoked by the organism's need to adapt to its environment. Darwin himself did not rule out this source of change, saying at the close of the introduction to the first edition of the *Origin*: "I am convinced that natural selection has been the main but not the exclusive means of modification." And, far from this being an early idea that Darwin rejected on the basis of further study, it became more pronounced in later years. In the last (sixth) edition, he reminded readers of that earlier statement (OS, 442). And, after saying that evolution has taken place "by the preservation or the natural selection of many successive slight favourable variations," he added: "aided in an important manner by the inherited effects of the use and disuse of parts; and in an unimportant manner . . . by the direct action of external conditions" (OS, 442). In *Varieties of Animals and Plants under Domestication,* published in 1868, Darwin gave several examples of what later came to be called "Lamarckian inheritance."

The idea that there was something un-Darwinian, even perverse, in the idea that inheritable changes could be induced by the behavior of the organism, perhaps in response to external conditions, was due primarily to attacks on the idea led by August Weismann beginning in the 1880s. These attacks provoked an advocacy of use-inheritance by some scientists who, although they did not adopt all or even the most central ideas of Lamarck himself, came to be called neo-Lamarckians (Cronin AP, 36–37; Bowler E, 58–140). In what Cronin calls Weismann's "most famous and influential experiments" (AP, 39), he cut off the tails of mice, pointing triumphantly to the fact that the mutilations were not inherited. This experiment, how-

---

10. My use of books by Denton, Wesson, and other critics of Darwinism, incidentally, does not imply an endorsement of all their criticisms. For example, at least most of the criticisms of the efficacy of natural selection based upon apparently maladaptive developments, such as the tail of the peacock and the antlers of the "Irish Elk," seem to ignore the power of sexual selection, which was a central part of Darwin's own theory and is now, after a long period of neglect, a central part of neo-Darwinism (Cronin AP, 113–249).

ever, was irrelevant to what the advocates of so-called Lamarckian inheritance were urging, because the "acquired characteristic" (the shortened tail) had in no way been brought about internally by the mice themselves.

Although Darwinists typically allege that the issue has been settled, it clearly has not. There are several reasons why some thinkers believe Lamarckian inheritance to occur, or at least to be worthy of further (and more serious) tests. I will here largely follow Wesson's discussion. A first argument in favor of Lamarckian inheritance is a purely negative one: It has never been disproved. Besides the fact that the experiments of Weismann were irrelevant, it is true more generally, Wesson says, that the "impossibility of input from environment to the genome has never been proved" (BNS, 227).

A second consideration concerns the "Weismann barrier" preventing any information from passing from the somatic cells to the germ or reproductive cells. The information is said to go only in the other direction, from reproductive cells to somatic cells. Inheritance of acquired characteristics is impossible because information about changes that have occurred in the body during the individual's lifetime cannot get transmitted to the reproductive cells. Weismann based this inviolability of the reproductive cells on the claim that "reproductive cells are already set apart in the early embryo, and they continue their segregated existence into maturity." However, Wesson claims, this segregation occurs only in some animals: It "does not occur in plants and a large majority of animal phyla." The "Weismann barrier," in other words, is "mostly nonexistent" (BNS, 10). Although this statement may be too strong, there seems to be a growing acknowledgment that the barrier is not absolute.

More recently, this idea that information goes only in the one direction—from DNA to protein, never from protein to DNA—has been called the "central dogma" of molecular biology.[11] The Weismann barrier and this central dogma have been part of a more general argument against Lamarckian inheritance, which is that there is no imaginable mechanism or route through which the environment or the organism itself could induce changes in the genome. This argument from our ignorance has surely never been a strong one. And today it is even weaker, because new knowledge and ideas have made the idea of such changes intuitively less implausible. For example, a well-known study by Howard Temin showed that viruses could carry genetic material into host cells, embedding it in the host's DNA (Taylor GEM, 51). Also, says Wesson, molecular biology "has raised doubts that information can flow in only one direction" (BNS, 227). One of the arguments Wesson gives is based on the fact that cells have proved to be enormously sensitive.

---

11. Although the term "dogma" now usually suggests an idea that cannot be doubted, Francis Crick, one of the authors of this idea, has reportedly said that he thought the word "meant a hypothesis, some arbitrary thing which was laid down for no particularly good reason" (Shapiro OSG, 291). We should not be surprised, therefore, to find exceptions.

Cells have to respond with great accuracy and sensitivity to instructions to become just what is needed at the right time in the correct relation to other specializing cells. . . . Not only do cells receive a set of instructions telling them how to build the new organism; throughout life, instructions change. . . . Dysentery-causing amoebae change from benign to malignant at least partly because of diet. In the immune system . . . , lymphocytes, in reaction to an indefinite variety of antigens, use a limited amount of genetic information to produce appropriate immunoglobins . . . [T]rypano-somes, the protozoa causing African sleeping sickness, alter their protein coat as rapidly as their host develops antibodies. . . . Cancer cells unfortunately raise their resistance to chemotherapy by gene amplification. (BNS, 228–29)

"It is no great leap," concludes Wesson, "for cells to react genetically to signals from the environment" (BNS, 228).

Another reason why hereditary changes involving responses to the environment should no longer be considered *a priori* unlikely is based on a new understanding of the functioning of the genome. Genetics has been transformed, Wesson points out, "by the discovery that the great majority of genes are regulatory. Clearly, a mutation in a regulatory gene could bring about a substantial change in the appearance of an organism at one blow. No longer is there any need to wait for the simultaneous appearance of several different structural mutations" (BNS, 175–76). Regulatory genes, according to this new view, determine which other genes are to be activated at a given time, which to be repressed.

The differentiation of cells in the body . . . means the activation of one or another part of the whole genome . . . , not the formation of new genes. However, the faculty of giving expression to certain genes and not to others in response to external cues is not very different from a faculty for changing heredity. Only a small fraction of DNA . . . is actually used. . . . [S]ome of it may well constitute a reservoir of potential changes. There is a huge storehouse of potential changes on which the body could draw without producing any new structural genes. (BNS, 231)

Wesson's point, in other words, is that Lamarckian, need-based changes would not necessarily involve a need-based mutation in the old sense. The inheritable change might result merely from the activation of previously dormant genes.

Besides these reasons not to consider Lamarckian changes impossible, there are empirical reasons to suspect that they actually occur. One of these is evidence supporting the old argument from disuse, summarized by Wesson: "Animals seem to lose unneeded structures and instincts much more rapidly than can be accounted for by an selective advantage of not having them. . . . Insects, crustaceans, salamanders, and so forth lose eyes in caves, in some cases rather quickly" (BNS, 232). Darwinists, incidentally, usually explain the loss of unused eyes in terms of economy: Not needing to expend energy on the eye gives a selective advantage. However, this argument, besides being weak in itself, is evidently in tension with various facts. For example, with regard to moles, whose underground habit led to their loss of eyes, Gordon Rattray Taylor says that "the presence of even fully functional eyes could hardly handicap the . . . mole, which does occasionally surface. The muscle which fills up the eye-socket actually takes more energy to maintain than the eye itself" (GEM, 41).

The above considerations are, of course, merely suggestive. The issue cannot be settled apart from extensive experimentation. Although there have not yet been enough experiments with positive results to claim definitively that need-based inheritable variations occur, some of the early experiments have evidently been promising. For example, Frederick Griffiths placed rats on slowly revolving turntables for many months. "When the wretched animals were freed," reports Taylor, "their heads constantly flicked in the direction in which they had been rotated, and their eyes flicked also. This flicking automatism reappeared in their progeny" (GEM, 49). Taylor also reports that Alan Durrant, by using fertilizers, made some flax plants heavier and larger, others lighter and smaller, and that the trends persisted through several generations of offspring (GEM, 49).

More recently, Harvard biologist John Cairns and colleagues, working with a strain of *E.* coli unable to digest lactose, found that these bacteria became able to do so when lactose was the only available nutrient, and that this was done not by restoring a lost gene but by bringing forth another one (OM, 142–45). In a letter written in response entitled "Is Bacterial Evolution Random or Selective?", Spencer Benson reported on experiments in which, when bacteria were placed in a medium with nutrient molecules too large to pass through the pores of its membrane, a mutation resulted in the membrane's becoming more permeable (IBE, 21–22). Shortly thereafter, Barry Hall, in an article entitled "Adaptive Evolution that Requires Spontaneous Mutations," reported that E. coli, when placed in a solution of salicin, which it could not metabolize, underwent two otherwise rare mutations together, after which it was able to metabolize the salicin. Another experiment involved a strain of E. coli unable to synthesize the amino acid tryptophan. The bacteria were given tryptophan for a few days, then deprived of it. The number mutating so as to produce tryptophan jumped as much as thirtyfold. Control bacteria, starved on different amino acids, did not undergo the change needed to synthesize tryptophan (AE, 887–97). John Endler and Tracy McLellan, on the basis of experiments showing that a bacterial transposon responds to an antibiotic so as to generate resistance, conclude that "the direction of evolution may be influenced by the environment not only via natural selection but by appropriately directed mutations" (PE, 413).

Another experimental fact that undermines the notion that mutations are random in every possible sense is that, when a mutation is needed for it to adapt to its environment, an organism's mutations become more frequent. It is as if the organism knows that *something* new is needed, even if it does not know exactly what. Wesson concludes: "If it is true that mutations are much more frequent where they are needed than where they are virtually certain to be harmful, they cannot be held to be random" (BNS, 239).

To summarize: If need-based mutations, which now should be understood to include activations of previously dormant genes, would be confirmed, this development would be important in several respects. Most obviously, by showing that there *are* factors involved other than natural selection of random variations, it would demonstrate the "evolutionary constriction" known as neo-Darwinism to be false.[12] It would also provide a way to understand how purposes can directly bring about structural changes. More generally, it would suggest that all living things, rather than being passive results of forces acting on them, are self-determining organisms,[13] helping to shape their own destinies and that of the future in general. Insofar as we cannot help thinking of ourselves in these terms, it would thereby overcome the dualism between humans and the rest of life, therefore between cultural and biological evolution. Also, it could help explain how evolution might have proceeded in a much more radically saltational manner than conceivable within current orthodoxy (which is to be discussed next). In all these ways, it would show that evolutionary theory needs a philosophical framework larger than the mechanistic, reductionistic one provided by neo-Darwinism.[14]

---

12. John A. Endler and Tracy McLellan (PE, 417), arguing the need for a "newer synthesis," point out that the major figures behind that earlier synthesis were involved in research programs that did not expose them "to the interactions between natural populations and their environment." They also add: "Other gaps in our knowledge have resulted from polarized discussion where ideology has become more important than understanding evolution. Evolutionary biology would benefit from . . . fewer instances of assuming that only one process causes evolution."

13. Lynn Margulis, famous for her now widely accepted theory that eukaryotic cells originated from prokaryotes by symbiosis, has argued that the mechanistic ideas of neo-Darwinism need to be replaced by the notion that living things are "autopoietic" (KA, 874), with "autopoiesis" referring to "the self-making and self-maintaining properties of living systems" (865). In an essay advocating "Physiological Autopoiesis versus Mechanistic Neo-Darwinism," she argues that "life does not *adapt* to a passive physicochemical environment, as the Neo-Darwinists assume. Rather, life actively *produces and modifies* its surroundings" (BTB, 230).

14. It is interesting to learn that mechanistic views, more than empirical evidence, evidently lay behind the neo-Darwinian endorsement of the micromutationist view of adaptation, according to which it always involves genes of small effect. Having reviewed the evidence, Jerry Coyne of the University of Chicago, who had previously accepted the standard view, and Allen Orr say: "We conclude—unexpectedly—that there is little evidence for the neo-Darwinian view: its theoretical foundations and the experimental evidence supporting it are weak, and there is no doubt that mutations of large effect are sometimes important in adaptation" (Orr and Coyne GA, 726). The primary theoretical support for micromutationism, they report, was the mechanistic view that each organism is a finely tuned machine, so that any large change would probably worsen its functioning (726–28).

The Clash Between Gradualism and the Fossil Record

The other empirical problem to be examined is the apparent clash between the fossil record and Darwin's commitment to gradualism. As we saw in the discussion of Darwinian evolutionism$_{10}$, Darwin's rejection of evolutionary jumps or saltations was criticized from the outset by Huxley and paleontologists, because the facts seemed to imply saltationism, not gradualism.[15] The first problem was that extant species seem to embody just the clearly defined forms or types that Darwinism considers unreal. Darwin himself posed this problem, asking,

> why, if species have descended from other species by fine gradations, do we not everywhere see innumerable transitional forms? Why is not all nature in confusion, instead of the species being, as we see them, well defined? (OS, 158)

This problem had, Darwin said, an easy answer: The transitional forms, being less well adapted for survival, would have become extinct. That answer, however, raised a truly serious problem: The fossil record shows little if any trace of these "innumerable transitional forms" (OS, 159). Because of this fact, Darwin said, the fossil record provided "the most obvious and serious objection" to his theory (OS, 287).

This problem has not disappeared in our century. It was raised most clearly by George Simpson, who wrote in 1944 that "the line making actual connection with [common] ancestry is not known in even one instance" and that the "regular absence of transitional forms . . . is an almost universal phenomenon" (TM, 106, 107). Although Darwin tried to dismiss this problem by speaking of the insufficiency of research and the imperfection of the record, today, after another century and a half of paleontological research dedicated to finding transitional species, this claim is much more difficult to maintain. Although we now have the remains of some 250,000 extinct species, most of which were unknown in Darwin's time, "the fossil record," says Denton, "is about as discontinuous as it was when Darwin was writing the *Origin*" (E, 162). In fact,

---

The contention that *a priori* considerations have been primary is supported by Ernst Caspari, who says (EGW, 22) that when Richard Goldschmidt's theory that speciation is a sudden event produced by a large mutation was proposed, it was rejected by most evolutionists because "it went too strongly against the customary way of thinking."

15. In discussing the meaning of "gradualism," Niles Eldredge says: "Gradual, of course, means 'by imperceptible degree.' But . . . it has also meant 'slow' and even 'steady' " (RD, 63). Everyone now agrees that evolution is not steady and that it is not as slow as previously thought. The big questions are whether these revisions imply that the first meaning of gradualism must be rejected (which Eldredge says) and, if so, it must be replaced by saltationism (which he rejects).

says Norman Newell, former curator at the American Museum of Natural History: "Many of the discontinuities tend to be more and more emphasized with increased collecting" (NFR, 267). Insofar as better knowledge of the record has changed the picture at all, in other words, it has made the tension between the record and the gradualistic theory even worse.[16]

The absence of transitional forms means that, in the words of Eldredge (RD, 95), evolutionary novelty "usually shows up with a bang." In Denton's words, "the first representatives of all the major classes of organisms known to biology are already highly characteristic of their class when they make their initial appearance in the fossil record." For example,

> the sudden appearance of the angiosperms [flowering plants] is a persistent anomaly which has resisted all attempts at explanation since Darwin's times. . . . The first representatives of [the various] fish groups were already so highly differentiated and isolated at their first appearance that none of them can be considered even in the remotest as intermediate with regard to other groups. (E, 162, 164)

Wesson extends the litany of unheralded appearances:

> [T]he oldest known turtle had a complete shell like turtles today. . . . There are no fossils leading to primitive chordates or linking them with the vertebrates to which they must have given rise. . . . The first known insect looked much like a modern bug. . . .When fossils of land plants appeared, without recorded ancestry, about 450 million years ago, major lines had already formed, with no evident linkage among them. . . . [T]he earliest known bat was almost indistinguishable from modern bats. . . . . [T]he earliest land animals appear with four good limbs, shoulder and pelvic girdles, ribs, and distinct heads. . . . [A] dozen orders of amphibians suddenly appear in the record, none apparently ancestral to any other . . . The earliest whale fossils . . . are about as specialized as modern whales. (BNS, 41, 44, 48–49, 50)

Besides not confirming the theory, the record even portrays the opposite of what it predicts. According to the theory, as we go higher up the taxonomical hierarchy from species to genera, to orders, to classes, and to phyla, we should find increasingly more transitional forms. But we find increasingly fewer. As Simpson has put it:

---

16. Referring to "ultra-Darwinians," such as Richard Dawkins, who "have tenaciously clung to the original Darwinian vision of gradualism," Eldredge says that, when their imaginative picture of how evolution occurred is "held up against the light of the fossil record," it is "stunningly out of whack" (RD, 58, 59).

> [T]he appearance of a new genus in the record is usually more abrupt than the appearance of a new species; the gaps involved are generally larger. . . . Gaps among known orders, classes, and phyla are systematic and almost always large. (HL, 149)

It is true that the record will probably always remain imperfect, because many organisms are unlikely to leave fossils. But this feature of the problem—involving the *systematic* character of the gaps—cannot be attributed to the imperfection of the record, because, says Denton:

> [T]here are fewer transitional species between the major divisions than between the minor. . . . And this rule applies universally throughout the living kingdom to all types of organisms, both those that are poor candidates for fossilization such as insects and those which are ideal, like molluscs. (E, 192)

The problem, Denton emphasizes, is that "this is the *exact reverse* of what is required by [Darwinian] evolution" (E, 192).

One way to describe the tension between theory and record is that the theory predicts a *sequential* pattern, with multiple derivations from a common ancestor, while the record suggests a *hierarchical* pattern, which is characterized by Denton thus:

> The only sequence implied is a theoretical or abstract logical programme whereby a very general concept is successively subdivided into the more specific subcategories. The nodes and branches of the tree signify concepts in the mind of the logician. (E, 122)

This view is threatening to the nontheistic, nominalistic metaphysics of neo-Darwinism, because this hierarchical, typological view fits well, as Denton points out, with "the creative derivation of all the members of a class from the hypothetical archetype which existed in the mind of God" (E, 132).[17]

Correlative with the fact that new types seem to appear suddenly, rather than evolving gradually through "innumerable transitional forms," is the fact that, once they appear, they generally remain about the same. Stressing "species stasis," Eldredge says that, "once they appear, species tend not to accumulate much anatomic change through the remainder of their existence" (RD, 75). Wesson simply says: "Species usually stand still" (BNS, 207). This fact is also problematic for Darwinian gradualism: If the evolutionary process was to bring about elephants, whales, and humans from bacteria in a few billion years through a series of tiny steps—tiny enough to have been produced by chance—then directional changes

---

17. It is interesting that Denton makes this point although he himself evidently is not a theist.

must have been occurring most of the time. The fossil record, however, suggests a different picture:

> Species are usually static, or nearly so, for long periods, species seldom and genera never show evolution into new species or genera but replacement of one by another, and change is more or less abrupt. (BNS, 45)

To be more concrete, the present evolutionary picture of life on Earth, in most general terms, goes something like this. First, life seems to have emerged around 3.5 billion years ago, about as soon as conditions on earth made it possible. Then for almost 2 billion years after emerging, life remained basically the same in the sense that only prokaryotic (pre-nucleate) cells existed.[18] Then, evidently about 1.2 billion years ago, eukaryotic cells, with a true nucleus, emerged. Virtual stasis then apparently reigned for 600 million years, followed about 540 million years ago by the most dramatic event on our planet, the "Cambrian explosion," during which most of the basic designs of multicellular life were created. According to science-writer Madeleine Nash's summary of this event, sometimes known as "biology's Big Bang,"

> within the span of no more than 10 million years, creatures with teeth and tentacles and claws and jaws materialized with the suddenness of apparitions. In a burst of creativity like nothing before or since, nature appears to have sketched out the blueprints for virtually the whole of the animal kingdom. . . . Since 1987, discoveries of major fossil beds in Greenland, in China, in Siberia, and now in Namibia have shown that the period of biological innovation occurred at virtually the same instant in geologic time all around the world. (WLE, 68)

In Darwin's time, this "instant" was thought to have consisted of about 75 million years, which was already impossibly short from the viewpoint of Darwin's gradualism. His solution was to postulate a long history of early animal life far back into the Precambrian period. The fossil record, however, was again unsupportive, leading Darwin to say: "The case at present is inexplicable; and may be truly urged as a valid argument against the views here entertained" (OS, 310). Darwin's hope that future discoveries would vindicate him have only been partly fulfilled. On the one hand, recent evidence does

---

18. Stephen Gould, in referring to this period, has said that "nothing much happened for ever so long" (WL, 309). Lynn Margulis, who has specialized in this period, has replied that "nothing is more inaccurate than this statement" (KA, 864). In an earlier work (EL), she had shown that a lot was happening, as many types of bacteria emerged, with the most fateful developments being the emergence of bacteria that give off oxygen, followed by the emergence of bacteria that could *use* oxygen. It is true, nonetheless, that life remained prokaryotic during this long period.

indeed suggest that various forms of metazoan life appeared in the Vendian period, some tens of millions of years prior to the Cambrian. On the other hand, it is not yet known whether some of these earlier forms provided links with the Cambrian explosion, or whether they were all simply replaced. This period of time, in any case, is far too brief for Darwinian gradualism, especially in the light of the evidence that new forms, once they arise, tend to remain static. Steven Stanley says:

> The rapid adaptive radiation that is apparent today confronts gradualism with a seemingly insoluble problem. We now know that for many groups of marine invertebrates an average species lasts for five or ten million years without evolving enough to be given a new name. How, then, are we to explain the origin of advanced groups, like arthropods and mollusks, from primitive ancestors in a few tens of millions of years? (NET, 90)

Besides the fact that the period of Precambrian multicellular life seems too brief, the "instant in geologic time" during which the Cambrian explosion occurred has also continued to be whittled down, with some saying that it may be closer to 5 than to 10 million years. One scientist involved in establishing these dates, Samuel Bowring, has been quoted by Nash as saying: "We now know how fast fast is. And what I like to ask my biologist friends is, How fast can evolution get before they start feeling uncomfortable?" (WLE, 70). His meaning, of course, is that the faster the major developments occurred, the less evolution seems to be compatible with the gradualistic framework presupposed by most biologists.[19] This issue, Michael Behe points out, has gone full circle: Darwin was originally confronted with the problem that, according to nineteenth-century physicists, the Earth was only about a hundred million years old, which was not long enough for evolution as conceived by Darwin. The discovery that the Earth is really over four billion years old seemed at first to provide the needed time. "With the discovery of the biological Big Bang, however, the window of time for life to go from simple to complex has shrunk to much less than nineteenth-century estimates of the earth's age" (DBB, 28).

The problem does not cease, furthermore, with the Cambrian explosion. The fossil record also suggests a rapid appearance of the flowering plants (which Darwin called "an abominable mystery"), the land animals, and the birds. It also suggests a rapid radiation of mammals. The latter fact, Stanley says, raises a problem that Darwin himself did not have to face:

---

19. Criticizing those who believe that macroevolution can be understood by simply extrapolating from the working of natural selection upon random variations within populations, Eldredge says that "the little progressive within-species change we see in the fossil record is simply too slow to account for the great adaptive changes wrought by evolution" (RD, 77).

Darwin was spared a confrontation with the extraordinarily rapid origins of the modern groups of mammals. . . . Today, our more detailed knowledge of fossil mammals lays another knotty problem at the feet of gradualism. Given a simple little rodentlike animal as a starting point, what does it mean to form a bat in less than ten million years, or a whale in little more time? . . . [In that period, gradual evolution] might move us from one small rodentlike form to a slightly different one, perhaps representing a new genus, but not to a bat or a whale! (NET, 93)

The most recent of the discontinuous appearances is the human form, with the earliest known hominid fossil being, in Wesson's words, "fully bipedal, with legs and pelvis very different from those of apes and fully of hominid type" (BNS, 41). The tension between neo-Darwinian gradualism and the fossil record, accordingly, runs the gamut of the latter, from beginning to end.

The major effort to deal with this problem within a still fundamentally neo-Darwinian framework is the theory dubbed "punctuated equilibria" by Niles Eldredge and Stephen Jay Gould (PE). According to this theory, a new species develops quickly, perhaps within a few thousand years, in a small, isolated group, in which the new characteristics can spread quickly. This new species then moves into the territory of the unchanged species, quickly replacing it. The result is that the fossil record suggests a saltation. Because the fossils of the transitional species are both so few and so far removed from the major fossil sites, we will be unlikely ever to discover them.

Although this theory has sometimes been likened to (heretical) saltational theories, it is, as Dawkins has stressed, still Darwinian in the sense of saying that major changes always occur through a series of small changes. Eldredge and Gould still affirm gradualism in the sense that, in Dawkins' words, "each generation is only slightly different from the previous generation" (BW, 241). They are only denying that it always occurs at a constant speed, saying instead that long periods of stasis are punctuated by "brief episodes of rapid gradual change" (BW, 243). Their theory of punctuated equilibria, accordingly, is not saltational:

> The jumps that it postulates are not real, single-generation jumps. They are spread out over large numbers of generations. . . . In the sense of the word in which Darwin was a passionate gradualist, Eldredge and Gould are also gradualists. (BW, 244, 250)

---

20. Eldredge has strongly affirmed this fact. Responding to those who had accused him and Gould "of abandoning the Darwinian camp altogether, promulgating instead a form of saltationism," he says (RD, 98, 100): "Whatever we are, we are not saltationists!" For a classic statement of the case for saltationism, see Richard Goldschmidt MBE.

In (rightly) stressing that Eldredge and Gould's theory is gradualistic in the sense of being non-saltational,[20] Dawkins has also (inadvertently) pointed out that it provides no basis for reconciling neo-Darwinism with the empirical facts. In Denton's words:

> While Eldredge and Gould's model is a perfectly reasonable explanation of the gaps between species . . . , it is doubtful if it can be extended to explain the larger systematic gaps.[21] The gaps which separate species . . . are utterly trivial compared with, say, that between a primitive terrestrial mammal and a whale . . . ; and even these relatively major discontinuities are trivial alongside those which divide major phyla such as molluscs and arthropods. Such major discontinuities simply could not, unless we are to believe in miracles, have been crossed in geologically short periods of time through one or two transitional species occupying restricted geological areas. Surely, such transitions must have involved long lineages including many collateral lines. . . . To suggest that the hundreds, thousands, or possibly even millions of transitional species which must have existed in the interval between vastly dissimilar types were all unsuccessful species occupying isolated areas[22] and having very small population numbers is verging on the incredible! (E, 193–94)

In sum, Darwinism, with its gradualism based upon nominalism, seems to be disconfirmed by the fossil record. This empirical problem, furthermore, is matched by a conceptual correlate, to which we now turn.

A CONCEPTUAL CORRELATE TO THE FOSSIL PROBLEM

Correlative to the empirical problem created by gaps in the fossil record is the difficulty of conceiving, in Denton's words, of "functional intermediates through which the gap might have been closed" (E, 213). The word "functional" stresses the point that each of the intermediate species had to be viable. As Gordon Taylor puts it, "the task is to improve a machine while it keeps running"

---

21. Although Eldredge has tried to suggest how this might be possible (RD), I cannot see that he has provided a plausible account. The apparent inadequacy is increased, furthermore, by the fact that he addresses only the question of the circumstances under which major changes might have occurred, not the conceptual problem, to be discussed below, of *how* they might have occurred.

22. The hypothesis in question is that we do not see the transitional forms because the new developments occurred in areas other than those in which we now find fossils. But, as Eldredge himself has said, "Evolution cannot forever be going on someplace else" (RD, 95).

(GEM, 113). This conceptual problem is so difficult because, in many cases, several changes would have to occur at once if the new organism is to be viable, and the possibility that these coordinated changes could occur by chance is remote.

One of the steps difficult to conceive is that from fish to amphibia. At least the following new structures were needed: legs to prevent the lungs from being compressed; a strengthened spine plus a pelvic girdle to support it; a new suspension system, so that the head would not move side to side with each step; impervious skin (in place of scales) to prevent the body from drying out; a new kind of eyes, complete with eyelids and tears to protect them from dust and drying; mucus for the nose; and ears (formed from the lateral line of the fish). Reflecting on these and other needed changes, Taylor asks "whether such an impressive array of coordinated changes could have taken place by chance" (GEM, 60).

Besides the conceptual problem created by the transition to amphibia in general, there are unique problems associated with particular types. Wesson reflects on the reproductive habit of an Australian frog:

> The female swallows its eggs and broods the tadpoles, about 20 in number, 6 or 7 weeks in the stomach, finally burping froglets, which she may reswallow. . . . [I]t is not easy to contemplate how this custom could have started; there could be no progressive stages in ingestion of eggs or the reduction of digestion in the stomach, which is rather radically altered. (BNS, 76–77)

Wesson then quotes M. T. Tyler, a specialist in gastric brooding frogs, who says that this habit must have come about "by a single, huge quantum leap."

Regarding the next transition, from amphibia to reptiles, Denton discusses the development of the amniotic egg:

> Every textbook of evolution asserts that reptiles evolved from amphibia but none explains how the major distinguishing adaptation of the reptiles, the amniotic egg, came about gradually as a result of successive accumulation of small changes. . . . There are hardly two eggs in the whole animal kingdom which differ more fundamentally. . . . The evolution of the amniotic egg is baffling. . . . Altogether at least eight quite different innovations were combined to make the amniotic revolution possible. (E, 218–19)

Taylor agrees, saying:

> From the shell, constructed of crystals of hydroxyapatate and waxed over, to the altered chemistry, based on fat rather than protein, the amniote egg was in a different class altogether, a stunning advance on the simple blob

of jelly that constituted the egg of frogs and fishes—a saltation if ever there was one. (GEM, 64)

Even more mysterious, perhaps, is the transition from reptiles to birds. One unsolved problem is how flight could have originated. Neither of the two major theories—that it arose from gliding or from leaping—is sufficiently free from problems to have created a consensus (E, 205–07). Another problem is the origin of feathers, which Denton discusses:

> It is not easy to see how an impervious reptile scale could be converted gradually into an impervious feather without passing through a frayed scale intermediate which would be weak, easily deformed and still quite permeable to air. (E, 209)

The problem is that the feathers, until they were impervious, would provide no survival advantage—indeed, they might even be a handicap—so would not be preserved by natural selection. Even if this problem could be solved, furthermore, there is the fact that many other changes had to occur simultaneously: The bones had to become hollow, the skull had to become very thin, and the tooth-studded jaw had to be replaced by a light beak, all of which reduced body weight. The body also had to become more compact, shedding the reptilian tail and snout and reducing the legs and feet to a minimum. Correlative with the need for greater balance and vision, the brain had to develop a larger cerebellum and a larger visual cortex. One other needed development, a system of thermo-regulation, is considered the most baffling by Taylor (GEM, 71). Still another development, the new lung and respiratory system, is considered the most remarkable by Denton, who says:

> Just how such an utterly different respiratory system could have evolved gradually from the standard vertebrate design is fantastically difficult to envisage. . . . In attempting to explain how such an intricate and highly specialized system of correlated adaptations could have been achieved gradually through perfectly functional intermediates, one is faced with the problem of the feather magnified a thousand times. (E, 211–12)

Darwin, we recall, had said: "If it could be demonstrated that any complex organ existed which could not possibly have been formed by numerous, successive, slight modifications, my theory would absolutely break down." Denton suggests that the avian feather and lung at least come close to filling the bill (E, 213). And, far from considering these examples unique, he adds that "practically every group of organisms possess complex adaptations equivalent in many ways to the feather" (E, 227).

The Rise of Life Itself

The conceptual difficulties, furthermore, do not exist only in the later stages of evolution. The rise of life itself is considered to be the greatest mystery by many, including Denton, who says:

> We now know . . . that it represents the most dramatic and fundamental of all the discontinuities of nature. . . . Molecular biology has shown that even the simplest of all living systems on earth today, bacterial cells, are exceedingly complex objects. Although the tiniest bacterial cells are incredibly small . . . , each is in effect a veritable micro-miniaturized factory containing thousands of exquisitely designed pieces of intricate molecular machinery. . . . The complexity of the simplest known type of cell is so great that it is impossible to accept that such an object could have been thrown together suddenly by some kind of freakish, vastly improbable, event. Such an occurrence would be indistinguishable from a miracle. (E, 249–50, 264)

Although Denton draws no theistic conclusions from this problem (see note 17), such conclusions are drawn by Michael Behe in *Darwin's Black Box: The Biochemical Challenge to Evolution.* The "black box" to which he refers is the cell, which biochemistry has enabled us to understand only recently, since the formulation of neo-Darwinism (DBB, 15, 25). The "challenge" to which Behe's subtitle refers involves the discovery that cells are not only *extremely* complex, making "the complexity of a motorcycle or television set look paltry in comparison" (DBB, 46–47), but also, more importantly, *irreducibly* complex, by which Behe means

> a single system composed of several well-matched, interacting parts that contribute to the basic function, wherein the removal of any one of the parts causes the system to effectively cease functioning. (DBB, 39)

Spelling out the negative implications for neo-Darwinism, Behe adds:

> An irreducibly complex system cannot be produced . . . by slight, successive modifications of a precursor system, because any precursor to an irreducibly complex system that is missing a part is by definition nonfunctional. . . . Since natural selection can only choose systems that are already working, then if a biological system cannot be produced gradually it would have to arise as an integrated unity, in one fell swoop, for natural selection to have anything to act on. (DBB, 39)

Much of Behe's book is devoted to describing a number of irreducibly complex "molecular machines," such as the cilium, the bacterial flagellum, the protein transport system, and the immune system, showing why each one involves a "chicken-and-egg" problem, because the functioning of each part of the system presupposes the functioning of all the other parts. Behe also argues that the biological literature, although it contains many articles on these systems, *contains not one that shows how any of these systems could have evolved* and very few that even try (DBB, x, 68, 72, 114, 138). He believes, furthermore, that this lack does not simply reflect present ignorance that will be overcome, but that "there are compelling reasons—based on the structure of the systems themselves—to think that a Darwinian explanation for the mechanisms of life will forever prove elusive" (DBB, x). One inference to be drawn, Behe argues in a chapter titled "Publish or Perish," is that "the theory of Darwinian molecular evolution has not published, and so it should perish" (DBB, 186).[23] The substantive conclusion to be drawn from all this, he argues, "is that many biochemical systems were designed. . . . Life on earth, at its most fundamental level, in its most critical components, is the product of intelligent activity" (DBB 193).

Although Denton rejects Darwinism and Behe rejects its atheism as well, it would be a mistake to assume that the origin of life is seen to be a problem only by those with a theistic, or at least an anti-Darwinian, ax to grind. In *Origins: A Skeptic's Guide to the Creation of Life on Earth,* Robert Shapiro, who says that he would never adopt a religious answer (OSG, 130), concludes from his survey that "no adequate scientific explanation to the problem has emerged" (7). And Francis Crick, whose acceptance of mechanistic reductionism in general and neo-Darwinism in particular is well known, has said:

> An honest man, armed with all the knowledge available to us now, could only state that in some sense, the origin of life appears at the moment to be almost a miracle, so many are the conditions which would have had to have been satisfied to get it going. (LI, 88)

Crick goes on, to be sure, to deny that this appearance means that we have good reasons to believe that life "could *not* have started on earth by a perfectly reasonable sequence of fairly ordinary chemical reactions." He regards that problem as

---

23. The response I have received from repeating Behe's claim about the evolutionary literature—which simply brings out the point being made implicitly by many others, such as Crick, Denton, Shapiro, Stanley, Taylor, and Wesson—is that I obviously have not read the right books. There are, I am assured, evolutionists who have described how the transitions in question could have occurred. When I ask in which books I can find these discussions, however, I either get no answer or else some titles that, upon examination, do not in fact contain the promised accounts. *That* such accounts exist seems to be something that is widely known, but I have yet to encounter anyone who knows *where* they exist.

serious enough, nevertheless, to suggest that life may have been sent to our planet from a higher civilization (LI, Chs. 8–13; Crick and Orgel DS)—which would, of course, simply push the problem back. There is, in any case, a growing recognition that at present a neo-Darwinian solution to the problem of the origin of life is not even in sight, with many saying that it seems impossible in principle.

Implications for Understanding Evolution

The difficulty of giving a naturalistic explanation of the origin of life was, of course, no problem for Darwin himself, because he referred the creation of the first form(s) of life to God. This difficulty, however, constitutes a serious problem for the theory that now bears Darwin's name. The implications for the completeness of the neo-Darwinian account are brought out by Denton, who says:

> The failure to give a plausible evolutionary explanation for the origin of life . . . represents yet another case of a discontinuity where a lack of empirical evidence of intermediates coincides with great difficulty in providing a plausible hypothetical sequence of transitional forms. . . . [T]he seemingly intractable difficulty of explaining how a living system could have gradually arisen as a result of known chemical and physical processes raises the obvious possibility that factors as yet undefined by science may have played some role. (E, 271)

This possibility raises, in turn, the more general possibility, Denton points out, that "these unidentified processes [may] have been involved in other problematical areas of evolution" (E, 271).

Denton's suggestion fits with a growing view in some evolutionary circles that neo-Darwinian evolution, with its natural selection of random variations, provides a good account of one kind of evolutionary change but not for the most interesting kind. Natural selection, Taylor says, "accounts brilliantly for the minor adaptations but it is by no means clear that it explains the major changes in evolution" (GEM, 13). One paleontologist is quoted by Nash as expressing this attitude in less guarded terms:

> What Darwin described in the *Origin of Species* was the steady background kind of evolution. But there also seems to be a non-Darwinian kind of evolution that functions over extremely short time periods—and that's where all the action is. (WLE, 74)

The question to be explored in the remainder of this chapter is whether one of the "other factors" posited by Denton might be divine influence naturalistically construed, and whether evolution thus influenced could be the "non-Darwinian kind of evolution that functions over extremely short time periods" and explains "the major changes in evolution."

Of course, there will be great resistance in the present-day scientific community to considering "other factors" that could account for a "non-Darwinian kind of evolution." Especially strong will be the resistance to the suggestion that one of these other factors is divine influence (even if this influence does not involve the supernatural interventionism that Behe seems to share with Plantinga and Johnson). There will be a strong hope that neo-Darwinism can be saved by simply modifying gradualism to allow for punctuated equilibria. However, the saltations needed to overcome the empirical and conceptual problems would have to be far greater than any that are compatible with a theory still meaningfully called Darwinian. As Denton (E, 229) and Dawkins (BW, 72–73) both agree, successful organisms resulting from purely fortuitous saltations are extremely unlikely. In Denton's words:

> [T]he total space of all combinatorial possibilities is so nearly infinite and the isolation of meaningful systems so intense, that it would truly be a miracle to find one by chance. Darwin's rejection of chance saltations as a route to new adaptive innovations is surely right. (E, 319)

The fact that a viable theory cannot be based upon purely fortuitous saltations, however, does not mean that a viable theory cannot be based upon saltations: What is evidently needed is an adjustment that allows the saltations not to be purely accidental.

One necessary condition for such a theory would be the rejection of the nominalism of Darwinism—its denial that forms are real and efficacious. Stressing the importance of forms, Taylor (GEM, 243) quotes Joseph Needham's statement that "[t]he central problem of biology is the form problem." Taylor also quotes (GEM, 141–42) the following statement of William Thorpe:

> What is it that holds so many groups of animals to an astonishingly constant form over millions of years? This seems to be *the* problem now—the problem of constancy, rather than that of change.

The problem, however, is more general, involving not only that of maintaining a certain form, once it has been attained, but also that of first attaining it. Wesson focuses on this side of problem, saying: "Innovation is the central problem that has troubled evolutionists ever since Darwin" (BNS, 53).

Even challenging the nominalism of Darwinism, however, would not get to the root of the problem. To affirm the reality of forms in the nature of things would beg the further question of where such forms could exist and how they could be efficacious. Realism about forms, as mentioned earlier, has usually gone hand in hand with theism. If an adequate framework for understanding evolution requires the rejection of nominalism, it would seem also to require the rejection of atheism. Given the assumption of a form-inducing divine power, the major transformations within biological evolution would not be so mysterious, because the idea of genuine saltations would not be unthinkable. That same assumption would render more intelligible developments both prior to biological evolution—the origin of our universe and of life—and those after it—such as the fact that the human mind's logical, mathematical, aesthetic, and moral notions seem to involve responses to normative forms somehow existent in the nature of things, not pure creations corresponding to nothing. In the final two sections of this chapter, I will develop these ideas by briefly indicating how the theistic naturalism developed in previous chapters could provide a more adequate framework for thinking about evolution.

A WIDER NATURALISTIC FRAMEWORK FOR EVOLUTIONARY THEORY

Although the inadequacies of neo-Darwinism have been increasingly recognized, even by many of those who have continued publicly to support it, there has been a strong tendency to ignore or belittle these inadequacies. Even most of those who have faced certain inadequacies squarely have not been thereby led to suggest the need for a fundamentally new framework for thinking about evolution. One reason for the hesitancy to discuss the failures of neo-Darwinism in a global way has surely been the perceived need to "keep the faith" in the face of the creation science movement. As Wesson says: "Biologists, under attack, do not want to admit doubts that might undermine their central theory" (E, 20). The more general problem is the assumption that the neo-Darwinian view of evolution provides the only naturalistic alternative to supernaturalistic creationism. The final portion of this chapter is not the place to try to develop a full-blown alternative theory of evolution. I will attempt, however, to show how the wider naturalism articulated in this book provides a more helpful framework than the hitherto dominant one for looking open-mindedly at all the relevant evidence and developing more plausible hypotheses.

This wider naturalism provides three new elements: a panexperientialist ontology, a nonsensationist doctrine of perception, and a naturalistic theism. I will show how these three elements, especially when taken in combination, provide a basis for affirming evolutionism$_{1-4}$ more adequately and coherently while rejecting or modifying the remaining doctrines (evolutionism$_{5-14}$) constituting neo-Darwinism. In the present section, I treat this wider naturalism as a philosophical framework for evolutionary theory, discussing the implications for more strictly scientific issues in the final section.

Real Chance

With regard to Darwinian evolutionism$_1$, this wider naturalism provides no reason to doubt the assumption that at least much microevolution occurs by means of natural selection of changes resulting from mutations that are random$_{na}$ and even random$_{eps}$. The adoption of the panexperientialist ontology would, in fact, allow these mutations to be considered "random" or "chance" events in an even stronger sense, for two reasons. In the first place, every individual, meaning most precisely each individualized event, is an occasion of experience with at least some degree of spontaneity. In the second place, each such event is causally influenced, to at least some slight degree, by all prior events. By putting these two factors together, we can speak of "chance" in the true sense of the term.

Real Evolution

With regard to Darwinian evolutionism$_2$, which affirms macroevolution as such, panexperientialism allows for real evolution, in the sense of *internal* changes in actualities and the emergence of *higher-level actualities*. By contrast, materialism, which allows for only external relations between things, has to think of evolution as mere rearrangement of things. Whitehead stressed this point, saying:

> The aboriginal stuff, or material, from which a materialistic philosophy starts is incapable of evolution. . . . Evolution, on the materialistic theory, is reduced to the role of being another word for the description of the changes of the external relations between portions of matter. There is nothing to evolve, because one set of external relations is as good as any other set of external relations. There can merely be change, purposeless and unprogressive. (SMW, 107)

By understanding all enduring individuals to be comprised of momentary experiences, this wider naturalism can regard them as internally influencing each other and also as capable of embodying forms, hence as capable of embodying novel, more complex forms. This type of naturalism can thereby do justice to, in Whitehead's words, "the whole point of the modern doctrine," which is "the evolution of the complex organisms from antecedent states of less complex organisms" (SMW, 107).

Naturalism and Uniformitarianism

Darwinian evolutionism$_3$, which stipulates that evolution of every type occurs without benefit of any supernatural interruptions of the normal causal processes of the universe, is affirmed. This wider naturalism, of course, considers some kinds of causality to be perfectly natural that the materialistic version of naturalism considered supernatural (and therefore impossible). But it is precisely this wider construal of the nature of nature that, by allowing for the naturalness of more types of occurrences, precludes the need to choose between denying the evident or resorting to supernaturalism to account for it. This point implies that Darwinian evolutionism$_4$, which is now identical with the doctrine of ontological uniformitarianism, is enthusiastically affirmed.

Theistic Guidance Without Supernaturalism

I turn now to the various dimensions of Darwinian evolutionism that are, from the viewpoint of this wider naturalism, to be rejected or at least modified. The first such dimension is the insistence that biological evolution occurs without any theistic influence (Darwinian evolutionism$_5$). Actually, there is agreement on this point insofar as what is usually meant by "theism" is supernaturalistic theism. Because most Darwinists have evidently been unaware of the distinction between supernaturalistic and naturalistic theism, they have assumed that the rejection of supernaturalism required the rejection of theism of all sorts. Although strongly ideological Darwinists have had other reasons for rejecting theistic influence, making the distinction between the two kinds of theism seem unimportant, it is, in fact, the distinction between a theism that seeks to provide a naturalistic account of evolution and one that does not. It is thereby the distinction between a worldview that could in principle be accepted by the scientific community and one that could not. We can hope that the next generation of evolutionary scientists and philosophers will be less ideologically committed to atheism, thereby more able to consider whether a naturalistic form of theism might help in constructing a more adequate theory of evolution.

Whiteheadian theism, it should be stressed, *is* fully naturalistic: Divine influence in the world is a regular, necessary part of the normal causal process, not an occasional interruption of this process, and it is consistent with uniformitarianism, because divine influence is said to occur in basically the same way always and everywhere: by providing possible forms for actualization. The divine influence does vary in content, in that different forms are relevant for different occasions. But variability in this sense does not violate uniformitarianism any more than does the fact that, although formally speaking my mind always influences my body in the same way—by providing aims for its various parts—my mind provides different aims for different parts of my body at the same time and different aims for the same part of my body at different times.

This uniform mode of operation is not simply a matter of divine decision, which could in principle be rescinded. As explained in Chapter 4, God is not a being external to the universe, in the sense of one who, being able to exist apart from any universe of finite beings whatsoever, created our universe *ex nihilo.* That view of God implied that, because the universe's causal processes and principles had been freely created, they could be freely interrupted. In this naturalistic theism, by contrast, God is *essentially* the soul of the universe, which entails that God and a universe, meaning a multiplicity of finite events, exist with equal necessity, being coeternal. Although our particular world is a contingent creation, it was (by hypothesis) created out of a multitude of finite events exemplifying the same basic causal principles that now obtain. These causal principles, not being contingent, cannot be interrupted. They, like God, exist naturally, belonging to the very nature of things (because, as I suggest elsewhere [RWS, Chs. 4 and 5], they belong to the divine essence). Belief in God in this sense does not threaten the scientific community's naturalistic assumption that there are no interruptions of the basic cause-effect relations.

Evil

Because of this crucial difference from theism of the supernaturalistic type, the return to speaking of God does not bring back the problem of evil. All the creatures have some power of their own, which cannot be canceled or overridden. Each creature has the twofold power to exert at least some iota of self-determination (assuming the creature to be an individual rather than an aggregational society) and some power to influence others. Although the divine power is unique in many ways, being omnipresent and the home of all previously unactualized forms, it is nonetheless one of the multitude of powers within the all-inclusive system influencing every finite event, not a power outside the system that can interrupt its systemic causal patterns. The divine power, in other words, is persuasive, not coercive or unilaterally determining.

Cosmic Support for Truth, Beauty, and Morality

Thanks to its inclusion of this soul of the universe, Whiteheadian naturalism need not be nominalistic—so that Darwinian evolutionism$_{11}$, which is responsible for so much of the inadequacy of neo-Darwinism, is overcome. The soul of the whole can provide a home for the eternal forms (which Whitehead called "eternal objects"), be they logical, mathematical, geometric, moral, or aesthetic forms. And the divine appetition for these forms to be actualized in the world can explain how these forms, being mere possibilities, can have causal efficacy, so that their presence can be felt by the creatures. This dimension of the soul of the whole provides a ground from which forms of all types can pervade the universe. The eternal forms are the material of the divine persuasion: The soul of the universe, with its appetitive vision for various forms to become incarnate in the world in due season, influences us by whetting *our* appetites for these forms.

The divine aim (by hypothesis) is most fundamentally aesthetic. God seeks beauty of every type. The most basic distinction is between inner and outer beauty. Inner beauty, which is beauty in the primary sense, occurs insofar as experience itself is both intense and harmonious. Outer beauty is the beauty of appearance—the tendency of the outer appearance of something to contribute to the inner beauty of its beholders. This outer beauty, called by Whitehead the "beautiful" (AI, 255), is obviously a somewhat relative quality, as that which is beautiful to one species may not be so to others. For a particular species, however, there tends to be agreement, so that things can be called beautiful with some objectivity: They are objectively beautiful in relation to a particular species insofar as they tend to contribute to the beauty of the experiences of the members of that species (AI, 256). This combination of objectivism and relativism seems to correspond with Darwin's own beliefs.

Beauty in the primary sense, the beauty of experience, can involve various types of forms. At the human level, we can enjoy aesthetic experience in the normal sense of the term, as when we enjoy beautiful music, poetry, scenery, food and drink. We can enjoy the intellectual beauty of seeking and discovering truth, whether in mathematics, philosophy, the physical sciences, or any other area. We can appreciate moral beauty, whether in ourselves or others, even the "beauty of holiness." We are attracted to Beauty, Truth, and Goodness because these values are entertained appetitively by the Eros of the universe (AI, 11), whose appetites we feel. This point means the rejection of Darwinian evolutionism$_{13}$, according to which the universe as the all-inclusive evolutionary process is amoral, providing no ground for moral norms.

The importance of this point can be illustrated by reference to Thomas Huxley and Richard Dawkins. Huxley was at first an advocate of evolutionary ethics, saying that it was liberal education's task to create in students a "desire to move in harmony with" the evolutionary laws of nature (Greene SIW, 159). Later, however, he became a critic, arguing in *Evolution and Ethics* that ethical progress "depends, not on imitating the cosmic process . . . , but in combating it" (EE, 83). Ethical goodness, said Huxley,

> involves a course of conduct which, in all respects, is opposed to that which leads to success in the cosmic struggle for existence. In place of ruthless self-assertion it demands self-restraint; in place of thrusting aside, or treading down, all competitors, it requires that the individual shall not merely respect, but shall help his fellows. . . . Laws and moral precepts are directed to the end of curbing the cosmic process. (EE, 81–82)

Herbert Spencer criticized Huxley's position that "we have to struggle against or correct the cosmic process," pointing out that this position "involves the assumption that there exists something in us which is not a product of the cosmic process" (LL, 336). Spencer presumably would see the same inconsistency in the position of Richard Dawkins, who says:

> We have the power to defy the selfish genes of our birth. . . . We can even discuss ways of deliberately cultivating and nurturing pure, disinterested altruism—something that has no place in nature, something that has never existed before in the whole history of the world. We are built as gene machines . . . but we have the power to turn against our creators. We, alone on earth, can rebel against the tyranny of the selfish replicators. (SG, 215)

Dawkins is right to say that we have this power for cultivating disinterested altruism. But Spencer would also be right to point out that Dawkins' affirmation of this power involves a sudden introduction of a dualism between humanity and nature that contradicts the otherwise wholly materialistic, reductionistic worldview that Dawkins has enunciated. Like Huxley, Dawkins assumes that we human beings (alone) have something in us—a genuine moral sense and the power to act upon it—that is not rooted in the cosmic process. But that is to suppose a magical emergence *ex nihilo*. To avoid such an inconsistency while doing justice to our (limited but real) capacity for disinterested moral altruism (as well as our related capacities for the disinterested pursuit of truth and beauty), we need a cosmology according to which this set of capacities can be seen as "a product of the cosmic process."

Evolutionary Progress

In any case, whereas truth and moral goodness are evidently forms of beauty that can significantly enrich the experience of only the cognitively most developed creatures, beauty as such is completely democratic, having forms appropriate for creatures of every level. Every true individual is capable of positive intrinsic value, of having experience characterized by beauty. More complex creatures can enjoy higher, more complex forms of value, thereby having more intrinsic value, more value in and for themselves. Does anyone really seriously doubt that a human being has more intrinsic value than a chimpanzee, that a chimpanzee has more than a mouse, a mouse more than a fly, and a fly more than an amoeba? Given the divine aim at increasing the beauty or intrinsic value of experience, we have a standard for speaking of progress, and we can say that, insofar as the evolutionary process has tended over time to bring forth creatures with increasingly more capacity for realizing more complex forms and thereby experiencing greater intrinsic value, progress has occurred. We need not try, accordingly, to affirm Darwinian evolutionism[14]. The recovery, within a naturalistic context, of a Cosmic Purpose and Valuation allows us to affirm explicitly our sense, which it is virtually and perhaps actually impossible to eradicate, that the higher animals really are "higher" from a nonrelativistic viewpoint.

Cosmic Meaning

To speak of a Cosmic Valuation is to refer to the second dimension of the soul of the universe, called by Whitehead "the consequent nature of God." Just as our own minds or souls not only influence our bodies, but also are influenced by them in return, synthesizing the multitudes of cellular experiences into a unified experience with intrinsic value far surpassing the intrinsic value of any of the brain cells qualitatively, something similar (by hypothesis) occurs in the universe as a whole. From this perspective we can reject that aspect of Darwinian evolutionism$_{13}$ according to which the cosmic evolutionary process is meaningless. The universe will never be as if we had never been: Our lives will have immortality in the everlasting divine experience. In Chapter 7, furthermore, we looked at reasons to believe that one of the latest evolutionary breakthroughs achieved by the persuasive power of the divine Eros may be the capacity of the human soul to survive the death of its physical body. If that is correct, then our contribution to the universe will not be limited to that which we can make in the present life.

Radical Empiricism and Uniformitarianism

Having looked at the implications of the theistic dimension of this wider naturalism, we turn now to the implications of its epistemology, especially its doctrine of perception. As we saw in earlier chapters, sense-perception is not our only mode of perception. The clear and distinct perceptions that can arise from our physical sensory organs are a high-level, secondary form of perception, derivative from a nonsensory prehension. Unless the mind could prehend its brain cells in this direct way, for example, it would not be able to receive the data from the eyes and ears. Having recognized that we must be receiving information from this nonsensory mode of perception all the time, albeit unconsciously in most cases, and having learned from parapsychology that our nonsensory reception of information is by no means limited to information from our brains, we can reject the positivistic-materialistic dimension of Darwinism (Darwinian evolutionism$_6$). Causality, in other words, need not be limited to material objects and the constituents thereof. In particular, being aware through the evidence for telepathy that our minds can be directly influenced by other minds, we can attribute causal power to minds of all levels, including a mind of the universe.

Although this wider naturalism rejects sensate empiricism, it retains empiricism as such, according to which no unexperienced elements should be allowed into our ontologies. Affirming what William James called *radical empiricism*, which recognizes nonsensory experience and thereby the elements at the very fringes of awareness as well as the clear and distinct data, we can honor the deeper intent behind positivism, which is to limit meaningful talk to that which is experienced, while still talking about a soul of the universe through which forms become efficacious. We can do so because we directly experience this soul's power by virtue of experiencing cognitive, aesthetic, and moral forms, through which we feel the call of Truth, Beauty, and Goodness as normative ideals. In positing the influence of this form-inducing power in eras long gone, therefore, we can remain true to uniformitarianism, because we are talking about a causal power known to be active today.

Beyond Reductionism and Determinism

By rejecting sensationist positivism and thereby materialism, we are also set free from the apparent need to affirm reductionism and determinism (Darwinian evolutionism$_7$). Having agreed with Darwin's rejection of a dualism between humans and the rest of nature and having affirmed the actuality of human minds, which we know to have the capacity to exercise self-determination, we can generalize a degree of this power to other actualities. Also relevant here is another feature of the panexperientialist ontology, the distinction between compound individuals and mere aggregational societies of individuals. It was primarily from the example of astronomy and physical dynamics, the first successful modern sciences, that predictive determinism and reductionism became ideals for science. These sciences, however, deal with inorganic aggregational societies, such as huge planets and little steel balls, which have no self-determining power, so that their behavior is reducible to the behavior of their parts. Reductionism and predictive determinism are not appropriate ideals, however, when dealing with living organisms, especially the more complex compound individuals. Given the organizational duality between compound individuals and aggregational, nonindividuated things, we can recognize that the behavior of some things is predictable in principle, at least virtually, while that of other things is not. It should become an essential part of science to recognize this distinction.

The importance of having a philosophical context that allows for genuinely free decisions can be illustrated by reference to Ernst Mayr's statement that "the pacemaker of evolution" is behavioral change in animals. "In animals, almost invariably," says Mayr, "a change in behavior is the crucial factor in initiating evolutionary innovation" (TNP, 408). Wesson brings out the significance of Mayr's statement with the observation that "if behavior leads the way, the prime mover in innovation is not inheritable variation, as Darwin postulated, but the animal's choice" (BNS, 241). If animal choice is central to biological evolution, furthermore, then it is surely important to have a philosophy that allows this idea to be taken seriously.

IMPLICATIONS FOR SCIENTIFIC EVOLUTIONARY THEORY

Having dealt with various ways in which our wider naturalism overcomes not only some of the philosophical constraints on neo-Darwinism as a theory of macroevolution (Darwinian evolutionism$_{3-7}$) but also some its negative philosophical implications (Darwinian evolutionism$_{13-14}$), I turn now to the import of this wider naturalism for the more strictly scientific dimensions (Darwinian evolutionism$_{8-10}$).

Macroevolution Unconstricted

In the first place, we can reject Darwinian evolutionism$_8$, the dictum that macroevolution is to be understood entirely in terms of the principles operative in microevolution. Even if the doctrine of microevolution be enlarged so as to allow factors beyond random variations and natural selection, such as need-induced mutations and the inheritance of certain kinds of acquired characteristics, we need not rule out the possibility that in macroevolution, which somehow gives rise not only to new species but also to new genera, families, orders, classes, phyla, and kingdoms, something extraordinary happens from time to time, resulting in the well-defined character of nature, with its apparent absence of intermediates. We need not rule out this possibility, that is, if it can be shown to be compatible with ontological uniformitarianism and thereby with naturalism.

With regard to Darwinian evolutionism$_9$: As intimated above, we can reject this evolutionary constriction, which limits the causal factors to random variations and natural (including sexual) selection. For one thing, we can, if the evidence seems to imply it, return to Darwin's own conviction that acquired characteristics are sometimes inherited. Whiteheadian panexperientialism suggests that actualities of every type and level can influence those of every other type and level. Nothing has a completely impregnable wall around it. Whitehead himself alluded to the improbability of the Weismann barrier (MT, 138–39). Given the vast hierarchies of interacting actualities in an organism, there are numerous possible ways in which a phenotypical or behavioral change could get imprinted in the genome.

Inheritance of Novelty

Beyond this general point, Whitehead provides a way to understand conceptually how a form that is a novel addition in one occasion can become canalized as part of the tradition in a later moment (PR, 107–08). Each occasion of experience, we recall, has a physical pole and then a mental pole. The physical pole involves merely an incorporation and repetition of forms received from previous occasions of experience. When this physical pole prehends a previous occasion in terms of that previous occasion's physical pole, we have a *pure* physical feeling. Insofar as the relations from occasion to occasion in an enduring individual are dominant and consist of pure physical feelings, the individual remains the same, repeating virtually the same forms over and over. Habit reigns supreme. (The "virtually indestructible proton" evidently exhibits this mode of existence to the highest degree.) But there can also be *hybrid* physical feelings, in which the present occasion prehends an earlier one in terms of the earlier one's *mental* pole. If that mental pole contained a novel form, that once-novel form may be taken into the physical pole of the new occasion. It can then, by the subsequent occasion, be prehended by a *pure* physical prehension. It has then become *canalized,* being now part of the tradition that is passed along automatically. This explanation can apply to cells and macromolecules as well as human minds. Whitehead's panexperientialism, unlike mechanistic philosophies, can thereby explain how a pattern that is novel at one stage can become part of the normally inherited tradition at a later stage.

Furthermore, whether or not the inheritance of acquired characteristics sometimes results in some sense from desires or purposes, we can allow that desires and purposes do in some cases induce inheritable changes. In one sense, of course, this is already allowed. It is well accepted that organisms, by purposefully adopting a new pattern of behavior, can bring about genetic changes in their species, insofar as the new behavior means that certain random mutations, if and when they occur, will be selected for. Whiteheadian naturalism, however, can go beyond this Darwinian permission (to which Mayr's statement about animal choices refers). It can allow that an organism's need or desire might more directly lead to a change in the genome. It may well be that Darwinism has been *largely* correct on this issues—that genetic variation is usually random$_{na}$ and even random$_{eps}$, as much evidence suggests. If so, it is tempting, given a mechanistic, reductionistic philosophy in which there could be no change introduced by purposes and no downward causation, to go beyond the evidence and turn this "usually" into "always." Given a more permissive naturalism, there is no reason to go beyond the evidence in this way. There is certainly no reason to fly in the face of the evidence.

Divinely Induced Novelty

The previous paragraph described how the organism's own purposes could bring about change that is not random$_{eps}$ because it is not random$_{na}$. This form of naturalism, furthermore, allows for genetic change that is not random$_{eps}$ because it is influenced by the soul of the universe. This cosmic soul has (by hypothesis) an overall aim at beauty in the most general sense of the term,[24] so that those forms of beauty that are possible in a given context are encouraged. This cosmic aim encourages, in Whitehead's phrase, "a three-fold urge: (i) to live, (ii) to live well, (iii) to live better," with the latter meaning "to acquire an increase in satisfaction" (FR, 8). If an increased capacity for beauty of experience and thereby positive intrinsic value is generally correlated with increased complexity of structure, the general aim of the soul of the universe would provide a ground for the most pervasive trend evident in the evolutionary process. This explanation can apply at every level, because actualities at every level are occasions of experience originating with prehensions of their environments, which would always include the soul of the whole. If mutations do, at least in some circumstances, reflect this derandomizing bias to at least some slight extent, the divine influence could be understood to be operating on the genomes in various ways—perhaps indirectly, by influencing the organism's aims, which would then affect its own genome in the way discussed above, and perhaps even directly.

Evolutionary Progress and the Inner Criterion of Success

This kind of cosmic bias would mean only that not all mutations are random$_{eps}$. It would not necessarily mean that they are not random$_{na}$, because the cosmic aim, by hypothesis, is not toward adaptation to the immediate environment or even toward a value that would be positively correlated with it. Adaptation has to do with the phenotype, which is the only side of an organism the environment can judge. However, the cosmic aim, being for an increase in the capacity for beauty in the sense of intrinsically rewarding experience, is directed toward the inside of each actuality, which is hidden from the agents of natural selection. There need be, in general, no positive correlation between increased intrinsic value and increased survival power.

This distinction between inner and outer criteria of excellence correlates perfectly with the distinction between fitness for survival and our intuitions about higher and lower forms of life: On the one hand, we cannot help but believe that the evolutionary process from bacteria to diverse forms of life existing today, including human life, has involved progress. On the other hand, this

---

24. In a book (GAD) that I saw in manuscript form only after my own manuscript was ready for final editing, John Haught, likewise seeing the divine purpose to be such that "the cosmos is a restless aim toward ever more intense configurations of beauty," quotes in support physicist Freeman Dyson's suggestion (IAD, 298) that "the laws of nature and initial conditions are such as to make the universe as interesting as possible."

progress does not mean an increase in survival power. If we can ask about progress only in terms of this externalist criterion, then those who deny progress are clearly right—dolphins and humans will not outlast bacteria! But the assumption that survival is the only valid criterion is a reflection of the positivist assumption that a "scientific" perspective can deal only with features knowable through our sensory perception, thereby only with external features. This assumption leads to what Whitehead called the "evolutionist fallacy," which is not the truism that the fittest survive, which is obvious, but "the belief that fitness for survival is identical with the best exemplification of the Art of Life" (FR, 4). If the overall goal were survival, there would be no explanation for the rise of life itself:

> [L]ife itself is comparatively deficient in survival value. The art of persistence is to be dead. . . . The problem set by the doctrine of evolution is to explain how complex organisms with such deficient survival power ever evolved. (FR, 4, 5)

The explanation offered by this theistic naturalism is that life and then more complex forms of life emerge because the universe lives within an appetitive, form-inducing soul, which whets the appetites of the creatures for life and then more abundant forms of life, a criterion that is at right angles, at it were, to the criteria for survival.

This Whiteheadian criterion for judging evolutionary progress—greater capacity for experience that is intrinsically valuable—is positively correlated with greater capacity to include more feelings and objective data from the environment in one's experience. This standard is similar, therefore, to that employed by Francisco Ayala, "the ability of an organism to obtain and process information about the environment" (CP, 90). Ayala's wording, however, stresses the purely cognitive side of the experience, whereas Whitehead's criterion is primarily aesthetic. The other difference is that Ayala, *qua* Darwinian scientist, must say that nothing in the evolutionary process makes his criterion of progress "best or more objective than others" (CP, 95). Within Whitehead's philosophy, however, the criterion can be thought to be objective, in that it is believed to be the criterion of the soul of the universe itself, which has inspired the whole evolutionary process. Within this wider naturalism, furthermore, reference to a soul of the universe would not be ruled out of "scientific" discourse any more than reference to any other agents believed really to exist and to exercise explanatory causal influence.[25]

---

25. We, of course, cannot perform experiments to see whether there really is such a soul of the universe. But we also cannot perform experiments to show that there really are logical, mathematical, and aesthetic principles, or that all events occur within a universal, unbroken causal nexus, or that numerous other ideas that we presuppose are really true.

In any case, Ayala's distinctions with regard to this criterion for progress are helpful. To say that progress can and does occur, it is not necessary to mean *general* progress, "which occurs in all historical sequences of a given domain of reality and from the beginning of the sequences until their end." It is sufficient to mean *particular* progress, "which occurs in one or several but not all historical sequences, or . . . during part but not all of the duration of the sequences" (CP, 80). Also, Ayala points out, it is obvious that *uniform* progress has not taken place. The question is only whether *net* progress has occurred (CP, 84). These distinctions are important, because the idea of evolutionary progress is often rejected by pointing out that there is no criterion in terms of which uniform and general progress has occurred. In terms of Whitehead's criterion, which incorporates Ayala's, we can say that *net* progress has occurred in the evolutionary process as a whole by virtue of having occurred in *particular* sequences. Darwinism$_{14}$ is thereby rejected.

Inner Gradualism and Outer Saltationism

The fact that evolution, especially progressive evolution, is *not* uniform, according to all the evidence, brings us to the need to replace gradualism (Darwinian evolutionism$_{10}$) with a saltational view of macroevolutionary change. Because of its nominalism, neo-Darwinism could not accept saltational mutation: The probability that all the parts of an organism would *by chance* make simultaneous leaps in just the directions needed to result in a viable organism was rightly considered virtually nil, and this improbability was increased exponentially in sexually reproducing organisms by the unlikelihood that male and female "hopeful monsters" (Goldschmidt MBE, 390–95) of the same sort would appear purely by chance at the same time and place so as to be able to mate.

As Wesson has stressed, however, such macromutations would not be so unlikely if there are ideal forms or archetypes serving as "attractors" for various forms of life, such as phyla, orders, and species, at least if this point is combined with the above points about the universal capacity to apprehend such forms, the capacity for purposes to exert causal influence, the capacity for downward as well as upward and horizontal causal influence, and the capacity for the genome in particular to be subject to causal influences from a variety of quarters. Given an unconscious apprehension of a new possible form to be actualized, an apprehension that is itself appetitive through responsive sympathy to a cosmic appetitive envisagement of this form, the mutation would *not* be purely fortuitous. Also, because the forms would be serving as attractors for *all* the members of a particular species that had attained a similar level in their internal lives, we would not assume that the mutation would occur only in a single individual. All would be called, and perhaps many would respond, more or less simultaneously.

This idea of saltations in the phenotype can even be combined with the general intuition that change, if naturalistic, must be gradualistic. In terms of the previous distinction between inner and outer, we can see that outer change could be saltational even though inner change is gradualistic. This possibility is based upon the distinction, discussed above, between pure and hybrid physical prehensions. The cosmic soul, by hypothesis, influences the world by whetting the appetites of the various creatures for novel forms, ones that are related to but go beyond forms that have already been canalized into the identity of the various individuals. Various steps, each of which may take a more or less lengthy period of time, must occur before a new form can appear in the phenotypical world upon which natural selection makes its judgment.

The first step, after a creature has reached the stage at which a new gestalt or archetype is a real possibility to be embodied, is for the creature's appetite to be whetted. This occurs only when the new form is prehended conceptually in the mental pole of the successive occasions of experience constituting the career of the enduring individual. The new form, therefore, is realized only in a *restricted* way (PR, 291). It is "an end realized in imagination but not in fact" (FR, 8). Each occasion, in prehending its predecessors, continues their appetitions for the new form merely as appetitions, without physically realizing the form, instead simply reiterating the old forms, received from the physical poles of their predecessors. As long as the prehension of the new form remains purely mental, it results in no change in the shape of the individual.

This change of shape can occur only if the mental or merely appetitive prehension of the form in one occasion is prehended, by means of a *hybrid* physical prehension, in a subsequent occasion. Through this kind of prehension, the new form is *unrestrictedly* and thereby *physically* realized in the individual (PR, 291). Once this has occurred, the new form can then be passed along to successors in the normal way: by *pure* physical prehensions. The new form has now become canalized; it is a regular part of the tradition.

In compound individuals, however, even this development will not suffice to bring about a phenotypical change. If the canalization just described occurred in the psyche of an animal, two more steps would need to occur before any change would be apparent to the outside world: A similar process would need to occur in the relation between the psyche and the living occasions of its bodily cells, at least those involved in reproduction. Then a similar process would have to occur between the cell as a whole, or its living occasions, and some of the enduring individuals constituting the genome. After that last step occurs, the offspring may be significantly different from their parents.

In the event that all of this occurs, the resulting saltation may be so great as to seem to betoken supernatural intervention. That supposition could well appear necessary within a philosophy of nature that identified things with their outsides, their outer appearance. Within such a philosophy, the only conceivable kind of causation is external efficient causation, in which the change produced in one enduring individual must be attributed to the manifest powers of some other enduring individual or individuals. Because all causal power is based upon external, manifest qualities, the changes in the manifest properties of individuals must be very small to seem naturalistic. Within a panexperientialist naturalism, by contrast, there is more to an individual than what appears. An enduring individual oscillates, many times per second, between subjectivity, which is hidden to others, and objectivity, which is manifest to, in the sense of being causally efficacious upon, others. Changes of appetition within the mentality of the individual can accumulate for a long time within its hidden subjectivity, adding up to a quite new gestalt, which can then all at once be made manifest. *Although there is a saltational change in the outer world, the change that has been going on behind the scenes, which is the only change that the divine appetitions can directly influence, has been gradualistic.* Any one of the steps might take dozens, hundreds, thousands, or millions of years. The whole process, from the time the novel form becomes relevant to the time that it is phenotypically incarnated, could take any length of time. In this way, Whitehead's theistic naturalism allows us to do justice to the apparent need for phenotypical saltationism.

Whitehead and Punctuated Equilibria

It might be assumed that, although Whitehead's philosophy allows for this kind of saltationism, Whitehead himself assumed the truth of gradualism, so that this use of his philosophy, while possible, would run counter to his own beliefs. In the first place, however, Whitehead clearly connected his doctrine of eternal forms, and thereby his reaffirmation of Platonic realism, with "the sharp-cut differences between kinds of things." Having discussed this feature of nature with regard to quantum theory—which, he says, would have surprised Newton but not Plato—Whitehead adds:

> It is well to remember that the modern quantum theory, with its surprises in dealing with the atom, is only the latest instance of a well-marked character of nature, which in each particular instance is only explained by some *ad hoc* dogmatic assumption. The theory of biological evolution would not in itself lead us to expect the sharply distinguished genera and species which we find in nature. There might be an occasional bunching of individuals round certain typical forms; but there is no explanation of the almost complete absence of intermediate forms. (PR, 95)

Whitehead's point is that, just as the well-marked character of physical and chemical objects was surprising from the perspective bequeathed by Newton, the well-marked character of biological species and genera has been surprising from the viewpoint of the perspective fostered by Darwin, "the Newton of biology," but that, from the viewpoint of Whitehead's more Platonic outlook, this well-marked character is what should be expected in biology as well as in physics and chemistry.

Also, several of Whitehead's statements show that he affirmed a strong version of the view that is now called "punctuated equilibria." In a general discussion, meant to apply to the rise of both new organic species and new scientific methodologies, he says:

> With a happy choice, the new method quickly reaches its meridian stage. . . . On the whole, the evidence points to a certain speed of evolution from a nascent methodology into the middle stage which is relatively prolonged. (FR, 19)

That Whitehead anticipated the idea of punctuated equilibria within the history of science, which has been articulated by Thomas Kuhn in terms of occasional "paradigm shifts," is shown by a passage that might have indirectly inspired the Kuhnian distinction between "ordinary science," which works within the reigning paradigm, and "revolutionary science," which brings a new paradigm into being. In this passage, Whitehead says:

> The advance of any reasonably developed science is twofold: There is the advance of detailed knowledge within the method prescribed by the reigning working hypothesis; and there is the rectification of the working hypothesis dictated by the inadequacies of the current orthodoxy. (AI, 223)

With regard to natural evolution in particular, Whitehead shows that he affirms not only that equilibrium reigns for long periods, but also that progress, insofar as it occurs, is not universal:

> [I]f we survey the universe of nature, mere static survival seems to be the general rule, accompanied by a slow decay. The instances of the upward trend are represented by a sprinkling of exceptional cases. (FR, 29)

Whitehead on Divinely Induced Ideals

The idea that the occasional rise of radically new forms of order is inspired by the Divine Reality, which provides novel ideals serving as attractors for new achievements, is also clearly expressed by Whitehead. Having said that our experience at least implicitly presupposes that the universe includes "a source of ideals" and that the "effective aspect of this source is deity as immanent in the present experience," Whitehead continues:

Thus there is an essential relevance between deity and historic process. For this reason, the form of process is not wholly dependent upon derivation from the past. As epochs decay amid futility and frustration, the form of process derives other ideals involving novel forms of order.... [A]s the present becomes self-destructive of its inherited modes of importance, then the deistic influence implants in the historic process new aims at other ideals.[26] (MT, 103)

The fact that this implantation of novel forms into the evolutionary process was central to Whitehead's idea of the deity of the universe is shown by a subsequent passage:

We are now discussing an alternative rendering of Descartes' notion of perfection. It is the notion of that power in history which implants into the form of process, belonging to each historic epoch, the character of a drive towards some ideal, to be realized within that period. (MT, 120)

Although in this passage Whitehead is talking primarily about human history, his nondualistic philosophy means that this idea of a divinely inspired drive towards an ideal applies to all epochs and all levels of existence. For example, referring to the fact that the "material universe has contained ... some mysterious impulse for its energy to run upwards," Whitehead says that "there must have been some epoch in which the dominant trend was the formation of protons, electrons, molecules, the stars" (FR, 24). The fact that the idea of a divinely inspired "drive towards some ideal" was, in Whitehead's perspective, relevant to biological evolution in particular is shown by the following passage, which follows upon a denial that there is one ideal order for all:

It is notable that no biological science has been able to express itself apart from phraseology which is meaningless unless it refers to ideals proper to the organism in question. This aspect of the universe impressed itself on that great biologist and philosopher, Aristotle. His philosophy led to a wild overstressing of the notion of "final causes" during the Christian middle ages; and thence, by a reaction, to the correlative overstressing of the notion of "efficient causes" during the modern scientific period. (PR, 84)

Although this passage indicates that Aristotle, or at least Christian Aristotelians, exaggerated the importance of formal causes, which can serve as final (in the

---

26. Although Whitehead used the term "deistic" here, he was not endorsing "deism" in the sense of a Creator who, after creating, no longer exerts influence in the evolutionary process. Rather, the passage was a commentary on Samuel Alexander's *Space, Time, and Deity,* and Whitehead was simply using the adjectival form of Alexander's term "deity."

sense of attracting) causes, it also suggests that modern Darwinian biology has *underestimated* their importance. By combining this passage with the previous ones, we can see that in Whitehead's opinion the process of biological evolution will never be intelligible apart from the idea of the occasional implantation of new ideal forms into the process by the divine power of the universe, in which all previously unactualized forms have their home until they are needed.

The language of the *occasional* implantation of new forms, to be sure, may seem to contradict ontological uniformitarianism (Darwinian evolutionism₄). If the Divine Reality occasionally acts in an extraordinary way, rather than acting in one and the same way always and everywhere, has not our purportedly naturalistic theism become supernaturalistic? The contradiction, however, is only apparent. The Divine Reality acts in the same way in every period and in relation to every individual event: by providing possible forms for actualization. The fact that certain periods of the evolutionary process are extraordinary, in the sense that radically new forms quite suddenly get incarnated in the world, implies no new type or even intensity of divine activity. Rather, what makes the epoch extraordinary is that some of the creatures have become ready to incarnate new forms of order. Exactly what makes them thus "ready" will probably always be largely beyond our understanding. The suggestion made above, however, is that the divine appetition for new forms has finally evoked a sympathetic appetition for these forms in the creatures in question, an inner appetition that eventually results in a saltational change at the phenotypical level. The divine activity is constant. Only the dramatic responses to this activity are occasional.

No Supersaltations

It might be wondered whether this version of naturalistic theism renders superfluous, as does supernaturalistic theism, the very idea of an evolutionary process. Some supernaturalists, as we saw in Chapter 3, have achieved harmony with both the accepted time table and the evidence for occasional saltations by affirming progressive creationism, according to which each new species is created by God. The problem, as we saw, is that this view makes us wonder why God, being able to create new species *ex nihilo,* would take so long to get to the higher forms of life, with their greater intrinsic value. Some might suppose that the version of theistic naturalism articulated here, by virtue of allowing for saltations, creates the same problem. The crucial difference, how-ever, is that although this view does allow for saltations, it does not allow for *super*saltations. It does not, in other words, allow for leaps of any size what-soever. Besides not allowing for creations out of *nothing*, it also does not allow for creations out of *inadequate* materials. Humans, for example, could not have been formed directly out of bacteria, or the first multicelled organisms, or even the first mammals. This wider naturalism does allow for larger steps than did the nontheistic, nominalistic version of naturalism. But it still entails that the creation of the present state of the world had to involve a step-by-step process. Large steps, in fact, are involved only in the external side of the process. Internally, in line with Darwin's intuition that naturalism requires gradualism, the steps are still small.

The Origin of Our Universe

At this point, however, the defender of supernaturalism may retort that, although this naturalistic form of theism may be compatible with the world's evil and the slow, step-by-step nature of the process through which our world was created, it is not adequate to the new account of how our universe began. That is, Whiteheadian theism portrays itself as consistent with evil and evolution by virtue of its doctrine that divine power is persuasive, not coercive, an essential dimension of which is the affirmation that our universe was created not *ex nihilo* but out of chaos. However, the critic may contend, the evidence now supports the idea that our universe arose from a big-bang singularity, through which time itself arose—which contradicts the pantemporalism of the Whiteheadian worldview. This initiating event of our universe, furthermore, surely required omnipotence as traditionally affirmed: The establishment of all the finely tuned "cosmic constants" could not have been effected by a God with merely persuasive power. In order to get these constants so precisely obeyed, Fred Hoyle's "supercalculating intellect" must also be an omnipotent agent with coercive power (see Corey *BD*).

I can here only indicate the direction my response would take. First, I would argue that we have no evidence whatsoever that the beginning of our particular universe was an absolute beginning of finite existence as such. The virtual consensus that there was a Big Bang, therefore, provides no support for the idea that it involved the beginning of time or for the idea of a deity that, having existed all alone, had a monopoly on power.

Second, with regard to the fine tuning, my argument would be that although divine power is and always has been persuasive, not coercive, in a situation approaching absolute chaos the divine power could have coercive-like effects. The crucial issue is whether the realm of finite actualities has enduring objects, in the sense of temporally ordered societies of actual occasions. When an occasion is a member of an enduring object, it is still, like all occasions, influenced by all prior occasions, but it is *primarily* influenced by a dominant line of inheritance. Each new electronic occasion, for example, primarily repeats the form of the electronic occasions in the enduring electron to which it belongs. In the world as we know it, divine power cannot unilaterally effect its will because of competition from the power of trillions of such enduring objects. In this Whiteheadian naturalism, it will be recalled, the so-called laws of nature are really its most long-standing habits. Enduring objects, such as protons, atoms, molecules, and cells, represent various forms of long-standing habits, each of which tends to impose itself upon future actualities. Each actual occasion, in coming into existence, receives causal influence from the entire past universe, with its habitual ways of being. Each new occasion is also influenced by the initial aim received from God, but this influence is, in one sense, merely one more influence among many. So, even apart from relatively high-level occasions, which have a significant mental pole and thereby a significant degree of self-determining power, the divine influence cannot unilaterally bring about states of affairs.

A state approaching absolute chaos, however, would by definition be a state in which there are no enduring objects. There would be a plenum of finite actual occasions, to be sure, but they would occur randomly, in the sense that none of them would be members of enduring individuals with a dominant line of inheritance. In such a situation, there would be no habits. There would, of course, be the *metaphysical* principles, which would necessarily be exemplified by all actual occasions. But there would be no contingent, cosmological habits. There would, accordingly, be no significant competition to the divine suggestions. In that situation, a powerful thought—"Let there be such and such!"—might be instantiated with a high degree of conformity. From then on, given the existence of enduring individuals embodying the proposed patterns, the divine purposes will never again be so efficacious (at least not until this universe has wound completely down, so that absolute chaos is again approached). In that one moment, however, the Creator's persuasive power could simulate coercive omnipotence, so that a finely tuned universe could result.

Summation

In previous chapters, we have seen that, on the basis of the wider naturalism articulated in this book, we can have a worldview that, besides supporting the basic presuppositions of Christianity and other theistic religions, also provides a more adequate framework for science than did the materialistic version of naturalism. In the present chapter, we have seen that this is also true for what has arguably been the basic and most difficult issue, that of reconciling the religious belief in the world as divine creation with the scientific community's evolutionary naturalism. What we have found is not merely that belief in evolution does not rule out belief in creation and vice versa. We have also found that the scientific community, to develop an adequate evolutionary theory, will evidently need a worldview involving freedom, purposive causation, and even a naturalistic theism, and that the religious community, to have a form of theism that is not falsified by the world's evils and other facts, requires an evolutionary naturalism, according to which a world such as ours could only have come about by a long, slow, step-by-step process. We need not, therefore, rest content with merely showing that there is no outright contradiction between science and our religious and moral intuitions. The results of some two centuries of scientific research and reflection on the evolutionary development of our world seem to be best interpretable in terms of a integral worldview that, while fully naturalistic, is also fully religious.[27]

---

27. It has been widely assumed that any doctrine, such as Whiteheadian naturalistic theism, that affirms creation out of chaos, instead of creation *ex nihilo*, would be contrary to the biblical tradition and thereby inadequate for Judaism, Christianity, and Islam. Recent studies by Jon Levenson (CPE) and Gerhard May (CEN), however, have shown conclusively that the doctrine of creation *ex nihilo* is postbiblical. Levenson shows that the idea of creation out of chaos was central to Hebrew cultic life, while May shows that the doctrine of creation *ex nihilo* does not even, as widely thought, appear in the inter-testamental period, such as in 2 Maccabees (so that it could be presupposed in the New Testament), but was first developed in the second half of the second century C.E., in response to Marcion's position. I have summarized their arguments in "Creation Out of Nothing, Creation Out of Chaos, and the Problem of Evil" (in Stephen Davis, ed., EE).

9

Epilogue

In this brief epilogue, I will, besides summarizing the main points of the book, place these points in a somewhat broader context than provided earlier.

As suggested by many of the "isms" that have shaped both cultural and political events—such as Voluntarism (or Nominalism), Protestantism, Roman Catholicism, Imperialism-and-Colonialism (by both Protestant and Roman Catholic nations), Marxism, Nazism, Positivism, Scientism, Materialism, Darwinism, and Modernism—the history of the modern Western world could be written largely in terms of two types of movements: those reflecting a supernaturalistic Christian worldview and those reacting against this worldview and its effects. Such a history, especially if it included experiences in the depths of human souls as well as large-scale social and political developments, would provide a record of enormous tragedy, ranging from disappointment, despair, and desperation to massive exploitation, enslavement, and extermination.

The negative message of the present book is that much of this tragedy, sad to say, has been based on a mistake—or, really, a double mistake: the equation of theism with supernaturalism and the equation of scientific naturalism with a sensationist epistemology and an atheistic, materialistic worldview. It is appropriate for philosophers of religion and theologians to try to rectify this double mistake because both aspects of it originated primarily in religious beliefs and motives, with naturalism$_{sam}$ being derivative from doctrines originally formulated to support a supernaturalistic version of Christianity.

At the root of this supernaturalistic religion was the doctrine of *creatio ex nihilo*. The Hebrew Bible, Plato, and Aristotle provided resources for another way of thinking, according to which a realm of nondivine entities has always existed, the Divine Reality influences these entities persuasively by attraction, and the creation of our particular world involved persuading the nondivine realm to embody new forms of order, so that creation was out of chaos, not out of absolute nothingness. In what became orthodox Christian doctrine, however, the Divine Reality was said to be the sole necessary existent and thereby essentially the sole repository of power. Our world, therefore, was said to be contingent in all respects. All of its causal principles, having been freely created by God, could be interrupted at will. This doctrine was perhaps motivated primarily by eschatological concerns, as it guaranteed that, at the end, Divine Power could vanquish demonic power. But this doctrine of creation, by allowing for a unique incarnation of God in Jesus, infallible inspiration of the Christian Scriptures, and supernatural miracles to attest to God's stamp of approval, also allowed Christianity's theologians to portray it as the One True Religion. Although it was important to show the superiority of a religious view of reality to a materialistic, atheistic view, the more important concern was to declare the superiority of Christianity over other religions.

Although the doctrine that all power essentially belonged to God would have allowed theologians to declare that God brings all events about directly, various considerations, such as the problem of evil, the growing awareness of regularities, the desire for an autonomous science of the finite world, and simple common sense, led to the attribution of real (albeit derivative) power to the creaturely realm. This idea that the Creator and the creatures both exert causation could have led to a doctrine of divine-creaturely cooperation, according to which each event is brought about partly by divine causation and partly by finite causal power.

The doctrine of creation *ex nihilo*, however, suggested that divine causation is different in kind from creaturely causation, and the doctrine of divine omnipotence suggested that God's causal power wholly, not just partly, brings things about. For these and related reasons, the view that became dominant involved the primary-secondary scheme, according to which God is the sufficient cause of all things and events—their whatness as well as their thatness—but brings about the whatness of most things and events by means of secondary or natural causes. This means that, as long as attention is focused solely on the nature or whatness of events, as distinct from the question of why they exist at all, most things and events can in principle be given a sufficient causal explanation without referring to divine causation. This doctrine implied that if the category of divine causation is thought to be especially significant for the whatness of some event, an *interruption* of the normal causal processes of the world would be involved. This implication was, in fact, embraced wholeheartedly, as "miracles" were defined as events in which God as primary cause brought about the events immediately, without using *any* secondary causation.

This supernaturalistic view was intensified and undergirded at the outset of the modern world. Whereas some doctrines of God in the medieval period regarded the world as importantly derivative from the divine *nature* and *reason*, the fourteenth- and fifteenth-century movement known as Nominalism or Voluntarism, stressing the divine freedom from any constraints, regarded all things as derivative from the divine *will*. This intensification of focus on the divine omnipotence, in both Roman Catholic and Protestant thought, was then employed to develop a new framework for science not only overagainst medieval Aristotelianism but also overagainst a third movement, consisting of Neoplatonic, spiritualist, and magical ideas, which provided a threat to the idea of Christianity as the One True Religion. The most significant aspect of the development usually called "the rise of modern science" is simply the victory of this new framework over its rivals to become the accepted framework for scientific thinking.

This new framework, with its mechanistic doctrine of (physical) nature, its dualistic doctrine of the human being, and its sensationist doctrine of perception, undergirded the belief that our world has an Omnipotent Creator who acts supernaturally in the world. The mechanistic doctrine of matter, besides being used to prove the existence of a Creator responsible for its motion and apparent gravitational attraction, was also used to support, against those who held influence at a distance to be a natural capacity, the genuinely supernatural character of Christianity's "physical miracles." This mechanistic doctrine of matter was also used against mortalists, to support the idea that the human mind or soul, being self-moving, is different in kind from the bodily constituents, therefore arguably immortal. The fact that mind and body were different in kind was, far from an embarrassment, further evidence for the existence of an omnipotent deity, who can make totally unlike things interact, or at least seem to. The sensationist doctrine of perception, besides being used against "enthusiasts" who believed they could be directly inspired by God, perhaps in ways that would supersede the teachings of the institutional church, was also used to support the genuinely supernatural character of Christianity's "mental miracles." This sensationist doctrine of perception meant, furthermore, that our moral knowledge, being incapable of naturalistic explanation, provides yet another basis for believing in a supernatural deity. With these doctrines, the early modern worldview undergirded the belief both (1) that the normal course of events excludes any variable divine influence in the world, and (2) that the causal processes involved in this normal course of events was freely established by a Supernatural Creator who can, and sometimes does, interrupt these causal processes.

In adopting sensationism and mechanism for religious reasons, the advocates of the early modern worldview in effect adopted an all-or-nothing strategy, undercutting the basis for any form of religious belief except a wholly supernatural form, through which they could support their form of Christianity as the

One True Religion. This strategy soon backfired. The supernaturalism of the early modern worldview was rejected by the scientific community in particular and the intellectual community in general, with the result that the "scientific worldview" came to be equated with naturalism$_{sam}$, which ruled out not only supernaturalistic Christian faith but any significantly religious beliefs whatsoever. Devoid of the supernaturalistic framework, the mechanistic doctrine of matter meant that there could be no divine influence in nature. The sensationist doctrine of perception and the atheistic doctrine of the universe provided mutually supporting epistemological and ontological reasons for rejecting the possibility of authentic religious experience and for accepting moral relativism, even nihilism. The collapse of mind-body dualism into materialistic identism implied the impossibility not only of life after death but also of genuine freedom. The combination of the sensationist epistemology and the mechanistic-materialist ontology meant, furthermore, that Christianity and other historic religions, by virtue of the central role played in them by paranormal events, were regarded as hopelessly superstitious.

Because of these developments, the present worldviews of the scientific and religious communities are necessarily in conflict. Although this is especially true insofar as the religious communities still retain a supernaturalistic worldview, it is true even if this form of theism is eschewed: Given the equation of naturalism with naturalism$_{sam}$, any worldview that is religiously robust enough to allow a religious community to perpetuate itself will *seem* supernaturalistic.

The moral of this analysis is that none of the hitherto dominant responses has any chance of overcoming this conflict at the heart of modern culture. The conservative approach, which uses the inadequacies of naturalism$_{sam}$ to urge the scientific community to return to a supernaturalistic framework, whether of an interventionist or a deistic type, is surely quixotic. The modern liberal approach, insofar as it tries to accommodate religious beliefs to naturalism$_{sam}$, redefines these beliefs so drastically that religious communities could not adopt these theologies without committing suicide.

Whereas these heretofore dominant approaches share the equation of scientific naturalism with naturalism$_{sam}$, the present book suggests that we henceforth equate scientific naturalism, meaning the naturalism properly presupposed by the scientific community, only with naturalism$_{ns}$. This equation would make room for a new version of naturalism that might not only be more adequate for the scientific community but also more adequate than supernaturalism for the religious communities. I have suggested that Whitehead's philosophy, being both a scientific and a religious naturalism, could play this role. Someone else may, to be sure, come up with a position that would provide an even more satisfactory basis for a shared worldview. For now, however, Whitehead's philosophy seems to be the best candidate. I have, accordingly, employed it to illustrate how a less restrictive form of naturalism not only could help scientists deal more adequately with a range of important issues, but also could provide an adequate basis for articulating religious convictions.

One thing that this philosophy cannot do, to be sure, is provide the basis for holding Christianity (or some other faith) to be the One True Religion, through which alone people can come into harmonious relationship with the Divine Reality of the universe. But why, egotistical and other sinful desires aside, should we want to hold *this*? It was the desire to defend this supernaturalistic, exclusivistic form of Christianity, as we have seen, that led to the widespread reaction against Christianity altogether. And what should be important to us, surely, is that our own religion is basically true, not that it alone is true. Put otherwise, the crucial contrast today is that between a fundamentally religious view of the universe, which all the religions share, and the wholly nonreligious view of naturalism$_{sam}$, which has been spreading over the globe in the guise of the "modern scientific worldview." The various religious traditions share the view that there is a Holy Reality, that we can experience it and thereby be empowered and transformed, that we have a significant degree of freedom, that we should exercise our freedom in the light of moral and aesthetic norms rooted in the nature of things, that our present life is a part of a larger journey continuing beyond bodily death, and that our lives have an ultimate meaning. The currently dominant form of scientific naturalism, with its sensationism, atheism, and materialism, denies all these beliefs. Rather than focusing on their differences in a competitive way, therefore, the various religions should focus on what they have in common. For one thing, the fact that the various religions of the world share these beliefs provides a significant witness to their truth. For another thing, the exclusivistic idea that the Divine Reality chose only one of the religions of the Earth to be a vehicle of saving knowledge would actually contradict the most fundamental of all religious beliefs, which is that the Divine Reality is a loving, compassionate reality. Liberal religious thinkers have made this point at least since the 18th century. Unfortunately, however, liberal religious thought from deism to the present has been so emaciated by its acceptance of distinctively modern presuppositions that this liberal attitude toward other religions has been associated with an extremely thin, virtually vacuous, theology. Given the shift from a modern to a postmodern form of liberalism, however, this need no longer be the case. In this context, each religion can regard all the others not as challenges to its basic truth but as so many confirmations of this truth.

This epilogue is not, of course, the place to develop a doctrine of the relations among the various religions. I perhaps should say, however, that my prior comments should not be taken as an endorsement of the old view that all the religions share a common essence and that this essence is all that is really important about the various religions. I believe, instead, that the various traditions give witness to somewhat different truths and values, so that each could be enriched by learning from the others, somewhat as John Cobb has suggested in

Beyond Dialogue: Toward a Mutual Transformation of Christianity and Buddhism. Beyond this statement, however, I will have to leave this topic for the future.[1] For now, the point is that the absence in Whitehead's philosophy of any basis for declaring the exclusive truth of one religious tradition is, far from being a handicap, one more of its strengths.

In any case, what is important is that, whatever philosophy provides the basis, the scientific community—for its own sake, for the sake of truth, and for the sake of the world—leave naturalism$_{sam}$ behind. As I argue elsewhere—most fully in a work in progress (DC)—it is equally important that the religious communities—for their own sake, for the sake of truth, and for the sake of the world—leave supernaturalism behind. If this double transcendence occurs, the conflict between our religious intuitions and our scientific intuitions, which has been such a destructive feature of the late modern world, could itself be transcended, and a new era of harmony and cooperation could be ushered in.

---

1. I have made a first effort in dealing with the relations among the religions, especially the relation between theistic and nontheistic religions, in Chapter 7 of my RWS.

REFERENCES

Adler, Julius, and Wing-Wai Tse (DMB). "Decision-Making in Bacteria." *Science* 184 (21 June 1974): 1292–94.

Alcock, James E. (PSM). *Parapsychology: Science or Magic? A Psychological Perspective.* Oxford and New York: Pergamon Press, 1981.

Alcock, James E. (SS). *Science and Supernature: A Critical Appraisal of Parapsychology.* Buffalo: Prometheus Books, 1990.

Alston, William (PG). *Perceiving God: The Epistemology of Religious Experience.* Ithaca: Cornell University Press, 1991.

Asimov, Isaac (IB). *In the Beginning.* New York: Crown Books, 1981.

Atkinson, Rita, Richard C. Atkinson, Edward E. Smith, and Daryl J. Bem (IP). *Introduction to Psychology,* 10th ed. San Diego and New York: Harcourt Brace Jovanovich, 1990.

Ayala, Francisco J. (CP). "Can 'Progress' Be Defined as a Biological Concept?" In *Evolutionary Progress.* Edited by Matthew H. Nitecki, 75–96. Chicago and London: University of Chicago Press, 1988.

Baker, Gordon, and Katherine J. Morris (DD). *Descartes' Dualism.* London and New York: Routledge, 1996.

Barbour, Ian (RAS). *Religion in an Age of Science.* San Francisco: Harper and Row, 1990.

Baumer, Franklin (RRS). *Religion and the Rise of Scepticism.* New York: Harcourt, Brace, 1960.

Behe, Michael J. (DBB). *Darwin's Black Box: The Biochemical Challenge to Evolution.* New York: Free Press, 1996.

Beloff, John (EM). *The Existence of Mind.* New York: Citadel, 1965.

Benacerraf, Paul (MT). "Mathematical Truth." In *Philosophy of Mathematics.* Edited by Paul Benacerraf and Hilary Putnam, 2nd ed., 403–20. Cambridge: Cambridge University Press, 1983.

Benor, Daniel J. (HR). *Healing Research.* Munich: Helix Verlag, 1993.

Benson, Herbert (BRR). *Beyond the Relaxation Response.* New York: Times Books, 1984.

Benson, Spencer A. (IBE). "Is Bacterial Evolution Random or Selective?" *Nature* 336 (3 November 1988): 21–22.

Berman, Morris (RW). *The Reenchantment of the World.* Ithaca: Cornell University Press, 1981.

Birch, Charles (OP). *On Purpose.* Kensington: New South Wales University Press, 1990.

Bock, Walter J. (SE). "The Synthetic Explanation of Macroevolutionary Change: A Reductionistic Approach." *Bulletin of the Carnegie Museum of Natural History* 18 (1979): 20–69.

Bohm, David (PS). "Postmodern Science and a Postmodern World." In *The Reenchantment of Science: Postmodern Proposals.* Edited by David Ray Griffin, 57–68. Albany: State University of New York Press, 1988.

Bohm, David, and B. J. Hiley (UU). *The Undivided Universe: An Ontological Interpretation of Quantum Theory.* London and New York: Routledge, 1993.

Bowler, Peter J. (E). *The Eclipse of Darwinism: Anti-Darwinian Evolution Theories in the Decades Around 1900.* Baltimore: Johns Hopkins University Press, 1983.

Boyle, Robert (W). *The Works of the Honourable Robert Boyle.* London: A. Millar, 1744.

Braude, Stephen E. (LI). *The Limits of Influence: Psychokinesis and the Philosophy of Science.* New York: Routledge and Kegan Paul, 1986.

Broad, C. D. (RPPR). *Religion, Philosophy and Psychical Research.* New York: Humanities Press, 1969.

Broad, William, and Nicholas Wade (BT). *Betrayers of Truth: Fraud and Deceit in the Halls of Science.* New York: Simon and Schuster, 1983.

Brooke, John Hedley (SR). *Science and Religion: Some Historical Perspectives.* Cambridge: Cambridge University Press, 1991.

Broughton, Richard S. (PCS). *Parapsychology: The Controversial Science.* New York: Ballantine Books, 1991.

Cairns, John, Julie Overbaugh, and Stephan Miller (OM). "The Origin of Mutants." *Nature* 335 (8 September 1988): 142–45.

Campbell, Keith (BM). *Body and Mind,* 2nd ed. Notre Dame: University of Notre Dame Press, 1984.

Čapek, Milič, ed. (CST). *The Concepts of Space and Time* (Boston Studies in the Philosophy of Science, Vol. 22). Dordrecht: Reidel, 1976.

Čapek, Milič (NAT). *The New Aspects of Time: Its Continuity and Novelties: Selected Papers in the Contemporary Philosophy of Science,* ed. Robert S. Cohen. Dordrecht/ Boston: Kluwer Academic Publishers, 1991.

Čapek, Milič (UIF). "The Unreality and Indeterminacy of the Future in the Light of Contemporary Physics." In *Physics and the Ultimate Significance of Time: Bohm,*

Prigogine, and Process Philosophy. Edited by David Ray Griffin, 297–308. Albany: State University of New York Press, 1986.

Caspari, Ernst W. (EGW). "An Evaluation of Goldschmidt's Work After Twenty Years." In *Richard Goldschmidt: Controversial Geneticist and Creative Biologist.* Edited by Leonie K. Piternick. Basel and Boston: Birkhäuser Verlag, 1980, 19–23.

Cassirer, E., P. O. Kristeller, and John H. Randall, Jr. (RPM). *The Renaissance Philosophy of Man.* Chicago: University of Chicago Press, 1948.

Chihara, C. (AGT). "A Gödelian Thesis Regarding Mathematical Objects: Do They Exist? And Can We Perceive Them?" *Philosophical Review* 91 (1982): 211–17.

Churchland, Paul (OS). "The Ontological Status of Intentional States: Nailing Folk Psychology to Its Perch." *Behavioral and Brain Sciences* 11/3 (1988): 507–08.

Cobb, John B., Jr. (RS). "The Resurrection of the Soul." *Harvard Theological Review* 80/2 (1987): 213–27.

Cobb, John B., Jr. (SCE). *The Structure of Christian Existence.* Philadelphia: Westminster Press, 1967.

Collins, James (DPN). *Descartes' Philosophy of Nature.* London: Blackwell, 1971.

Corey, Michael A. (BD). *Back to Darwin: The Scientific Case for Deistic Evolution.* Lanham, Md.: University Press of America, 1994.

Crick, Francis (AH). *The Astonishing Hypothesis: The Scientific Search for the Soul.* London and New York: Simon and Schuster, 1994.

Crick, Francis (LI). *Life Itself.* New York: Simon and Schuster, 1981.

Crick, Francis, and Leslie E. Orgel (DP). "Directed Panspermia." *Icarus* 19 (1973). Reprinted in *The Quest for Extraterrestrial Life.* Edited by Donald Goldsmith. Mill Valley, Calif.: University Science Books, 1980.

Crombie, A. C. (M). "Marin Mersenne." C. G. Gillispie, ed., *Dictionary of Scientific Biography,* Vol. 9: 316–22. New York: Scribner's, 1974.

Cronin, Helena (AP). *The Ant and the Peacock: Altruism and Sexual Selection from Darwin to Today.* Cambridge and New York: Press Syndicates of University of Cambridge, 1991.

Darwin, Charles (A). *The Autobiography of Charles Darwin,* ed. Nora Barlow. New York: Norton, 1969.

Darwin, Charles (DLL). *The Life and Letters of Charles Darwin,* ed. Francis Darwin, 2 vols. New York: D. Appleton, 1896.

Darwin, Charles (DM). *The Descent of Man* (together with *Selection in Relation to Sex).* London: John Murray, 1871.

Darwin, Charles (MLD). *More Letters of Charles Darwin*, 2 vols. New York: D. Appleton, 1903.

Darwin, Charles (NS). *Charles Darwin's Natural Selection: Being the Second Part of His Big Species Book Written from 1856 to 1859,* ed. Robert C. Stauffer. Cambridge: Cambridge University Press.

Darwin, Charles (OS). *The Origin of Species.* New York: Mentor Books, 1958.

Darwin, Charles (VAP). *The Variation of Animals and Plants Under Domestication,* 2 vols. New York: Orange Judd and Co., 1868.

Davis, Stephen T. (EE). *Encountering Evil: Live Options in Theology,* 2nd ed. Atlanta: Westminster/John Knox, 2001.

Dawkins, Richard (BW). *The Blind Watchmaker: Why the Evidence of Evolution Reveals a Universe without Design.* New York and London: Norton, 1987.

Dawkins, Richard (SG). *The Selfish Gene,* 2nd ed. Oxford: Oxford University Press, 1989.

De Beer, Gavin (E). "Evolution." *The New Encyclopedia Britannica,* 15th ed. Vol. 7: 7–23. London: Encyclopedia Britannica, 1973–74.

Dennett, Daniel E. (CE). *Consciousness Explained.* Boston: Little, Brown and Co., 1991.

Dennett, Daniel E. (DDI). *Darwin's Dangerous Idea: Evolution and the Meaning of Life.* New York: Simon and Schuster, 1995.

Denton, Michael (E). *Evolution: A Theory in Crisis.* London: Burnett Books, 1991.

Descartes, René (PWD). *The Philosophical Writings of Descartes,* ed. John G. Cottingham, Robert Stoothoff, Dugald Murdoch, and (Vol. III) Anthony Kenny. Cambridge University Press, 1985: Vol. I and II; 1991: Vol. III.

Desmond, Adrian, and James Moore (D). *Darwin.* New York: Viking Penguin, 1991.

Dobbs, Betty Jo (FNA). *The Foundations of Newton's Alchemy.* Cambridge: Cambridge University Press, 1975.

Dobzhansky, Theodosius (CCE). "Chance and Creativity in Evolution." In *Studies in the Philosophy of Biology.* Edited by Francisco J. Ayala and Theodosius Dobzhansky, 307–38. Berkeley: University of California Press, 1974.

Dossey, Larry (HW). *Healing Words: The Power of Prayer and the Practice of Medicine.* San Francisco: HarperSanFrancisco, 1993.

Drees, Willem (RSN). *Religion, Science and Naturalism.* Cambridge: Cambridge University Press, 1996.

Duran, Jane (PD). "Philosophical Difficulties with Paranormal Knowledge Claims." In *Philosophy of Science and the Occult.* Edited by Patrick Grim, 196–206. Albany: State University of New York Press, 1982.

Dyson, Freeman (IAD). *Infinite in All Directions.* New York: HarperCollins, 1988.

Easlea, Brian (WH). *Witch Hunting, Magic, and the New Philosophy: An Introduction to the Debates of the Scientific Revolution 1450–1750.* Atlantic Highlands, N.J.: Humanities Press, 1980.

Eccles, John C. (HS). *How the Self Controls Its Brain*. Berlin, Heidelberg and New York: Springer-Verlag, 1994.

Eddington, Arthur (NPW). *The Nature of the Physical World*. Ann Arbor: University of Michigan Press, 1968.

Edel, Abraham (NHS). "Naturalism and Ethical Theory." In Krikorian, ed., NHS, 65–95.

Edge, Hoyt L., Robert L. Morris, John Palmer, and Joseph H. Rush (FP). *Foundations of Parapsychology: Exploring the Boundaries of Human Capability*. Boston and London: Routledge and Kegan Paul, 1986.

Einstein, Albert (MR). *The Meaning of Relativity*, 3rd ed. Princeton: Princeton University Press, 1950.

Eisenbud, Jule (PF). *Paranormal Foreknowledge: Problems and Perplexities*. New York: Human Sciences Press, 1982.

Eisenbud, Jule (PP). "Paranormal Photography." In *Handbook of Parapsychology*. Edited by Benjamin Wolman, 414–32. New York: Van Nostrand Reinhold, 1977.

Eisenbud, Jule (WTS). *The World of Ted Serios: "Thoughtographic" Studies of an Extraordinary Mind*. New York: Simon and Schuster, 1968.

Eldredge, Niles (RD). *Reinventing Darwin: The Great Debate at the High Table of Evolutionary Theory*. New York: John Wiley and Sons, 1995.

Eldredge, Niles, and Stephen Jay Gould (PE). "Punctuated Equilibria: An Alternative to Phyletic Gradualism." In *Models of Paleobiology*. Edited by T. J. M. Schopf, 82–115. San Francisco: Freeman, Cooper and Co., 1972.

Endler, John A., and Tracy McLellan (PE). "The Processes of Evolution: Toward a Newer Synthesis." *Annual Review of Ecology and Systematics* 19 (1988): 395–421.

Erickson, Millard J. (CT). *Christian Theology*. Grand Rapids: Baker Book House, 1985.

Flew, Antony (C). "Commentary." In Ray Hyman, *The Elusive Quarry. A Scientific Appraisal of Psychical Research*, 267–69. Buffalo: Prometheus Books, 1989.

Flew, Antony (PR). "Parapsychology Revisited: Laws, Miracles, and Repeatability." In *Philosophy and Parapsychology*. Edited by Jan Ludwig, 263–69. Buffalo: Prometheus Press, 1978.

Flew, Antony (PSP). "Parapsychology: Science or Pseudoscience?" In *A Skeptic's Handbook of Parapsychology*. Edited by Paul Kurtz, 519–36. Buffalo: Prometheus Press, 1985.

Forrest, Peter (GWS). *God Without the Supernatural: A Defense of Scientific Theism*. Ithaca: Cornell University Press, 1996.

Frazer, James (GB). *The Golden Bough: A Study in Magic and Religion*. London: Macmillan, 1923.

Futuyma, Douglas J. (EB). *Evolutionary Biology*. Sunderland, Mass.: Sinauer, 1979.

Futuyma, Douglas J. (ST). *Science on Trial: The Case for Evolution*. New York: Pantheon Books, 1983.

Gauld, Alan (FPR). *The Founders of Psychical Research.* New York: Shocken Books, 1978.

Gauld, Alan, and A. D. Cornell (P). *Poltergeists.* London: Routledge, 1979.

Gillespie, Neal C. (CD). *Charles Darwin and the Problem of Creation.* Chicago: University of Chicago Press, 1979.

Gödel, Kurt (WC). "What is Cantor's Continuum Problem? Supplement to the Second [1964] Edition." In *Collected Works,* Vol. II. Edited by Solomon Feferman et al., 266–69. New York: Oxford University Press, 1990.

Goldbeter, A., and D. E. Koshland, Jr. (SMM). "Simple Molecular Model for Sensing Adaptation Based on Receptor Modification with Application to Bacterial Chemotaxis." *Journal of Molecular Biology* 161/3 (1982): 395–416.

Goldschmidt, Richard (MBE). *The Material Basis of Evolution.* New Haven: Yale University Press, 1940.

Gould, Stephen Jay (ESD). *Ever Since Darwin.* New York: W. W. Norton, 1977.

Gould, Stephen Jay (GP). "A Geology Professor Answers Questions on Creationism." *Unitarian Universalist Word* 13/2 (15 February 1982): 1, 8.

Gould, Stephen Jay (HT). *Hen's Teeth and Horse's Toes.* New York: W. W. Norton, 1983.

Gould, Stephen Jay (ISJ). "Impeaching a Self-Appointed Judge." *Scientific American,* July 1992: 118–21.

Gould, Stephen Jay (MM). *The Mismeasure of Man.* New York: W. W. Norton, 1981.

Gould, Stephen Jay (PT). *The Panda's Thumb.* New York: W. W. Norton, 1982.

Gould, Stephen Jay (RA). *Rocks of Ages: Science and Religion in the Fullness of Life.* Ballentine, 1999.

Gould, Stephen Jay (RIP). "On Replacing the Idea of Progress with an Operational Notion of Directionality." In *Evolutionary Progress.* Edited by Matthew Nitecki, 319–38. London and Chicago: University of Chicago Press, 1988.

Gould, Stephen Jay (WL). *Wonderful Life: The Burgess Shale and the Nature of History.* New York: W. W. Norton, 1989.

Grad, Bernard (EO). "Experiences and Opinions of an Unconventional Scientist." In *Men and Women of Parapsychology: Personal Reflections.* Edited by Rosemarie Pilkington, 146–60. Jefferson, N.C., and London: McFarland and Co., 1987.

Greene, John C. (SIW). *Science, Ideology, and World View: Essays in the History of Evolutionary Ideas.* Berkeley: University of California Press, 1981.

Gregory, Frederick (IDE). "The Impact of Darwinian Evolution on Protestant Theology." In *God and Nature: Historical Essays on the Encounter between Christianity and Science.* Edited by David C. Lindberg and Ronald L. Numbers, 369–90. Berkeley: University of California Press, 1988.

Griffin, David Ray (CFSN). "Christian Faith and Scientific Naturalism: An Appreciative Critique of Phillip Johnson's Proposal." *Christian Scholar's Review* 28/2 (1998): 308–328.

Griffin, David Ray (DC). *The Divine Cry of Our Time.* Work in progress.

Griffin, David Ray (ER). *Evil Revisited: Responses and Reconsiderations.* Albany: State University of New York Press, 1991.

Griffin, David Ray (GPE). *God, Power, and Evil: A Process Theodicy*, 2nd edition with a new preface. Lanham, Md.: University Press of America, 1991.

Griffin, David Ray (HG). "Hartshorne, God, and Relativity Physics." *Process Studies* 21/2 (Summer 1992): 85–112.

Griffin, David Ray (IBI). "On Ian Barbour's *Issues in Science and Religion.*" *Zygon* 23/1 (March 1988): 57–81.

Griffin, David Ray (PAP). "Panexperientialism and Pantemporalism." In *The Textures of Time.* Edited by Paul A. Harris. Forthcoming.

Griffin, David Ray (PP). "Parapsychology and Philosophy: A Whiteheadian Postmodern Perspective." *Journal of the American Society for Psychical Research* 87/3 (July 1993): 217–88.

Griffin, David Ray (PPS). *Parapsychology, Philosophy, and Spirituality: A Postmodern Exploration.* Albany: State University of New York Press, 1996.

Griffin, David Ray, ed. (PUST). *Physics and the Ultimate Significance of Time: Bohm, Prigogine, and Process Philosophy.* Albany: State University of New York Press, 1986.

Griffin, David Ray (RE). "Religious Experience, Naturalism, and the Social Scientific Study of Religion." *Journal of the Academic Study of Religion* 68/1 (March 2000).

Griffin, David Ray (RWS). *Reenchantment Without Supernaturalism: A Process Philosophy of Religion.* Ithaca: Cornell University Press, 2000.

Griffin, David Ray (UW). *Unsnarling the World-Knot: Consciousness, Freedom, and the Mind-Body Problem.* Berkeley and Los Angeles: University of California Press, 1998.

Griffin, David Ray (WDEW). "Whitehead's Deeply Ecological Worldview." In *Worldviews and Ecology: Religion, Philosophy, and the Environment.* Edited by Mary Evelyn Tucker and John Grim, 190–206. Maryknoll: Orbis Books, 1994.

Griffin, David Ray, and Huston Smith (PT). *Primordial Truth and Postmodern Theology.* Albany: State University of New York Press, 1989.

Griffin, David Ray, John B. Cobb, Jr., Marcus P. Ford, Pete A. Y. Gunter, and Peter Ochs (FCPP). *Founders of Constructive Postmodern Philosophy: Peirce, James, Bergson, Whitehead, and Hartshorne.* Albany: State University of New York, 1993.

Griffin, Donald R. (QAA). *The Question of Animal Awareness: Evolutionary Continuity of Mental Experience.* New York: Rockefeller University Press, 1976.

Grim, Patrick, ed. (PSO). *Philosophy of Science and the Occult*. Albany: State University of New York Press, 1982.

Groos, Karl (PA). *The Play of Animals*. New York: D. Appleton, 1898.

Gruber, Howard (DM). *Darwin on Man: A Psychological Study of Scientific Creativity*, 2nd ed. Chicago: University of Chicago Press, 1981.

Gruenler, Royce Gordon (IG). *The Inexhaustible God: Biblical Faith and the Challenge of Process Theism*. Grand Rapids: Baker Book House, 1983.

Gunter, Pete A. Y. (HB). "Henri Bergson." In David Ray Griffin et al., *Founders of Constructive Postmodern Philosophy: Peirce, James, Bergson, Whitehead, and Hartshorne*, 133–64. Albany: State University of New York Press, 1993.

Gurney, Edmund (TQ). *Tertium Quid*. Vol. I. London: Kegan Paul, Trench, and Co., 1887.

Hall, Barry G. (AE). "Adaptive Evolution that Requires Spontaneous Mutations." *Genetics* 120 (1988): 887–97.

Hansel, C. E. M. (ESP). *ESP: A Scientific Investigation*. New York: Scribner's, 1966.

Hansel, C. E. M. (EP). *ESP and Parapsychology: A Critical Re-Evaluation*. New York: Buffalo: Prometheus, 1980.

Hansel, C. E. M. (ET). "Experiments on Telepathy in Children." *British Journal of Statistical Psychology* 13 (1960): 175–78.

Hardy, Sir Alister (LS). *The Living Stream: A Restatement of Evolution Theory and Its Relation to the Spirit of Man*. London: Collins, 1965.

Harman, Gilbert (NM). *The Nature of Morality: An Introduction to Ethics*. New York: Oxford University Press, 1977.

Harris, Marvin (CP). *Cows, Pigs, Wars and Witches*. New York: Random House, 1974.

Hartshorne, Charles (BH). *Beyond Humanism: Essays in the New Philosophy of Nature*. New York: Willett, Clark, 1937.

Hartshorne, Charles (CI). "The Compound Individual." In *Philosophical Essays for Alfred North Whitehead*. Edited by Otis H. Lee, 193–220. New York: Longmans Green, 1936. Reprinted in Charles Hartshorne, *Whitehead's Philosophy: Selected Essays 1935–1970*, 41–61. Lincoln: University of Nebraska Press, 1972.

Hartshorne, Charles (CSPM). *Creative Synthesis and Philosophic Method*. LaSalle, Ill.: Open Court; London: SCM Press, 1970.

Hasker, William (DTR). "*Darwin on Trial* Revisited: A Review Essay." *Christian Scholar's Review* 24/4 (May 1995): 479–88.

Hasker, William (MJP). "Mr. Johnson for the Prosecution." *Christian Scholar's Review* 22/2 (December 1992): 177–86.

Haught, John F. (GAD). *God After Darwin: A Theology of Evolution*. Boulder: Westview, 1999.

Haught, John F. (PN). *The Promise of Nature: Ecology and Cosmic Purpose.* Mahwah, N.J.: Paulist Press, 1993.

Hebb, D. O. (RNI). "The Role of Neurological Ideas in Psychology." *Journal of Personality* 20/1 (September 1951): 39–55.

Heidegger, Martin (WN). "The Word of Nietzsche: 'God is Dead.' " In Martin Heidegger, *The Question Concerning Technology: Heidegger's Critique of the Modern Age.* Translated by William Lovitt, 53–112. New York: Harper and Row, 1977.

Hersh, Reuben (WIM). *What is Mathematics, Really?* New York: Oxford University Press, 1997.

Hesse, Mary (FF). *Forces and Fields: The Concept of Action at a Distance in the History of Physics.* Totowa, N. J.: Littlefield, Adams, and Co., 1965.

Hick, John (DEL). *Death and Eternal Life.* San Francisco: Harper, 1976.

Hill, Christopher (WTU). *The World Turned Upside Down.* London: Temple Smith, 1972.

Hodge, Charles (ST). *Systematic Theology*, 3 vols. Grand Rapids: Eerdmans, 1982 (originally published 1872).

Hooykaas, R. (NL). *Natural Law and Divine Miracle: A Historical-Critical Study of the Principle of Uniformity in Geology, Biology, and Theology.* Leiden: E. J. Brill, 1959.

Hoyle, Fred (TU). "The Universe: Past and Present Reflections." *Engineering and Science* 45/2 (November 1981): 8–12.

Huxley, Thomas H. (D). *Darwiniana: Essays.* New York: D. Appleton, 1908.

Huxley, Thomas H. (EE). "Evolution and Ethics." James Paradis and George C. Williams, *Thomas Huxley's Evolution and Ethics: With New Essays on Its Victorian and Sociobiological Context,* 57–116. Princeton: Princeton University Press, 1989.

Huxley, Thomas H. (LL). Leonard Huxley, *Life and Letters of Thomas Henry Huxley*, 2 vols. London: Macmillan; New York: A. Appleton, 1901.

Hyman, Ray (EQ). *The Elusive Quarry: A Scientific Appraisal of Psychical Research.* Buffalo: Prometheus Books, 1989.

Jacob, James R. (BA). "Boyle's Atomism and the Restoration Assault on Pagan Naturalism." *Social Studies of Science* 8 (1978): 211–33.

James, William (ERE). *Essays in Radical Empiricism.* Edited by Ralph Barton Perry. New York: E. P. Dutton, 1971 (published in one volume with James's *A Pluralistic Universe*).

James, William (SPP). *Some Problems of Philosophy.* London: Longman and Green, 1911.

James, William (WJPR). *William James on Psychical Research,* ed. Gardner Murphy and Robert O. Ballou. Clifton, N. J.: Augustus M. Kelley, 1973.

Jay, Martin (DPC). "The Debate over Performative Contradiction: Habermas versus the Poststructuralists." In Martin Jay, *Force Fields: Between Intellectual History and Cultural Critique*, 25–37. New York and London: Routledge, 1993.

Johnson, Phillip E. (DT). *Darwin on Trial.* 2nd ed. Downers Grove, Ill.: InterVarsity Press, 1993.

Johnson, Phillip E. (RB). *Reason in the Balance: The Case Against Naturalism in Science, Law, and Education.* Downers Grove, Ill.: InterVarsity Press, 1993.

Johnson, Phillip E. (RH). "Response to Hasker." *Christian Scholar's Review* 22/3 (March 1993): 297–304.

Johnson, Phillip E. (RTH). "Response to Hasker." *Christian Scholar's Review* 24/4 (May 1995): 489–93.

Jung, Carl (MDR). *Memories, Dreams, Recollections.* Edited by Aniela Jaffe, translated by Richard and Clare Winston. New York: Random House, 1963.

Kant, Immanuel (CPR). *Critique of Pure Reason.* Translated by Norman Kemp Smith. New York: St. Martin's, 1965.

Kaufman, Gordon D. (IFM). *In Face of Mystery: A Constructive Theology.* Cambridge: Harvard University Press, 1993.

Kearney, Hugh (SC). *Science and Change 1500–1700.* New York: McGraw-Hill, 1971.

Keller, Evelyn Fox (FO). *A Feeling for the Organism: The Life and Work of Barbara McClintock.* New York: Freeman, 1983.

Keynes, J. M. (EB). *Essays in Biography.* Edited by Geoffrey Keynes. 2nd ed. London: R. Hart-Davis, 1951.

Kim, Jaegwon (SM). *Supervenience and Mind: Selected Philosophical Essays.* Cambridge: Cambridge University Press, 1993.

Kim, Stephen S. (JT). *John Tyndall's Transcendental Materialism and the Conflict between Religion and Science in Victorian England.* Lewiston: Mellen, 1996.

Klaaren, Eugene M. (RO). *Religious Origins of Modern Science: Belief in Creation in Seventeenth-Century Thought.* Grand Rapids: William B. Eerdmans, 1977.

Kocher, Paul (SR). *Science and Religion in Elizabethan England.* San Marino, Calif.: The Huntington Library, 1953.

Kors, Alan C., and Edward Peters (WE). *Witchcraft in Europe 1100–1700.* Philadelphia: University of Pennsylvania Press, 1972.

Koyré, Alexandre (FCW). *From the Closed World to the Infinite Universe.* Baltimore: The Johns Hopkins Press, 1957.

Krikorian, Yervant H., ed. (NHS). *Naturalism and the Human Spirit.* Morningside Heights: Columbia University Press, 1944.

Krutch, Joseph Wood (MT). *The Modern Temper: A Study and a Confession.* New York: Harcourt, Brace and World, 1956.

Kurtz, Paul (IPS). "Is Parapsychology a Science?" In *A Skeptic's Handbook of Parapsychology.* Edited by Paul Kurtz, 503–18. Buffalo: Prometheus Press, 1985.

Kurtz, Paul (MC). "More Than a Century of Psychical Research." In *A Skeptic's Handbook of Parapsychology*. Edited by Paul Kurtz, xi–xxiv. Buffalo: Prometheus Press, 1985.

Kurtz, Paul (TT). *The Transcendental Temptation: A Critique of Religion and the Paranormal*. Buffalo: Prometheus Press, 1986.

Lamprecht, Sterling (NHS). "Naturalism and Religion." In Krikorian, ed., NHS, 17–39.

Laudan, Rachel, ed. (DSP). *The Demarcation between Science and Pseudo-Science*. Blacksburg, Va.: Center for the Study of Science and Society (Virginia Polytechnic Institute and State University), 1983.

Lenoble, Robert (MNM). *Mersenne ou la naissance du méchanisme*. Paris: Librairie Philosophique J. Vrin, 1943.

Levenson, Jon D. (CPE). *Creation and the Persistence of Evil: The Jewish Drama of Divine Omnipotence*. San Francisco: Harper and Row, 1988.

Lewis, H. D. (EM). *The Elusive Mind*. London: George Allen and Unwin, 1969.

Lewis, H. D. (ES). *The Elusive Self*. London: Macmillan, 1982.

Lindberg, David C., and Ronald L. Numbers, eds. (GN). *God and Nature: Historical Essays on the Encounter between Christianity and Science*. Berkeley: University of California Press, 1986.

Lorimer, David (S). *Survival? Body, Mind, and Death in the Light of Psychic Experience*. London: Routledge and Kegan Paul, 1984.

Lovejoy, A. O. (AOE). "The Argument for Organic Evolution Before the *Origin of Species,* 1830–1858." In *Forerunners of Darwin: 1745–1859*. Edited by B. Glass, O. Temkin, and W. S. Strauss, 356–414. Baltimore: Johns Hopkins Press, 1959.

Lycan, William G. (C). *Consciousness*. Cambridge: MIT Press, 1987.

Mackie, John (E). *Ethics: Inventing Right and Wrong*. New York: Penguin, 1977.

Maddy, Penelope (RM). *Realism in Mathematics*. Oxford: Clarendon Press, 1990.

Madell, Geoffrey (MM). *Mind and Materialism*. Edinburgh: The University Press, 1988.

Manuel, Frank (PIN). *A Portrait of Isaac Newton*. Cambridge: Harvard University Press (Belknap Press), 1968.

Margenau, Henry (ESP). "ESP in the Framework of Modern Science." *Journal of the American Society for Psychical Research* 60 (1966): 214–28.

Margulis, Lynn (BTB). "Big Trouble in Biology: Physiological Autopoiesis versus Mechanistic Neo-Darwinism." In *Doing Science: The Reality Club*. Edited by John Brockman, 211–35. New York: Prentice-Hall, 1991.

Margulis, Lynn (EL). *Early Life*. Boston: Science Books International; New York: Van Nostrand Reinhold, 1982.

Margulis, Lynn (KA). "Kingdom Animalia: The Zoological Malaise from a Microbial Perspective." *American Zoologist* 30/4 (1990): 861–75.

Margulis, Lynn, and Dorion Sagan (WIL). *What Is Life?* New York: Simon and Schuster, 1995.

May, Gerhard (CEN). *Creatio Ex Nihilo: The Doctrine of 'Creation out of Nothing' in Early Christian Thought.* Translated by A. S. Worrall. Edinburgh: T and T Clark, 1994.

Mayr, Ernst (PSE). *Populations, Species and Evolution.* Cambridge: Harvard University Press, 1970.

Mayr, Ernst (TNP). *Toward a New Philosophy of Biology: Observations of an Evolutionist.* Cambridge and London: Harvard University Press, 1988.

McClenon, James (DS). *Deviant Science: The Case of Parapsychology.* Philadelphia: University of Pennsylvania Press, 1984.

McClenon, James (WE). *Wondrous Events: Foundations of Religious Belief.* Philadelphia: University of Pennsylvania Press, 1994.

McGinn, Colin (CM). *The Character of Mind.* Oxford: Oxford University Press, 1982.

McGinn, Colin (PC). *The Problem of Consciousness: Essays Toward a Resolution.* Oxford: Basil Blackwell, 1991.

McGuire, J. E., and P. M. Rattansi (NPP). "Newton and the Pipes of Pan." *Notes and Records of the Royal Society* 21 (1966): 108–41.

McMullin, Ernan (NSBC). "Natural Science and Belief in a Creator: Historical Notes." In *Physics, Philosophy, and Theology: A Common Quest for Understanding.* Edited by R. J. Russell, W.R. Stoeger, and G.V. Coyne, 49–79. Vatican City State: Vatican Observatory, 1988.

McMullin, Ernan (PDSC). "Plantinga's Defense of Special Creation." *Christian Scholar's Review* 21/1 (1991): 55–79.

Mebane, Alexander (DCM). *Darwin's Creation Myth.* Venice, Fl.: P and D Printing, 1994.

Merchant, Carolyn (DN). *The Death of Nature: Women, Ecology and the Scientific Revolution.* San Francisco: Harper and Row, 1980.

Mill, John Stuart (SL). *A System of Logic Rationcinative and Inductive.* Toronto: University of Toronto Press, 1973.

Monod, Jacques (CN). *Chance and Necessity: An Essay on the Natural Philosophy of Modern Biology.* New York: Vintage Books, 1972.

Moody, Raymond (LAL). *Life after Life.* New York: Bantam, 1975.

Mosse, George L. (PRE). "Puritan Radicalism and the Enlightenment." *Church History* 29 (1960): 424–39.

Mundle, C. W. K. (DCP). "Does the Concept of Precognition Make Sense?" In *Philosophy and Parapsychology.* Edited by Jan Ludwig, 327–40. Buffalo: Prometheus Book, 1978.

Mundle, C. W. K. (EE). "The Explanation of ESP." In *Science and ESP.* Edited by J. R. Smythies, 197–207. London: Routledge and Kegan Paul, 1967.

Murphy, Michael (FB). *The Future of the Body: Exploration into the Further Evolution of Human Nature.* Los Angeles: Jeremy Tarcher, 1992.

Murphy, Nancey (PJOT). "Phillip Johnson on Trial: A Critique of His Critique of Darwin." *Perspectives on Science and Christian Faith* 45/1 (March 1993): 26–36.

Myers, Frederick (HP). *Human Personality and Its Survival of Bodily Death.* Edited and abridged by Susy Smith. New Hyde Park, N. Y.: University Books, 1961.

Myers, Frederick (PA). "Presidential Address." *Proceedings of the Society for Psychical Research* 15 (1900): 110–27.

Nagel, Ernest (NHS). "Logic Without Ontology." In Krikorian, ed., NHS, 210–41.

Nagel, Thomas (MQ). *Mortal Questions.* London: Cambridge University Press, 1979.

Nagel, Thomas (VN). *The View from Nowhere.* New York: Oxford University Press, 1986.

Nash, J. Madeleine (WLE). "When Life Exploded." *Time,* 4 December 1995: 66–77.

Newell, Norman D. (NFR). "The Nature of the Fossil Record." *Proceedings of the American Philosophical Society* 103/2 (1959): 264–85.

Nitecki, Matthew H., ed. (EP). *Evolutionary Progress.* Chicago and London: University of Chicago Press, 1988.

Orr, A. Allen, and Jerry A. Coyne (GA). "The Genetics of Adaptation: A Reassessment." *American Naturalist* 140/5 (November 1992): 725–42.

Ospovat, Dov (DDT). *The Development of Darwin's Theory: Natural History, Natural Theology and Natural Selection 1838–1859.* Cambridge and New York: Cambridge University Press, 1981.

Otto, Rudolf (DR). "Darwinism and Religion." In Rudolf Otto, *Religious Essays: A Supplement to 'The Idea of the Holy.'* Translated by Brian Lunn, 121–39. Oxford: Oxford University Press; London: Humphrey Milford, 1931.

Otto, Rudolf (NR). *Naturalism and Religion.* London: Williams and Norgate; New York: G. P. Putnam's, 1907.

Pagels, Heinz R. (DR). *The Dreams of Reason: The Computer and the Rise of the Sciences of Complexity.* New York: Simon and Schuster, 1988.

Papineau, David (PN). *Philosophical Naturalism.* Oxford: Blackwell, 1993.

Peters, Ted (ST). *Science and Theology: The New Consonance.* Boulder: Westview Press, 1998.

Plantinga, Alvin (WFR). "When Faith and Reason Clash: Evolution and the Bible." *Christian Scholar's Review* 21/1 (1991): 8–32.

Plantinga, Alvin (ENAP). "Evolution, Neutrality, and Antecedent Probability: A Reply to McMullin and Van Till." *Christian Scholar's Review* 21/1 (1991): 80–109.

Popper, Karl R. (OCC). *Of Clocks and Clouds.* St. Louis: Washington University Press, 1966.

Popper, Karl R., and John C. Eccles (SB). *The Self and Its Brain: An Argument for Interactionism.* Heidelberg: Springer-Verlag, 1977.

Pratt, James Bissett (N). *Naturalism.* New Haven: Yale University Press, 1939.

Preus, J. Samuel (ER). *Explaining Religion: Criticism and Theory from Bodin to Freud.* New Haven and London: Yale University Press, 1987.

Price, George (SS). "Science and the Supernatural." In *Philosophy and Parapsychology.* Edited by Jan Ludwig, 145–71. Buffalo: Prometheus Books, 1978.

Prigogine, Ilya, and Isabelle Stengers (OOC). *Order Out of Chaos: Man's New Dialogue with Nature.* New York: Bantam Books, 1984.

Prior, Moody E. (JG). "Joseph Glanvill, Witchcraft, and Seventeenth-Century Science." *Modern Philosophy* 30 (1932–33): 167–93.

Provine, William (PE). "Progress in Evolution and Meaning in Life." In *Evolutionary Progress.* Edited by Matthew H. Nitecki, 49–74. Chicago and London: University of Chicago Press, 1988.

Randall, John Herman, Jr. (NHS). "Epilogue: The Nature of Naturalism." In Krikorian, ed., NHS, 354–82.

Randall, John L. (PNL). *Parapsychology and the Nature of Life.* New York: Harper, 1975.

Rao, K. Ramakrishna (C). "Comments." *Zetetic Scholar* 6 (1980): 107–09.

Rhine, J. B. (CP). "A Century of Parapsychology." In *The Signet Handbook of Parapsychology.* Edited by Martin Ebon, 7–16. New York: New American Library, 1978.

Rhine, J. B. (RM). *The Reach of the Mind.* New York: William Sloane, 1972.

Richards, Robert J. (MF). "Moral Foundations of the Idea of Evolutionary Progress." In *Evolutionary Progress.* Edited by Matthew Nitecki, 129–48. Chicago and London: University of Chicago Press, 1988.

Robinson, William S. (BP). *Brains and People: An Essay on Mentality and Its Causal Conditions.* Philadelphia: Temple University Press, 1988.

Roll, William G. (PP). "The Problem of Precognition." *Journal of the Society for Psychical Research* 41 (1961): 115–28.

Rolston, Holmes, III (SR). *Science and Religion: A Critical Survey.* Philadelphia: Temple University Press, 1987.

Russell, Bertrand (ML). *Mysticism and Logic.* London: Allen and Unwin, 1917.

Russell, Bertrand (OK). *Our Knowledge of the External World.* London: Allen and Unwin, 1921.

Russell, Colin A. (SSC). *Science and Social Change 1700–1900.* London: Macmillan, 1983.

Russell, Jeffrey Burton (WMA). *Witchcraft in the Middle Ages.* Cornell University Press, 1972.

Russell, Robert John (SP). "Special Providence and Genetic Mutation: A New Defense of Theistic Evolution." In *Evolutionary and Molecular Biology: Scientific Perspectives on Divine Action.* Edited by Robert John Russell, William R. Stoeger, S.J., and Francisco J. Ayala, 191–223. Vatican City: Vatican Observatory Publications, Berkeley: Center for Theology and the Natural Sciences, 1998.

Sabom, Michael (RD). *Recollections of Death: A Medical Investigation.* New York: Harper and Row, 1982.

Schleiermacher, Friedrich (CF). *The Christian Faith.* Edited by H. R. Mackintosh and J. S. Stewart. New York: Harper, 1963.

Schwartz, Joel S. (DW). "Darwin, Wallace, and the *Descent of Man.*" *Journal of the History of Biology* 17 (1984): 271–89.

Scott, Christopher (WPD). "Why Parapsychology Demands a Skeptical Response." In *A Skeptic's Handbook of Parapsychology.* Edited by Paul Kurtz, 497–501. Buffalo: Prometheus Books, 1985.

Seager, William (CIP). "Consciousness, Information, and Panpsychism." *Journal of Consciousness Studies* 2/3 (1995): 272–88.

Seager, William (MC). *Metaphysics of Consciousness.* London and New York: Routledge, 1991.

Searle, John R. (MBP). "The Mind-Body Problem." In *John Searle and His Critics.* Edited by Ernest Lepore and Robert van Gulick, 141–46. Cambridge, Mass., and Oxford: Basil Blackwell, 1991.

Searle, John R. (MBS). *Minds, Brains, and Science: The 1984 Reith Lectures.* London: British Broadcasting Corporation, 1984.

Searle, John R. (MBW). "Minds and Brains Without Programs." In *Mindwaves: Thoughts on Intelligence, Identity, and Consciousness.* Edited by Colin Blakemore and Susan Greenfield, 209–33. Oxford: Basil Blackwell, 1987.

Searle, John R. (RM). *The Rediscovery of the Mind.* Cambridge: MIT Press, 1992.

Sellars, Wilfrid (SPR). *Science, Perception, and Reality.* New York: Humanities Press, and London: Routledge and Kegan Paul, 1963.

Setzer, J. Shoneberg (PRBS). "Parapsychology: Religion's Basic Science." *Religion in Life* 39 (1970): 595–607.

Shapiro, Robert (OSG). *Origins: A Skeptic's Guide to the Creation of Life on Earth.* New York: Summit Books, 1986.

Shea, William M. (NS). *The Naturalists and the Supernatural.* Mercer University Press, 1984.

Sheils, Dean (CCS). "A Cross-Cultural Study of Belief in Out-of-the-Body Experiences, Waking and Sleeping." *Journal of the Society for Psychical Research* 49 (1978): 697–741.

Sheldrake, Rupert (NSL). *A New Science of Life: The Hypothesis of Formative Causation.* 2nd ed. London: Blond, 1985.

Siegel, Bernie S. (LMM). *Love, Medicine, and Miracles: Lessons Learned about Self-Healing from a Surgeon's Experience with Exceptional Patients.* New York: Harper and Row, 1986.

Simpson, George G. (HL). "The History of Life." In *The Evolution of Life.* Edited by Sol Tax, 117–80. Chicago: University of Chicago Press, 1960.

Simpson, George G. (TM). *Tempo and Mode in Evolution.* New York: Columbia University Press, 1944.

Skinner, B. F. (BFD). *Beyond Freedom and Dignity.* New York: Free Press, 1965.

Skinner, B. F. (SHB). *Science and Human Behavior.* New York: Free Press, 1965.

Spencer, Herbert (LL). *The Life and Letters of Herbert Spencer.* Edited by David Duncan. London: Methuen, 1908.

Sperry, Roger (SMP). *Science and Moral Priority: Merging Mind, Brain, and Human Values.* New York: Columbia University Press, 1983.

Stanley, Steven M. (NET). *The New Evolutionary Timetable.* New York: Basic Books, 1981.

Stapp, Henry (ET). "Einstein Time and Process Time." In Griffin, ed., PUST, 264–70.

St. James-Roberts, Ian (CS). "Cheating in Science." *New Scientist,* 25 November 1976: 466–69.

Stevenson, Ian (TI). *Telepathic Impressions.* Charlottesville: University Press of Virginia, 1970.

Strauss, David Friedrich (LJ). *The Life of Jesus Critically Examined.* Translated by George Eliot. Edited by Peter C. Hodgson. Philadelphia: Fortress Press, 1972.

Suchocki, Marjorie (IGP). *In God's Presence: Theological Reflections on Prayer.* St. Louis: Chalice Press, 1996.

Swinburne, Richard (EG). *The Existence of God.* Oxford: Clarendon, 1979.

Swinburne, Richard (ES). *The Evolution of the Soul.* Oxford: Clarendon, 1986.

Tanagras, A. (PE). *Psychophysical Elements in Parapsychological Traditions.* New York: Parapsychology Foundation, 1967.

Tart, Charles (PSI). *PSI: Scientific Studies of the Psychic Realm.* New York: E. P. Dutton, 1977.

Taylor, Gordon Rattray (GEM). *The Great Evolution Mystery.* London: Secker and Warburg, 1983.

Thalbourne, M. A. (CFP). "The Conceptual Framework of Parapsychology: Time for a Reformation" (unpublished 1984 ms.).

Thayer, H. S., ed. (NPN). *Newton's Philosophy of Nature.* New York: Hafner, 1953.

Thomas, Keith (RDM). *Religion and the Decline of Magic.* New York: Charles Scribner's Sons, 1971.

Trevor-Roper, H. R. (EWC). *The European Witch Craze of the Sixteenth and Seventeenth Centuries and Other Essays.* New York: Harper, 1956, 1969.

Truzzi, Marcello (SL). "A Skeptical Look at Paul Kurtz's Analysis of the Scientific Status of Parapsychology." *Journal of Parapsychology* 44 (March 1980): 35–55.

Tyndall, John (FS). *Fragments of Science,* 5th ed. London: Longmans, Green, and Co., 1876.

Tyrrell, George (A). *Apparitions.* New Hyde Part, N.Y.: University Books, 1961 (published in one volume with Tyrrell's *Science and Psychical Research*).

Uttal, William R. (PM). *The Psychobiology of Mind.* Hillsdale, N. J.: L. Erlbaum Associates, 1978.

Van Till, Howard J. (BA). "Basil, Augustine, and the Doctrine of Creation's Functional Integrity." *Science and Christian Belief* 8/1 (1996): 21–38.

Van Till, Howard J. (SC). "Special Creationism in Designer Clothing: A Response to *The Creation Hypothesis.*" *Perspectives on Science and Christian Faith* 47/2 (June 1995): 123–31.

Van Till, Howard J. (WFRC). "When Faith and Reason Cooperate." *Christian Scholar's Review* 21/1 (1991): 33–45.

Van Till, Howard J., Davis A. Young, and Clarence Menninga (SHH). *Science Held Hostage: What's Wrong with Creation Science and Evolutionism.* Downers Grove, Ill.: InterVarsity Press, 1988.

Vickers, Brian, ed. (OSM). *Occult and Scientific Mentalities in the Renaissance.* Cambridge: Cambridge University Press, 1984.

Weaver, Richard M. (IHC). *Ideas Have Consequences.* Chicago: University of Chicago Press, 1948.

Weinberg, Steven (DFT). *Dreams of a Final Theory: The Scientist's Search for the Ultimate Laws of Nature,* 2nd ed. New York: Vintage Books, 1994.

Weinstein, Deena (FS). "Fraud in Science." *Social Science Quarterly* 59 (1979): 639–52.

Weiss, P. (TP). "Time Proves Not Reversible at Deepest Level." *Science News* 154 (31 October 1998): 277.

Wesson, Robert (BNS). *Beyond Natural Selection.* Cambridge: MIT Press, 1991.

Westfall, Richard J. (IAN). "The Influence of Alchemy on Newton." In *Science, Pseudo-Science, and Society.* Edited by Marsha P. Hanen, Margaret J. Osler, and Robert G. Weyant, 145–70. Waterloo, Ont.: Wilfrid Laurier University Press, 1980.

Westfall, Richard J. (NR). *Never at Rest: A Biography of Isaac Newton.* Cambridge: Cambridge University Press, 1980.

Westman, Robert S., and J. E. McGuire, eds. (HSR). *Hermeticism and the Scientific Revolution.* Los Angeles: Clark Memorial Library (University of California), 1977.

Whitehead, Alfred North (AI). *Adventures of Ideas.* New York: Free Press, 1967 (originally 1933).

Whitehead, Alfred North (ESP). *Essays in Science and Philosophy.* New York: Philosophical Library, 1947.

Whitehead, Alfred North (FR). *The Function of Reason.* Boston: Beacon Press, 1968 (originally 1929).

Whitehead, Alfred North (MT). *Modes of Thought.* New York: Free Press, 1968 (originally 1938).

Whitehead, Alfred North (PR). *Process and Reality,* corrected edition, ed. David Ray Griffin and Donald W. Sherburne. New York: Free Press, 1978 (originally 1929).

Whitehead, Alfred North (RM). *Religion in the Making.* Cleveland: World Publishing Co., 1960 (originally 1926).

Whitehead, Alfred North (S). *Symbolism: Its Meaning and Effect.* New York: Capricorn, 1959 (originally 1927).

Whitehead, Alfred North (SMW). *Science and the Modern World.* New York: Free Press, 1967 (originally 1925).

Wilson, Edward O. (OHN). *On Human Nature.* New York: Bantam Books, 1979.

Yates, Frances (GB). *Giordano Bruno and the Hermetic Tradition.* Chicago: University of Chicago Press, 1964.

Yates, Frances (RE). *The Rosicrucian Enlightenment.* Boulder: Shambhala, 1978.

NOTE ON SUPPORTING CENTER

This series is published under the auspices of the Center for Process Studies, a research organization affiliated with the Claremont School of Theology and Claremont Graduate University. It was founded in 1973 by John B. Cobb, Jr., Founding Director, and David Ray Griffin, Executive Director; Marjorie Suchocki is now also a Co-director. It encourages research and reflection on the process philosophy of Alfred North Whitehead, Charles Hartshorne, and related thinkers, and on the application and testing of this viewpoint in all areas of thought and practice. The center sponsors conferences, welcomes visiting scholars to use its library, and publishes a scholarly journal, *Process Studies*, and a newsletter, *Process Perspectives*. Located at 1325 North College, Claremont, CA 91711, it welcomes new members and gratefully accepts (tax-deductible) contributions to support its work.

INDEX